Genders and Sexualities in History

Series Editors: **John H. Arnold, Joanna Bourke and Sean Brady**

Palgrave Macmillan's series, Genders and Sexualities in History, aims to accommodate and foster new approaches to historical research in the fields of genders and sexualities. The series promotes world-class scholarship that concentrates upon the interconnected themes of genders, sexualities, religions/religiosity, civil society, class formations, politics and war.

Historical studies of gender and sexuality have often been treated as disconnected fields, while in recent years historical analyses in these two areas have synthesised, creating new departures in historiography. By linking genders and sexualities with questions of religion, civil society, politics and the contexts of war and conflict, this series will reflect recent developments in scholarship, moving away from the previously dominant and narrow histories of science, scientific thought and legal processes. The result brings together scholarship from contemporary, modern, early modern, medieval, classical and non-Western history to provide a diachronic forum for scholarship that incorporates new approaches to genders and sexualities in history.

Between the end of the nineteenth and the beginning of the twentieth centuries, a number of women were diagnosed as sexual inverts in Britain and Europe. This book examines the ways in which female same-sex desires were represented across a wide range of Italian and British medical writings between1870 and 1920. Chiara Beccalossi has undertaken extensive archival research to bring to light new documents and give a detailed account of how the female invert was positioned alongside other figures of same-sex desires, such as the tribade-prostitute, the *fiamma* (flame), the nymphomaniac, and women with abnormal genitalia or bodily dysfunctions. In this way the book shows the richness of medical representations of female same-sex desires that has previously been unexplored in historical scholarship.

Titles include:

John H. Arnold and Sean Brady (*editors*)
WHAT IS MASCULINITY?
Historical Dynamics from Antiquity to the Contemporary World

Cordelia Beattie and Kirsten A. Fenton (*editors*)
INTERSECTIONS OF GENDER, RELIGION AND ETHNICITY IN THE MIDDLE AGES

Chiara Beccalossi
FEMALE SEXUAL INVERSION
Same-Sex Desires in Italian and British Sexology, c.1870–1920

Peter Cryle and Alison Moore
FRIGIDITY
An Intellectual History

Jennifer V. Evans
LIFE AMONG THE RUINS
Cityscape and Sexuality in Cold War Berlin

Kate Fisher and Sarah Toulalan (*editors*)
BODIES, SEX AND DESIRE FROM THE RENAISSANCE TO THE PRESENT

Christopher E. Forth and Elinor Accampo (*editors*)
CONFRONTING MODERNITY IN FIN-DE-SIÈCLE FRANCE
Bodies, Minds and Gender

Dagmar Herzog (*editor*)
BRUTALITY AND DESIRE
War and Sexuality in Europe's Twentieth Century

Andrea Mansker
SEX, HONOR AND CITIZENSHIP IN EARLY THIRD REPUBLIC FRANCE

Jessica Meyer
MEN OF WAR
Masculinity and the First World War in Britain

Jennifer D. Thibodeaux (*editor*)
NEGOTIATING CLERICAL IDENTITIES
Priests, Monks and Masculinity in the Middle Ages

Hester Vaizey
SURVIVING HITLER'S WAR
Family Life in Germany, 1939–48

Forthcoming titles:

Heike Bauer and Matthew Cook (*editors*)
QUEER 1950s

Matthew Cook
QUEER DOMESTICITIES
Homosexuality and Home Life in Twentieth-Century London

Rebecca Fraser
GENDER AND IDENTITY IN ANTEBELLUM AMERICA
From Northern Woman to Plantation Mistress

Julia Laite
PROSTITUTION AND REPRESSION IN THE METROPOLIS
Criminalization and the Shaping of Commercial Sex in London, 1885–1960

Melissa Hollander
SEX IN TWO CITIES
The Negotiation of Sexual Relationships in Early Modern England and Scotland

---

Genders and Sexualities in History Series
Series Standing Order 978–0–230–55185–5 Hardback
978–0–230–55186–2 Paperback
(*outside North America only*)

You can receive future titles in this series as they are published by placing a standing order. Please contact your bookseller or, in case of difficulty, write to us at the address below with your name and address, the title of the series and the ISBN quoted above.

Customer Services Department, Macmillan Distribution Ltd, Houndmills, Basingstoke, Hampshire RG21 6XS, England

---

Female Sexual Inversion

Same-Sex Desires in Italian and British Sexology, c.1870–1920

Chiara Beccalossi

Postdoctoral Research Fellow, Centre for the History of European Discourses, University of Queensland, Australia

First published 2012 by
PALGRAVE MACMILLAN

Palgrave Macmillan in the UK is an imprint of Macmillan Publishers Limited, registered in England, company number 785998, of Houndmills, Basingstoke, Hampshire RG21 6XS.

Palgrave Macmillan in the US is a division of St Martin's Press LLC, 175 Fifth Avenue, New York, NY 10010.

Palgrave Macmillan is the global academic imprint of the above companies and has companies and representatives throughout the world.

Palgrave® and Macmillan® are registered trademarks in the United States, the United Kingdom, Europe and other countries.

ISBN 978–0–230–23498–7

This book is printed on paper suitable for recycling and made from fully managed and sustained forest sources. Logging, pulping and manufacturing processes are expected to conform to the environmental regulations of the country of origin.

A catalogue record for this book is available from the British Library.

Library of Congress Cataloging-in-Publication Data
Beccalossi, Chiara, 1976–
    Female sexual inversion : same-sex desires in Italian and
    British sexology, c. 1870–1920 / Chiara Beccalossi.
       p.   cm.
    Includes index.
    ISBN 978–0–230–23498–7
    1. Lesbianism—Italy—History. 2. Lesbianism—
    Great Britain— History. I. Title.
    HQ75.6.I8B43 2011
    306.76′630946—dc23   2011029570

10   9   8   7   6   5   4   3   2   1
21 20 19 18 17 16 15 14 13 12

Printed and bound in Great Britain by
CPI Antony Rowe, Chippenham and Eastbourne

To Natalie

Contents

Acknowledgements

I began work on the PhD thesis that forms the basis of this book in 2004 at the Queen Mary University of London, and since then different institutions and many people have offered me their support, help, and advice. This book would not have been possible without the support of the Westfield Trust Scholarship during my PhD. The Central Research Fund of the Institute of Historical Research provided me with financial assistance to undertake archival research in Italy during the second year of my PhD. I have been very fortunate to have Daniel Pick and Julian Jackson as supervisors of my doctoral thesis. I owe special thanks to them for their enduring support and for all the times they have been present when I needed advice. This has meant a lot to me. I am also grateful to the Centre for the History of European Discourses at the University of Queensland, which allowed me to finalise the manuscript of this book in the course of my postdoctoral research fellowship.

Sally Alexander, Valeria Babini, Sean Brady, Ivan Crozier, Thomas Dixon, Colin Jones, Lucy Riall, and Jeffrey Weeks read my work in various forms and at various stages; their comments helped me to rethink and re-examine my project. I am grateful to Palgrave Macmillan and the editors of the Genders and Sexualities in History series, John Arnold, Joanna Bourke, and Sean Brady for their encouragement and patience. Carolina Caliaba Crespo deserves special thanks for her invaluable editing of the final version of this book, although any faults remain my own.

My research in the fragmented Italian archives has been facilitated by the staff of a number of libraries. Although it is impossible to list all the people and institutions across the country who have helped me, special mention must be made of the suggestive Biblioteca Scientifica 'Carlo Livi' in Reggio Emilia: I felt there as though I had entered a nineteenth-century asylum, with the good fortune to be able to enter and leave at will. The Centro di Studi e di Documentazione di Storia della Psichiatria e dell'Emarginazione Sociale 'Gian Franco Minguzzi', Biblioteca Archiginnasio, the Biblioteca della Società Medica Chirurgica in Bologna were all invaluable. The Biblioteca Alessandrina, the Bibioteca Nazionale, and especially the Biblioteca dei Lincei and Santa Maria della Pietà in Rome, the Biblioteca Nazionale in Florence, the Biblioteca Nazionale and Biblioteca Universitaria in Naples were all indispensable resources for this project. In Naples I found the friendliest people. I wish to thank the archivist Lucia Pollio for taking a day off to guide me through the archive of the psychiatric clinic formerly directed by Leonardo Bianchi. I was amazed by the richness of this archive and sad to see it was forced to close its doors to researchers due to lack of funding. The psychiatrist Francesco Catapano assisted me in the puzzling search for the history of Neapolitan psychiatry. Antonio Penta – grandson of Pasquale Penta, on whom Chapter 6 is based – shared his family history with me. I owe them both a debt of gratitude.

My enduring Italian friendships have always made my trips to Italy more pleasant: thank you to Caterina, Daniela, Gabri, Frenz, Ilaria, Luca, Raul, and especially Patrizia, for putting me up every time I needed to come to Bologna. My friends in London, Alex, Alessia, Carrie, Gemma, Matthew, Neil, Miguel, Tracey, and Zoe have been especially precious. All of you have shown me what true friendship is. My new friends in Australia, especially Ryan, have prevented me from losing my patience during the last stage of this book. My parents and my sister Katia have helped me to maintain my sense of perspective and proportion. Above all, you have instilled a sense of commitment.

But it is to Natalie that I owe the most. Thank you for your steady support, patience, and love. Above all, thank you for making me laugh when I thought my head might explode.

Note on Terminology

This work draws mainly on specialised medical literature written and published between c.1870 and 1920. Texts originally printed before 1870 have also been considered when successive editions continued to be used in this period of the nineteenth century, or have been influential in the Italian or the British context. Different editions of the same work, such as Cesare Lombroso's *L'uomo delinquente*, have also been taken into account because medical writers broadened the scope of their research in response to new scientific insights as they became available. Studying the transformation each text underwent with successive editions reveals how each discipline developed. A number of specialised journals have proved particularly useful in the reconstruction of sexological debates: the Italian *Archivio di psichiatria, scienze penali e antropologia criminale* [Archive of Psychiatry, Penal Sciences and Criminal Anthropology], first published in 1880; the *Archivio delle psicopatie sessuali* [Archive of Sexual Psychopathies], published in 1896 and then continued as *Rivista mensile di psichiatria forense, antropologia criminale e scienze affini* [Monthly Journal of Forensic Psychiatry, Criminal Anthropology and Related Sciences] from 1898 and 1904; and the British *Journal of Mental Science*, first published in 1857. In citing the titles of Italian books and articles, I have transcribed the original faithfully rather than adopting the English style of capitalisation. Translations from the Italian are mine unless specified otherwise in the notes.

My book aims to reproduce the language of the authors so as to chart as carefully as possible the debates of the time; hence the use of terms such as 'sexual invert', 'pederast', 'sodomist', 'urning', 'tribade', and 'sapphist'. Occasionally, sexual inversion referred to what we might call today transgender, but broadly speaking, in both Italian and British medical literature this term used to designate same-sex desires or practices. 'Urning' was used as an alternative to sexual invert, especially during the early years of sexological enquiry, whereas 'homosexual' was not commonly employed before the end of the nineteenth century, and it became more and more frequent in the twentieth century. 'Sapphic' and 'lesbian love' were used to refer to female same-sex desires. Unless used by the doctors analysed here, I have avoided the terms 'homosexuality' and 'lesbianism' because they have implicit political values that continue to motivate debates at the core of the current gay and lesbian community. This book is not concerned with such debates about politics of sexual identity.

I use the term 'sexology' to refer to a range of medical disciplines that engage with the study of sexual perversion, ranging from psychiatry to gynaecology. 'Sexology' is a retrospective term used by twentieth-century historians, largely for convenience. Although it can be found in texts published in the second half of the nineteenth century, its use is rare and does not always match modern definitions. For example, it can be traced at least to the 1860s, when the American feminist Elizabeth Willard entitled a book *Sexology*.[1] Willard applied it to refer to the relationship between the sexes rather than to a medical discipline, arguing that there were essential differences between men and women, that these were part of 'natural sexual law', and that if they embraced 'true scientific knowledge', men and women might harmonise their mutual relationships. Willard shared the belief of subsequent medical writers that the organisation of sexual relations had become an important scientific project.[2] By the early 1900s, individuals in fields as diverse as anthropology, biology, psychiatry, and various other medical disciplines had all contributed to the emergence of a specialised medical branch today called sexology. Other than in my concluding remarks, I have not discussed how psychoanalysis dealt with female same-sex desires in Italy and Britain because in both countries this discipline began influencing debates on sexual perversions after World War One. Perhaps the only exception is Pasquale Penta, who acknowledged Sigmund Freud's early works on same-sex desires, and regarded them as a continuation of the work of the French psychologist Alfred Binet. My book is predominantly concerned with information produced within the medical community, even though in the second half of the nineteenth century there were radical thinkers and feminists such as Willard, who engaged with the investigation of sexual knowledge. This book does not discuss them at any length, except when these non-medical experts had a considerable impact on the work of medical writers; John Addington Symonds's co-authoring of Ellis's *Sexual Inversion* is one such example.

Finally, in the second chapter, Italian history is described in more detail than British history as this book is designed for an Anglophone readership that might not be familiar with the history of Italian unification.

Abbreviations

AP	*Archivio di psichiatria, antropologia criminale e scienze penali*
APS	*Archivio delle psicopatie sessuali*
BL	*Papers and Manuscripts*, British Library
BMJ	*British Medical Journal*
JMS	*Journal of Mental Science*
RMPF	*Rivista mensile di psichiatria forense, antropologia criminale e scienze affini*
RSF	*Rivista sperimentale di freniatria e di medicina legale in relazione con l'antropologia e le scienze giuridiche e sociali*

Part I

1
Introduction: Female Sexual Inversion and other Medical Embodiments of Female Same-Sex Desires in Italy and Britain, circa 1870–1920

This woman is a cretin; she is fifty years old, and an inmate in the asylum of Pesaro, a small town in Italy. Her looks are rustic, and her appearance is mannish. She has a dolichocephalous skull, a wide forehead, badly implanted ears, dark skin, atrophic breasts, and abnormal genitals: at four times the normal size and nearly as hard to the touch as cartilage, her left lip is hypertrophic. Her clitoris is larger at the base [1885]. This woman is a spinster; she lives in Liverpool, England. Her body produces too much calcium and as a result her voice is low, man-like. She is also flat-chested. She has been masturbating since puberty, and her health has been bad ever since. The guilt over her own evil ways was so unbearable that her clitoris and labia were excised [1916]. This woman is a pretty middle-class teenage girl; she lives in a boarding school in Padua, Italy. Her face is pale. She is highly strung and has a restless temperament. Her attitude is masculine, and she is self-confident [1898]. This woman is single. She is forty-eight years old, and is an inmate at the Bethlehem asylum in London, England. She has no pubic hair and no breasts. The hands of this woman are like those of a man. She is very tall and broad, and has an overall masculine appearance. Her post-mortem examination revealed her uterus was like that of a child [1878]. This woman is a prostitute; she works in a luxurious brothel in Rome, Italy. Her hair is short, her clothes are fashionable, her preferred sports are typically masculine [1891]. This woman is in business. She is a prominent figure in professional and literary circles in London, England. The conformation of her body is overall feminine. She has a fine intelligence, and is perfectly healthy, but this is not what sets her apart from most women: her medical examination revealed that if she extends her arms before her palms-up, with the inner sides of her hands touching, she cannot bring the inner sides of her forearms together [1897]. These women have a common

3

feature: between the end of the nineteenth and the beginning of the twenti-eth centuries, they were diagnosed as sexual inverts. This book examines the often different but sometimes similar ways in which female same-sex desires were theorised and represented across a wide range of Italian and British medical writings in the 1870–1920 period.

The inception of the new medical category of sexual inversion set off a series of shifts that continue to have indirect effects on contemporary under-standing of human sexuality. This has determined that the story of how and when Western physicians began describing sexual inversion – especially male sexual inversion – has been written many times and from several dif-ferent perspectives, from literary and medical histories to queer studies and the philosophy of science. To a certain extent, it is possible to say that this medical category has enjoyed more success as a basis for new understand-ings of modern sexuality than as a basis for medical history. Although some of my interpretations build on the work of earlier scholars, I have refined the topic with special focus on how the sexual invert was positioned along-side other medical representations of female same-sex desires such as the tribade-prostitute, the *fiamma*, and gynaecological theories about same-sex desires. All of these configurations have at least one feature in common: the notion that women who engaged in same-sex practices did not conform to the sanctioned behaviour for their gender.

One of the limitations informing much of the debate about sexual inver-sion is that, by focusing on certain ruptures introduced by this medical category in the Western perception of same-sex desires, scholars often under-play the internal resistance, contradictions, continuities, and anticipations at the core of medical thought and practice. In doing so, they create an illu-sion of medicine as a monolithic, homogeneous, and stable field. In this work, I have tried to show the extent to which the medicalisation of female same-sex desires in the late nineteenth century was a process of simultan-eous continuity and change; such medicalisation was informed by conflict-ing explanations put forward by competing medical fields, and by different national medical traditions. Drawing too sharp a contrast between a falsely coherent and homogeneous notion of sexual inversion and other ideas of female same-sex desires leads to ignoring how nineteenth-century medi-cal debates operated. In order to understand how medical concepts were constructed, and how they reflected their intellectual milieu, I have endeav-oured to look at medical ideas about same-sex desires through the lens of more than one medical discipline. At times, as with gynaecology and psy-chiatry, these fields competed against each other; at other times, as with psychiatry and criminal anthropology, the fields were allied. Even though in the long term the psychiatric model of sexual inversion became received knowledge beyond medical circles, a careful analysis of medical writings about female same-sex desires shows that this new category was hotly contested at the time of its inception. Such an analysis also reveals that sexual inversion did not replace earlier explanations of same-sex desires, and that these formulations were not discredited as residual ideas from early modernity. Moreover, I have tried to resist portraying nineteenth-century medical writers who engaged in the study of female same-sex desires as either stigmatising agents of sexual non-conformity, or champions of sexual liberation. This dichotomy, which has informed many historical analyses, fails to take into account the intellectual and cultural context of the time, underestimates the controversies around same-sex desires, and ignores the fact that doctors would support one view over another based on considerations of professional interest. These attendant factors have framed my analysis. Although the conclusions of my historical research are consequently more revisionist than I had originally envisaged, my main aim is not to debunk old historiographical myths, but to bring to light new historical evidence on the subject of female same-sex desires. Contrary to widespread assumptions, late nineteenth-century physicians formulated detailed theories of love and sexual acts between women. Rather than mirroring more sophisticated explanations of male same-sex desires, medical ideas of female same-sex desires had their own rich narratives, and their own multilayered history.

The emergence of the medical category of sexual inversion: continuities and variations

By the mid-nineteenth century, a few medical writers including the French physician Claude-François Michéa and the German forensic doctor Johann Ludwig Casper had presented analyses of sodomy in which they noted that the preference for members of the same sex could be innate; Casper explicitly associated such inclinations in men with effeminacy.[1] This notion was more comprehensively formulated almost twenty years later: in 1868, a leading German psychiatrist practising at the Charité psychiatric clinic in Berlin, Wilhelm Griesinger, associated sexual desire for one's own sex with congenital 'neuropathic' conditions.[2] Griesinger believed that brain lesions caused psychiatric problems, some of which might be manifest only as behavioural problems such as sexual perversion. He stressed that such psychiatric problems often expressed themselves only in the psychological realm, and did not interfere with the individual's ability to follow logical processes in their daily life. A doctor, Griesinger noted, could easily underestimate the anomalous nature of these forms of neuropathy, with their 'abnormal vagaries, instincts, drives, desires'.[3] In the following year Carl Westphal, who was Griesinger's student and his successor at the Charité clinic, published the first psychiatric article dealing with what he called *conträre Sexualempfindung* [contrary sexual feeling]. In this article he reported two case-studies, the more important and influential of which was based on a woman referred to as N.[4] Although her examination revealed she had normal genitals, N. had desired other women from an early age, showed aversion for men, and held the 'feeling of having a man's nature'.[5] In 1877, Richard von Krafft-Ebing published an important article in the *Archiv für Psychiatrie und Nervenkrankheiten* in which he explained *conträre Sexualempfindung* as a 'functional sign of degeneration'. This would go on to become the dominant psychiatric view of sexual inversion until the 1890s, paving the way for further sexological studies on various sexual deviations.[6] The German term *conträre Sexualempfindung* was subsequently translated as *inversione sessuale* [sexual inversion] by the Italian forensic doctor Arrigo Tamassia in 1878; as *inversion sexuelle* by the French neurologist Jean-Martin Charcot and psychologist Valentin Magnan in 1882; and as sexual inversion in a series of English and American articles published in the early 1880s.[7] Influenced by Westphal's text, Continental doctors began compiling case-studies of sexual inverts whose main psychological characteristics were gender-inverted behaviour and sexual longings for individuals of the same sex. Most of these (male) case-histories remarked that the patients' genitalia were normal and bore no marks of same-sex practices. By the 1890s, well-established psychiatrists on both sides of the Atlantic had been addressing the topic of sexual inversion not only in articles in important medical journals, but also in entire monographs.[8]

Both late nineteenth-century physicians and twentieth-century historians agree that the introduction of sexual inversion to psychiatry played a pivotal role in the discussion of homosexuality because it posed the question of whether same-sex desires were pathological. In 1976, Michel Foucault famously drew attention to the different styles of sexual prohibitions typical of pre-modern legal codes and the nineteenth-century psychiatric approaches to sexual perversion, pointing to the emergence of sexual inversion in the 1870s. During this period, Foucault goes on to say, the homosexual became 'a personage: a past, a history, and a childhood, a character, a type of life; also a morphology, with an indiscreet anatomy and possibly a mysterious physiology'.[9] Foucault's historical observation on the emergence of homosexuality as a disposition, a psychological characteristic, was not unprecedented. In 1968, the sociologist Mary McIntosh published a pioneering work, 'The Homosexual Role', in which she argued that modern conceptions of homosexuality are not identical to earlier understandings of same-sex acts; she pointed out that the concept of homosexuality as both a 'condition' and as a 'type' emerged in Western culture at the end of the seventeenth century.[10] In the same years that Foucault drew attention to the emergence in the nineteenth century of sexual inversion as a category of psychopathology, scholars such as Vern L. Bullough and Jeffrey Weeks began to describe how the modern medical model of homosexuality was constructed in Western culture.[11]

From the 1980s onwards a plethora of studies have offered a more precise account of how homosexuality came to be regarded as a psychological condition. Building on Foucault's insight, Arnold Davidson has suggested that homosexuality's becoming a disease in the nineteenth century is in fact part of a major epistemological rupture in Western thought. His sophisticated framework sheds light on the conditions that made the emergence of the modern concept of sexuality possible, which came to be increasingly considered as a personality trait and a mode of sensibility. Sexology had a decisive role in this process.[12] Davidson discerns three stages in the history of sexual perversion, each of which is characterised by a different supposed cause of the disease: genitalia, brain, or sexual instinct. 'In the first stage, sexual perversion was thought to be a disease of the reproductive or genital organs.'[13] In the second stage, sexual perversion came to be understood in terms of neurophysiology and the anatomy of the brain. The genital and cerebral explanations of sexual perversion, he synthesises, shared a 'commitment to the anatomo-pathological style of reasoning'.[14] In the third stage, sexual perversion was held to be the result of functional deviations of the sexual instinct, and therefore not reducible to any cerebral or organic pathology. In this phase, perversion was viewed and treated at the level of psychology, 'not at the grander level of pathological anatomy'.[15] This shift represents a major change in the history of sexual perversion as a disease category because an 'anatomical mode of reasoning' was replaced by a psychologically based 'psychiatric style of reasoning which conceptualized new symptoms related to drives, inner states, and consciousness'.[16] Thus, the idea that knowing a person's sexuality was a way of knowing that person became available only with the rise of the psychiatric style of reasoning. According to Davidson, it was only from the late nineteenth century onwards that homosexuality came to identify an individual's personal characteristics rather than a mere sexual behaviour.[17]

Davidson claimed that pathological anatomy did not substantially influence the clinical description of sexual perversion.[18] As the second part of this book shows, accounts of female same-sex desires present a different picture. While anatomical theories might not have hugely influenced Continental sexological literature on male same-sex desires, British medical and gynaecological approaches to female sexual perversion continued to rely on anatomical explanations. When nineteenth-century British gynaecologists encountered cases of women who indulged in same-sex pleasures, they identified the cause of those sexual desires in an abnormal reproductive system or genitalia. These anatomical explanations of female same-sex desires were so powerful in the British medical community that a few British psychiatrists such as George Savage embraced bodily oriented arguments when discussing the topic. Although to a lesser extent than in Britain, anatomical explanations were also adopted in Italy. Most Italian psychiatrists remarked on sexual inverts' normal genitalia and searched for the origin of same-sex desires in the whole individual's degenerate constitution. However, some of these doctors, including Lombroso, resorted to clitoridectomies to contain the spread of female same-sex practices in asylums. Therefore, in medical treatments, same-sex desires were believed to have a genital origin. While it is certainly true that Italian psychiatry was committed to psychological interpretations of female same-sex desires, anatomical explanations continued to be a part of daily practice towards the end of the nineteenth century and beyond. In part, the popularity of these anatomical explanations derived from earlier medical formulations of female same-sex desires. Before the concept of sexual inversion emerged as such, Italian and British doctors believed female same-sex desires were caused by an enlarged clitoris. The figures of the tribade and the virago, with their excessive sexual desires and abnormal clitorises that could replace a penis, continued to circulate widely in medical writings and did not disappear when the sexual invert entered the scene.

While I recognise the great merit of Davidson's analytical framework, the tension between continuity and change in the medical representations of female same-sex desires is central to my analysis. Old medical explanations of female same-sex desires persisted throughout the nineteenth and well into the twentieth centuries, and contributed to shape the idea of sexual inversion itself. I do not wish to imply that there were no shifts in the perception of same-sex desires at the end of nineteenth century. By the 1890s, when the concept of sexual inversion became accessible to non-medical audiences, it had become a powerful notion that changed the meaning, subjectivity, and life experience of many individuals, and helped shape the modern idea of homosexuality. Nevertheless, as many historians whose research draws on empirical evidence have recently pointed out, modern homosexual identity was shaped by several factors – not just by the medical conceptualisation of sexual deviance.[19] These other social and cultural factors were class, gender, religion, ethnicity, and a growing psychological understanding of the self related to the rise of Romanticism. However important the interaction of these factors may have been, this book does not focus on how the modern lesbian identity emerged, or how medical theories might have influenced this process. Such investigation would require measuring the impact of nineteenth-century medical theories on the lives of women, rather than an analysis of medical ideas of female same-sex desires.

My preoccupation with continuity in the midst of change in cultural meanings is shared by many historians, including those who have contributed to the history of sexuality. Robert Nye has noted the suitability of Davidson's model for understanding the shifting concepts of sexual perversion in medical theories while pointing out the inadequacies of Foucauldian-informed historical accounts that favour discursive ruptures. On the whole, Nye agrees with Davidson's description of the conceptual changes in thinking about sexual perversion in psychiatry. In pointing out some of its shortcomings, however, Nye explains that Davidson's tri-partition does not easily fit the French situation, where the older anatomical style coexisted with the new psychiatric style of reasoning for many years. Moreover, according to Nye, undue emphasis on discontinuities obscures much that anticipates modern notions of sexuality, which continue to mobilise some of the language and deterministic implications of pathological anatomy.[20]

While Nye has addressed his critique from the perspective of a historian of medicine, the queer theorist Eve Kosofsky Sedgwick argued against the assumption that one model of understanding same-sex desires simply disappears when another is invented. She also noted that it is equally erroneous to conceive of these modes as a linear, chronologically ordered, sequence.[21] Regardless of such caveats, studies of same-sex desires in various periods of Western thought continue to emphasise the role of conceptual ruptures, which have grown to become a largely unexamined axiom. Scholars interested in the history of homosexuality routinely talk about changing paradigms, and it is not infrequent to come across contrasts between the 'acts paradigm' and the 'identity paradigm', or the 'paradigm of bodily structure' and the 'paradigm of desire'. Such a distinction is taken to coincide with a shift between the concept of sodomy as a criminal act, and homosexuality as a psychological type – the latter initiated by the introduction of the concept of sexual inversion into psychiatry.[22] Yet, medical accounts of same-sex desires are more complicated than such a clear-cut narrative of replacement of paradigms suggests.

For more than a decade David Halperin has been studying how the appearance of sexual inversion changed discourses on same-sex desires; while he originally emphasised ruptures, his later work offers a more nuanced balance between change and continuity.[23] Inspired by Sedgwick's critique, Halperin has rightly pointed out how the 'tension between interpretative emphases on continuity and discontinuity […] appears with almost painful intensity in the historiography of homosexuality'.[24] After re-examining the role of continuity and change in explanations of same-sex desires throughout history, Halperin offered four transhistorical explanatory models of same-sex desires: effeminacy, pederasty or active sodomy, friendship or male love, and passivity or inversion.[25] According to Halperin, homosexuality as we understand it today is a conjunction of these different models. Although my work has benefited from Halperin's most recent insights, I do not aim to identify transhistorical medical taxonomies of female same-sex desires. Nor do I intend to look for precursors to sexual inversion in fundamentally different epistemological domains, because this would inevitably lead to anachronism. This book is concerned with developments in a number of medical specialities, which rearticulate same-sex desires differently. I am concerned with how different knowledges managed to coexist within medicine, which was not a unified system of knowledge.[26] Wary of seductive theories, I wish to show the historical trajectories of medical ideas of female same-sex desires within their traditions and within a broader intellectual framework. Perhaps this will make future accounts of the topic mindful of historical detail, and thus more rigorous; perhaps this will indirectly further the debate and rethinking of the history of homosexuality. If it does, it will have exceeded its original goal, which is to map the multiple meanings of female same-sex desires within medicine. This book is, above all, an intellectual history of nineteenth-century medicine.

National traditions and competing medical fields

At the turn of the century, medical sexual knowledge circulated easily in Europe and was increasingly international in many aspects, yet it was formulated differently in each country. Historians who have explained the developments of sexology have relied heavily on the internal dynamics of national history. In doing so, they have shown that medical sexual knowledge responded to local concerns. Nye, for example, has explained the rise of French sexology in terms of the French government's demographic preoccupations during the last decades of the nineteenth century.[27] After the Franco-Prussian War, anxieties about a declining birth-rate led French sexologists to cast all perversions – especially homosexuality – as deviations from, and threats to, heterosexual norms that needed to be bolstered as a matter of national urgency.[28] Harry Oosterhuis has explained the rise of German sexology in terms of the emergence of a new homosexual identity.[29] Sexology in Germany and Austria was increasingly associated with movements of sexual reform, which were chiefly aimed at abolishing or revising laws against male same-sex acts.[30] Sexology in Britain combined a detailed assessment of European and American sexology with the political motivations of homosexual rights activists such as Edward Carpenter and John Addington Symonds; the work of the well-known sex psychologist Havelock Ellis illustrates this very clearly. Lesley Hall has argued that British sexological concerns were a response to the rise of feminism,[31] but Ivan Crozier's research convincingly shows that British sexological research developed out of traditional psychiatric concerns about moral insanity and degeneration.[32] In this book, I explain how Italian sexology grew out of criminal anthropology, political debates, and a strong anticlerical sentiment that ensued after the unification of Italy, with special emphasis on how Italian sexology developed a specific interest in female sexual deviation.

In this work I look at how two distinct medical communities articulated ideas of female same-sex desires, and how such formulations diverged and converged: while British medical writers were reluctant to endorse the Continental psychiatric model of sexual inversion, Britain and Italy represented female same-sex desires in much the same way, through the figures of the tribade-prostitute, the virago or masculine woman, and the *fiamma* or adolescent female same-sex crushes in school environments. These similarities in part depended on common readings and the influence of early nineteenth-century medical authors such as Alexandre Jean Baptiste Parent-Duchâtelet, who had written on the topic of love between women. The diverging trajectories of Italian and British medical writings, on the other hand, are explained by reference to the historical contingencies in which the two medical communities were embedded in the last decade of the nineteenth century. Following the unification of Italy, Italian physicians were committed to secularise Italian culture and thus had a relevant cultural role in the *post-Risorgimento*. When the time came to draft the civil and penal codes for the new kingdom, legal and scientific experts engaged in high-profile debates about the roles of family and of women in society, and what form of national law to adopt to regulate sexual behaviours that had been previously punished or allowed under different legal codes that followed the traditions of various foreign countries. The political urgency of these debates and the prominent cultural role of Italian scientists in the process of secularisation fostered a great number of studies on female sexual deviancy.

One of the aspects that makes the Italian and the British cases stand apart in their historical and political contexts is that sexological works such as Ellis's aimed to discredit laws against male same-sex acts on the grounds that they were not supported by scientific evidence. More traditional medical writers who engaged with the study of same-sex desires had been forced to be very cautious when addressing the topic in order to avoid conflicts with British law. Under the weight of legal constraints, it was easier for British physicians to regard same-sex desires as an immoral conduct, than to endorse Continental psychiatric theories that claimed homosexuality was an inborn condition. The discrepancies between the two countries can also be explained through the impact of different medical disciplines. Thus, at the turn of the century, criminal anthropology was a leading discipline, and one of the main proponents of the study of sexual psychopathologies in Italy. By this stage, however, the British medical community had grown sceptical about some of the deterministic assumptions favoured by Lombroso's school in its approach to deviant behaviours, and therefore dismissed its study of sexuality altogether.

Historians have often assumed that although national factors influenced the development of sexology in different countries, by the end of the 1890s the Continental psychiatric understanding of sexual inversion dominated most Western medical literature. The underlying assumption is that the process that led to different national communities and medical fields taking up this category must have been similar. Yet, at the turn of the twentieth century, medical thought around same-sex desires was complicated and contested by trends of professionalisation, national traditions, and the impact of new methods of enquiry. Since the inception of sexual inversion, this concept was rearticulated differently in distinct medical disciplines. The increasing specialisation of medicine in the nineteenth century resulted in the development of various disciplines that competed to extend their disciplinary domain over abnormal sexual behaviours. Certainly, these medical specialities had their own interest in supporting one view of sexuality over another, but different views also depended on how a practitioner approached first the diseased body, and then the sexual body, so to speak. Different views of sexuality rested on how a writer approached the body and its sexual function. Thus, a practitioner working in public hygiene was concerned with how the environment could affect the body and with the social dimension of the disease, and sexual behaviour. Gynaecologists argued that diseased ovaries could result in injuries to the nervous system, and ultimately lead to mental illnesses or abnormal sexual behaviours; psychiatrists, meanwhile, reversed this way of thinking. In other words, each medical discipline framed and, to a large extent, determined its practitioners' perspectives on sexuality. As I will illustrate throughout this book, different disciplines approached same-sex desires in various ways, and once psychiatry introduced the category of sexual inversion, competing disciplines redefined the subject to the point that, at the beginning of the twentieth century, a psychiatrist and a gynaecologist might have been looking at two different things even though they both used the term sexual inversion.

Differences amongst disciplines aside, clinical observations were complicated by local and historical circumstances. While the psychological dimension of sexual inversion made it fashionable in Continental psychiatric circles, British medicine was reluctant to adopt certain Continental psychiatric ideas, and continued to engage with sexual perversion in anatomical terms. It is not a coincidence, then, that British gynaecologists managed to influence even some British psychiatrists in their understanding of sexual behaviour. Bodily oriented explanations continued to be adopted in the early twentieth century, when British endocrinological theories were still denying there was a psychological explanation for same-sex desires. Even within the Italian scene psychological explanations of same-sex desires were contested, but in a different way. A number of criminal anthropological works, which took up Parent-Duchâtelet's social and cultural analysis of the phenomenon of love relationships amongst female prostitutes, explained same-sex practices in terms of environmental influence. Much late nineteenth-century medical literature proposes that rather than psychological factors, same-sex desires emerged as a consequence of the social context of prostitutes, prisoners, and even students living in boarding schools.

Narratives about the emergence of a homogeneous medical model of homosexuality, then, become more difficult to sustain when trying to pinpoint how national contexts and competing medical fields shaped medical texts. One of the reasons I decided to go beyond the boundaries of a single nation was practical and biographical in more than one way. When I began researching the topic, I was struck by what I perceived as a historiographical impasse in studies on female same-sex desires. Despite there being much literature on male sexual inversion, scholars working in the British context in particular have emphasised that in Victorian times medicine rarely addressed female same-sex desires.[33] Yet it was enough to leaf through one of the major Italian psychiatric journals, the *Archivio di psichiatria*, to conclude that this was not the case in Italy.[34] My research, then, was prompted by a single question: why were accounts of female same-sex desires so common in Italy but almost unknown in Britain? Historians who work on late-modern and contemporary subjects often remain within the boundaries of the nation-state, whereas early modernists effortlessly transcend national boundaries in their research. The mere fact that such a methodological division is in place confirms that the formation of nation-states in the late eighteenth century is a historiographical milestone. There are further practical considerations to take into account: early modernists cross national boundaries in an attempt to bring new primary sources to light. Having identified a number of sources in a specific geographical and temporal location, early modern historians favour comparative approaches so as to interrogate sources in a new way, to follow new paths, and to discover new sources when there might seem to be none in specific geographical areas. In the same way, I found that for an historian interested in the meanings and values attached to female same-sex desires, transcending national borders could be very rewarding, particularly when tracing new lines of enquiry and identifying sources where, at first glance, there was nothing. For example, in practical terms it is true that medical writings on the subject of love between women were not as abundant in Britain as they were in Italy. Yet, female same-sex practices were considered a deviant sexual behaviour in both countries. Conventionally, in mid-nineteenth-century psychiatry, deviant sexual behaviour implied perverted and unmanageable instincts, and these were found in mental disorders such as monomania, moral insanity, and degenerative disorders. It is amongst these disorders that the British provided examples and explanations of female same-sex desires. The difference between British and Italian psychiatrists was that the British did not generally adopt the term sexual inversion, and were very cautious, brief, or resorted to circumlocutions when addressing the issue of same-sex desires, whereas Italian physicians tended to be explicit and indulge in almost voyeuristic details. Similarly, in both Britain and Italy, debates on prostitution or women-only environments included analysis of female same-sex desires.

Working on two different national contexts helped me single out possible common sources and lines of enquiry, and look at well-known sources in a new light.[35] Moreover, looking at more than one national context has also enabled me to see how medical knowledge about female same-sex desires circulated amongst countries. For instance, a lot of British research, including Symonds's and Ellis's *Sexual Inversion*, grew out of an interest in Italian sexological knowledge. Italian and British periodicals of the time bear evidence that there was an active exchange of ideas between Britain and the Continent, and between Italy and the rest of Europe; in turn, this has led me to question whether and how different histories and models of scientific communities shaped diverging discourses on sexual perversion.

The hero/villain dichotomy

Drawing on past medical authorities to justify their enquiry into sexual perversion and to present it as a plausible enterprise in the wider field of medicine, sexologists such as Ellis fashioned themselves as champions of reform and treated the cause of sexology as the cause of sexual emancipation. Such narratives of progress were popular throughout the first half of the twentieth century because historians interpreted sexology – and later psychoanalysis – as intellectual achievements that helped liberate Western thought from Victorian repressive attitudes towards sex.[36] In the second part of the twentieth century, however, these narratives of scientific progress started to be challenged: Foucault famously interpreted the emergence of sexology as a determining factor in the imposition of standards of normalcy whereby new ways of regulating sexual behaviour replaced the authority of religion and the law.[37] From at least the 1970s, feminist historians have questioned the concept of value-free and neutral science, and showed how 'objective' findings were tarnished by the scientist's biases. Medical theories of an organic female nature have been interpreted as attempts to preserve the social status quo.[38] Numerous feminist historians have argued that sexology pathologised female same-sex relationships, created a climate of opinion that stigmatised single women and their relationships, and favoured heteronormativity.[39] Some feminist writers have also argued that nineteenth-century sexologists reinforced the idea of a clear boundary between 'normal' and 'abnormal' sexual behaviour, and have charged them with undermining feminism and female homosexuality.[40] In more recent years there have been substantial challenges to the assumption that sexologists contributed to stigmatise sexual minorities, and a few historians have warned against simply labelling sexological descriptions as misogynistic. Laura Doan, for instance, has suggested that for lesbians in particular, the blurring of categories of gender and the greater dissemination of sexual knowledge made possible new frameworks for self-understanding, which in turn paved the way for the formation of subcultures.[41]

In this work I have tried to overcome the dichotomy that presents medical writers as either heroes of sexual liberation or villainous agents of a culture that discriminated against same-sex desires. In nineteenth-century medical thought, sexual normality and pathology were notions that were more fluid than is generally acknowledged. While many nineteenth-century medical writers conceived of same-sex desires as cause and symptom of various diseases – especially mental disorders – or as a form of insanity in itself, such views were not uncontested. Especially in the context of European psychiatry, same-sex desires defied a stable categorisation and could be located across a complex and unsteady borderline between normal and pathological states. In my work I have endeavoured to trace the intellectual developments of such ideas, how such texts and theories were created, their intellectual debts and provenance, and not least, the wider medical debates in which they participated.

Depending on who the practitioners were, and on their professional, cultural, and intellectual context, sexology itself became a conservative or a radical endeavour. While nineteenth-century medicine had an established tradition of addressing problematic desires, with the rise of sexology same-sex longings became a central issue as sexologists started questioning the assumption that only reproductive sexual behaviour was normal. This, in turn, brought abnormal sexual behaviours to the fore. The third part of this book in particular shows how sexological research could serve a number of different professional and political purposes. My study of Italian and British sexology is organised around four medical writers, whose statements about same-sex desires and sexuality in general mark them as either 'conservative' or 'progressive' in regards to their own historical and cultural context. Cesare Lombroso's and William Blair-Bell's views on sexual inclinations might be interpreted as conservative, yet their beliefs were not monolithic and some aspects of their research were quite innovative in the fields in which they emerged. Pasquale Penta and Havelock Ellis, on the other hand, advanced radical positions and faced professional ostracism for their views on sexuality, yet some of their writings present a certain degree of ambiguity. The careers of these scientists also illustrate the trends that shaped their respective countries' medical fields. The eminent psychiatrist Lombroso was the founder and leader of criminal anthropology, which had a significant influence in the Italian medical community, whereas the psychiatrist Penta eventually specialised in sexology, establishing international networks and founding the first European sexological journal. Trained as a doctor, Ellis was an outsider to the traditional medical community in many ways, and British physicians' reactions to his work on sexual inversion tell much about the British rejection of Continental ideas. The famous gynaecologist Bell is representative of a field that influenced other British medical disciplines in shaping ideas of female same-sex desires.

The (in)visibility of female same-sex desires

While psychiatrists did not seem to encounter any shortage of case-studies of men, female inverts appeared more difficult to find, to the extent that in 1897 Ellis could still write that physicians knew 'comparatively little of sexual inversion in [women;] of the total number of recorded cases of this abnormality, now very considerable, but a small proportion are in women, and the chief monographs on the subject devote but little space to women'.[42] Indeed, some of the most important monographs on the topic confirm Ellis's observation: Krafft-Ebing's seventh edition of *Psychopathia Sexualis* (1892) contained forty-one male case-studies, but only eleven female case-studies.[43] Likewise, Ellis's *Sexual Inversion* contained twenty-seven male case-studies, in contrast to four female case-studies.[44] According to late nineteenth-century sexologists, female homosexuality remained much less known than male homosexuality due to the scarcity of case-histories on women. Historians have tended to take this remark as true, and have thus focused mainly on male sexual inversion; it is currently believed that almost all medical texts across a variety of specialisations ignored female same-sex desires.[45] Yet, as this book shows, this perception is wholly misguided, and derives more from a gap in the historical enquiry than from hard evidence. Such a misleading assumption has also been reinforced by a strong trend in lesbian studies to denounce male culture's attempts to refuse visibility to female same-sex desires.[46]

As this book demonstrates, late nineteenth-century medical writers' theories about female same-sex desires were fully articulated. While in the British context there were only a few case-histories of female sexual inverts, Italian medical journals regularly published them. But it is only when we move our attention away from the female sexual invert that we are struck by the complex and multiple representations of female same-sex desires. The female invert was positioned alongside other medical figures of same-sex desires such as the tribade-prostitute, the *fiamma*, the nymphomaniac, and women with abnormal genitalia or bodily dysfunctions. These figures might overlap, so it was possible for a physician to describe a tribade-prostitute who was also an invert, or to discuss a nymphomaniac with abnormal genitals who had sexual relations with women, or a sexually inverted *fiamma*. By the end of the nineteenth century, diagnosing a person with sexual inversion meant the physician would have been influenced by a combination of socio-cultural, physical, and psychological approaches to same-sex desires. Sexual inversion had become multilayered, and at the same time multiple images embodied different ideas of female same-sex desires. And yet, Italian and British medical communities remained widely different: while British physicians were cautious in approaching the topic, proceeded by circumlocutions and adopted an almost coded language, in Italy, female same-sex desires captured doctors' imaginations, and consequently studies on the topic multiplied. Regardless of the mode of engagement, one thing is clear: female same-sex desires were a complex reality, and a much debated topic in medical writings.

Despite the multiple guises of female same-sex desires unveiled in various medical texts, this book is a history of male medical perceptions of same-sex desires, rather than a work on different forms of love between women. Whatever else it may be, this book is not a lesbian history in the sense of uncovering women's homoerotic experiences; here, love between women is an object of medical knowledge. Medical case-histories certainly reveal aspects of the lives of real women; likewise, medical accounts of same-sex desires might betray cultural anxieties, biases, and the expected role of women in society. Medical accounts therefore represent just one of many possible starting points for a lesbian or queer history, as shown by recent brilliant readings of medical records in early modern and modern literary studies.[47] In the nineteenth century, medical texts were written by male physicians for other physicians, quite often serving professional purposes and according to medical codes and rules. Although a few of these treatises fell in women's hands, the authors did not intend these texts for a female audience.

It has been impossible to analyse female sexual inversion without unpacking medical theories of their male counterpart, because although female same-sex desires were not simply the mirror of the male invert, both formulations share their conceptual foundations. Ultimately, the multiple embodiments of female same-sex desires in different medical communities in the late nineteenth and early twentieth centuries urge historians to interrogate whether such accounts are specific to female same-sex desires. However, such an investigation would require a detailed account of medical representations of male same-sex desires, which is beyond the scope of this book.

In this history, I favoured a close and detailed reading of the texts in order to preserve the richness and complexity of the medical debates, and to recover various representations of female same-sex desires. Nonetheless, I have endeavoured to show the broader context in which such representations were situated; how the roles of institutions and associations, certain medical fields' ongoing struggle to secure professional prestige, and how editorial policies of journals had an impact on the way physicians spoke about female same-sex desires. This awareness of the intellectual milieu in turn made it possible to trace the trajectories of specific ideas in the work of individual thinkers. This, then, is a history of ideas that pays attention to the environment that nurtured them.

2
Sexuality in *Post-Risorgimento* Italy and Victorian Britain

The list of international contributors to the pioneering Italian sexological journal *Archivio delle psicopatie sessuali* featured, amongst others, the British sexologist Ellis and the Franco-Russian poet and journalist Marc-André Raffalovich.[1] Raffalovich had moved to Britain in the 1880s and, although he was not medically trained, he began writing on sexual inversion soon after. Unlike his literary works, which were written in English and printed in his new country, his material on sexology was never published in Britain. Rather, it had been published in French in the *Archives de l'Anthropologie Criminelle* [*Archives of Criminal Anthropology*] since the early 1890s, and in a monograph, *Uranisme et Unisexualité* [*Uranism and Unisexuality*], in 1895. In this monograph he argued that male sexual inverts who maintained their masculine characteristics were normal rather than diseased individuals; while it has never been translated into English, *Uranisme et Unisexualité* was promptly translated into Italian and reviewed in medical journals.[2] In 1896 he published an article in the *Archivio delle psicopatie sessuali* about the current state of research on psychopathologies in England. Raffalovich lamented how difficult it was to write about such a subject in Britain, and castigated the British medical community for the backwardness of its research on same-sex desires. Raffalovich denounced the hypocrisy of British physicians when handling sexual matters, pointing out that their scientific community lagged behind the Italian in its investigation into human sexual instinct. Unlike their British colleagues, who were portrayed as 'fearful ignorants and liars', Italian physicians working in the field of sexology were open minded and published extensively.[3] Raffalovich explained that the British 'caste-mentality' did not exist in the Mediterranean because social hierarchies were much less rigid than in Britain, where the middle classes had to maintain a level of (sexual) respectability. Therefore, it was easier for Italian physicians to debate the topic openly, than it was for their British colleagues. Raffalovich remarked that even rich and educated Englishmen who visited the Peninsula would perceive how much more relaxed the attitudes towards sexual matters were, and would take pleasure in meeting compliant fishermen, gondoliers, facchini, cocchieri, urchins, sailors and boulevard boys.[4]

Today's reader might find it strange to compare Britain and Italy; after all, were the Germans and the French not closer to Britain – at least geographically – than the Italians? Its economical backwardness, its unstable political situation, and its subjection to the Catholic Church are only some of the factors that mark Italy as the opposite of Britain. What, then, invited Raffalovich to make such a comparison? While Britain perhaps differed from Continental countries such as Germany and France in its approach to sexual matters, it was certainly dissimilar to Italy. To the intellectuals who reflected on the conditions of those who loved their own sex in either country, this contrast served to illuminate the differences between Italy and Britain.[5] Like Raffalovich, the literary scholar John Addington Symonds had compared the state of Italian and British medical studies on sexual inversion; he also had found that cultural differences in dealing with sexual matters enabled Italian physicians to speak about the topic more easily than in his own country.[6] Observers were certainly struck by the evident difference in Italian and British sexual customs, but this was not the only reason why British intellectuals of the Victorian period turned to Italy.

In the late nineteenth century, Italy was the favourite destination of British men looking for sexual intercourse with other men. Like Symonds, the writer Frederick Rolfe visited Venice regularly because there he could buy the companionship of gondoliers and sailors.[7] On the whole, however, the British were particularly taken by the southern parts of Italy. Oscar Wilde took pleasure in describing gigolos and adolescents who sold themselves in the streets of Naples.[8] This sexual tourism was silently accepted by the local population because it entailed financial advantages, both for the male prostitutes and the community as a whole. Even British female travellers were drawn to the charms and possibilities of the Peninsula. In Capri there was a small but renowned homosexual community made up of both male and female homosexuals from around the world.[9] In the middle of the century, the well-known American actress Charlotte Cushman travelled to Rome with her partner Matilda Hays and a group of friends, where they established a woman-centred community that stressed female solidarity and emancipation from men. From 1852 to 1868, her salon of friends and lovers attracted a number of unconventional American – and English – women.[10]

This homosexual tourism is part of a broader trend of reciprocal exchanges between Britain and Italy. Since the Renaissance, Italian culture had periodically left an impression on British, and especially English, life. English scholars travelled to Italy to study under the supervision of Italian humanists, and Italian scholars, merchants, and diplomats who went to England brought with them new learning that had repercussions on nearly every aspect of the nation's life – manners, commerce, scholarship, music, literature, art. During the Renaissance, the most important physicians and musicians working in London were Italian. This direct influence was in decline by the middle of the seventeenth century, but Italy continued to be a meaningful presence in the cultural life of the upper classes, who regularly visited Italy in the eighteenth and nineteenth centuries to round off their education, in the context of a Grand Tour.[11]

At the beginning of the nineteenth century Italy regained its influence, which has led the historian Peter Brand to talk about an 'Italo-mania' between 1800 and 1840.[12] In early nineteenth-century Britain, knowledge of Italian and some acquaintance with Italian literature were considered normal and even necessary, not only for scholars and poets, but also for well-educated young men and women from the upper classes. A journey to Italy was deemed an appropriate conclusion for the education of a young British man of good family, while an accomplished young lady was expected to be able to sing in Italian. Italy shaped the artistic tastes of Britain, especially in England, where families of good standing would customarily own a box at the Italian Opera. Painters based their education on Raphael, Michelangelo, and Titian; sculptors and architects were trained to know and admire the Roman and Italian Renaissance styles, and they would often base their work on the old masters.[13] While such Italo-mania reached its peak in the first forty years of the nineteenth century, in the subsequent decades Italy did not cease to occupy an important place in the minds of the educated British. Indeed, British intellectuals continued to be seduced by Italy well into the twentieth century. The novelist Edward Morgan Forster had travelled to Italy after graduating from Cambridge at the beginning of the twentieth century, and Italy remained one of his favourite landscapes for his novels such as *Where Angels Fear to Tread* (1905) and *A Room with a View* (1908).[14] The English historian George Macaulay Trevelyan wrote a number of books on Italian history, most notably his trilogy on Giuseppe Garibaldi, which was published in the first decade of the twentieth century.[15]

The British also showed a vivid interest in Italian politics throughout the nineteenth century. The outbreak of the French Revolution was met with fears that a similar event might take place in Britain, and the expansion of Napoleon's power made British elites take an active interest in the Italian political situation. If the expansion of Napoleon was going to be contained, it was important to stop France from descending onto the Peninsula. By defeating Napoleon's fleet at the Battle of the Nile (1798) and acquiring the island of Malta, British naval power increasingly played a key role in the Mediterranean, a region that had assumed critical strategic importance because it kept open the lines of communication with Britain's rapidly expanding empire in the east. British governments monitored the political situation in the unstable and fragmented Peninsula both in the interest of the empire, and of the burgeoning trade with Italian states.[16] Britain followed the Italian *Risorgimento* closely: by the 1840s, British public opinion was supportive of Italian liberals and nationalists, and in the 1860s it greeted the unification with enthusiasm. The British government was indeed the first to officially recognise the new kingdom.[17]

Britain gave asylum to eminent Italian exiles such as the renowned poet Ugo Foscolo, the lawyer Antonio Panizzi, the poet Gabriele Rossetti, and the economist Giuseppe Pecchio. The first wave of émigrés arrived in London in 1815; they were later joined by growing numbers of Italians who left the Peninsula following the local disorders of 1820–1 and 1831. These exiles had an important role in nourishing not only the British interest in Italian culture and politics, but also in contributing to European anglophilia.[18] In 1837, the well-known politician Giuseppe Mazzini moved to London after living in France and Switzerland. During his stay in England, Mazzini founded several organisations aimed at liberating Italy and other nations from foreign domination, which in practice meant he was running a radical network with links across Europe and America. Garibaldi aroused strong enthusiasm in Victorian England, and London received him as a hero in 1864.

As people travelled from Italy to Britain and vice versa, ideas and ideals crossed cultural boundaries, and although different in numerous aspects, the two countries shared many of the sexual values typical of nineteenth-century society. Despite Raffalovich's and Symonds's comments on the contrasting attitudes towards male homosexuality in Italy and Britain, in both countries the family was a symbol of order and stability, and the foundation of social order and the state; its role in social control was consequently crucial. The middle classes were equally mindful of sexual respectability in Victorian Britain and in the Italian *post-Risorgimento*. Although official discourses condemned extra-marital sexuality, men's transgressions were tacitly justified on the grounds that their natural sexual needs drove them to look for sexual intercourse outside marriage. Men who consorted with prostitutes and mistresses were silently tolerated, whereas women who satisfied their sexual desires outside marriage were harshly criticised. From labour to sexuality, all aspects of female and male experience were divided into gender-specific spheres: in both countries, medical-scientific theories posited the naturalness of this division by arguing that woman's passive nature left her ill-equipped for the rough and tumble world of economic competition and politics. Thus, for instance, women's excessively delicate nervous system, monthly 'illness', smaller brain, and specific reproductive organs made it unhealthy – indeed unnatural – for them to work, write, or vote. The rest of this chapter will provide an overview of some historical features, especially some Italian peculiarities, which will contextualise the different medical discourses about female same-sex desires analysed in detail in Parts II and III of this book.

Historical contexts

The unification of Italy in 1861 did not completely resolve the instability which had characterised the political life of the Peninsula throughout the nineteenth century, and the unified state's failure to achieve a real national unity was evident on many fronts. Firstly, the Venetian territories and Rome were initially excluded from the new kingdom. Secondly, there was the pressing *'questione meridionale'*: the government was unable to bridge the gap between the wealthy urban north and the impoverished rural south. The economic and cultural differences between the north and the south caused growing concern amongst the ruling elites and intellectuals, along with anxieties about southerners in general, who, like the Irish in Victorian Britain, were held responsible for their own misfortunes. The Italian ruling class was incapable of reaching a political consensus and because economic growth continued to be slow, Italy seemed unable to catch up with the more powerful European countries.

It was only in 1866 that the Venetian territories became a part of Italy, after the Prussian victory over Austria and Napoleon III's defeat at the Battle of Sedan. Yet Rome, the Papal territories, and the new state's relationship with the Church continued to be thorny problems at the end of the century and beyond. Medical attitudes towards sexual morality in the *post-Risorgimento* were inevitably shaped by this tension with the Roman Catholic Church. In 1861 the Church refused to recognise the new state, instructed Catholics in Italy to withdraw their support for the new government, and requested the Catholic powers in Europe to assist the Church militarily and diplomatically in its efforts to restore its lost temporal power. Initially at least, the government tried to compromise with the Vatican, maintaining the first article of the *Statuto Albertino*, which proclaimed that the state's religion was Roman Catholic, and that other religions would only be tolerated.[19] In 1864, the Vatican responded with the unaccommodating *Quanta cura*, one of the most famous encyclicals of modern times, with its accompanying Syllabus of Errors. This document listed the principal ideological misconceptions of the time, and called Catholics to reject supposedly widespread notions such as people's freedom to profess whatever religion they thought best, or that even non-Catholics might go to heaven. It also indicated that the Pope was entitled to exert temporal power over his own state, and no dissension on this matter would be tolerated; Church and state should not be separated; and that the Pope did not agree with progress, or modern liberal civilisation.

Finally, in 1867, as the government started to believe that it would not be possible to reach an agreement with Catholic authorities, it began taking actions to limit the Vatican's power. With the exception of parish lands and a few other buildings, all the Church's property was confiscated, and monasteries and convents were closed or forced to operate under restrictions. In 1868, the Vatican issued another famous encyclical, the *Non expedit*, which barred Catholics from being elected to Parliament, voting, and participating in politics. The *Non expedit* continued to be valid in the early twentieth century, and it was a constant reminder of the Church's continuing refusal to recognise the legitimacy of the Italian government. Only in 1870 did Italian troops break through the city walls at Porta Pia; as they conquered Rome and proclaimed it the capital of Italy, Pope Pius IX was forced to withdraw into the Vatican. In the following year the government approved the Law of Guarantees, which in practice gave the Pope the status of an independent sovereign in power, guaranteed his freedom to communicate with Catholics around the world, to keep diplomatic relations with other countries, and assigned a generous annual endowment to cover the expenses of the Vatican. The Law of Guarantees also abolished the requirement whereby bishops had to swear allegiance to the king.[20]

To a certain extent, the Law of Guarantees helped to improve the relationship between Church and state. In the long term, however, the Roman Catholic Church's hostile attitude towards unified Italy further weakened the state's legitimacy and unity itself. The Church's refusal to acknowledge the new nation caused a further problem: Italians continued to be Catholic, and the beliefs and rituals of the Church were amongst the few things that drew the people of the north and south together. If the cohesive force of the religious creed was going to be effectively ousted, the ruling class and the supporters of unification – mostly members of the middle classes – had to find a new cultural basis for national identity. Positivism thus emerged as the only compatible discourse with the liberal and secular aspirations of the new state. In Italy, science came to play a pivotal cultural role and became especially important as a weapon against the traditional cultural hegemony of Catholic thought.[21]

Rather than academics, Italian positivistic scientists were what we now refer to as public intellectuals, eager to popularise research, influence legislation, and engage with practical issues.[22] For instance, a number of Cesare Lombroso's followers were members of Parliament or bureaucrats at the Ministries of Internal Affairs and Justice, and thus pushed for the new institutions to reflect positivist understandings of how society works. They championed criminological positivism as a secular alternative to Catholicism, a 'lay faith' upon which to build the institutions of the young state.[23] Legal and penal authorities were moved by a reformist sentiment, which partially explains the appeal of criminal anthropologists in the 1880s; endorsed by a supposedly coherent programme that would allow them to defend society against criminality and social disorders, criminal anthropologists would solve the problems faced by the new government. Physicians, and in particular criminal anthropologists, endeavoured to find practical solutions to the increasingly chaotic situation of the new Italian kingdom. Italians had remained poor and strikes broke out in both the north and the south. The riots of *Fasci* workers in Sicily in 1894 and in Milan in 1898 were brutally put down, restrictions were placed on the freedom of the press, and the socialist party was dissolved. Additionally, it was believed that the *'questione meridionale'* was becoming progressively graver. In 1896, the defeat in Eritrea (Adowa) caused more deaths than all the *Risorgimento* wars put together. All of these factors culminated in the anarchist Bresci's assassination of King Umberto I in 1900.[24]

These events at the turn of the century shattered the beliefs of intellectuals and positivist physicians in historical progress. From an initial phase of faith in progress and universalism, militant positivists retreated into pessimism and abandoned their progressive cosmopolitanism. But until then, scientists had been actively involved in the politics and cultural life of the *post-Risorgimento*. Although Lombroso's followers were at the forefront of public life and established themselves as part of the ruling group after the unification of Italy, medical writers in general exploited their scientific credentials to express views on ethics and morality.[25] It is not a coincidence that two doctors, Luigi Carlo Farini and Giovanni Lanza, became prime ministers in the years after the unification, and others who became ministers or members of Parliament also exerted their influence on political matters. More than any other group, doctors contributed to set the kingdom's new moral guidelines in an attempt to replace the Church's influence on the Peninsula.[26]

The conflict between Catholic intellectuals and anticlerical medical writers was particularly intense until the first decade of the twentieth century. Debates about morality, sexual habits, and new lifestyles often became occasions for fierce conflict between the two camps. Supporters of a 'new sexual morality' were often identified with anticlericalism and socialism; opponents with Catholicism and conservatism. Many positivist medical practitioners and writers came from socialist ranks, and likewise most of the members of the *Lega italiana neomalthusiana* [Italian Neo-Malthusian League] were socialists. The Church perceived this neo-Malthusian association as an explicit threat because it promoted birth control and information about sexuality amongst the working classes; it also supported divorce and free-love unions. A substantial part of the urban population responded quite favourably to the activities of the *Lega italiana neomalthusina*, which, along with doctors' popularisation of sexual knowledge, aimed to educate, reform, and guide the nation in its attitude towards sexual matters.[27] The Church did not approve of such campaigns, but notwithstanding this antagonism, science and religion shared a prescriptive discursive structure, and their discussions of sexuality had, generally speaking, a distinctly moralistic overtone. Therefore, the point at which the Church's attitude towards sexual matters diverged the most from medical writers' was the latter's willingness to speak about sexuality.

In nineteenth-century Britain there was no political event comparable to Italian unification. In Victorian society there was no crisis of the governing class equivalent to the process that led to and ensued after Italian unification. This does not mean that Britain was not troubled by revolts. Indeed, the social upheavals of the 1830s and 1840s, and then again in the late 1860s and 1880s, unsettled Victorian political life, which was nevertheless a stable society if compared to Italy. In the first half of the nineteenth century, Britain had established itself as an industrial nation with significantly developed urban areas, a growing economy, and by the end of the nineteenth century, an expanding empire. The monarch was the head of the Anglican Church, and the link between Church and state was frequently held to be a vital pillar of the constitution, especially by conservatives in the first half of the nineteenth century.[28]

To a certain extent, in Britain as in Italy, there existed a multinational diversity. The separate identity of Ireland, Scotland, and Wales, as well as powerful regional and local consciousness, meant that political and cultural realities were underpinned by tensions between Englishness and Britishness. The long-term enmity with France was, however, a cohesive element, as was Britain's commitment to Protestantism and its powerful anti-Catholic animosity, with the clear exception of Catholic Ireland, which presented a problem for England.[29] Overall, however, since the 1830s Britain had put uniform policies in place through parliamentary reform: in 1832, the Reform Acts and the Irish Act unified the voting requirements for citizens of England, Scotland, Wales, and Ireland, and made the nation more uniformly British than it had previously been. In the 1830s, Catholic emancipation incorporated the Irish and other Catholics into British culture. Above all, as a political entity, Britain was defined by the institution of monarchy. The image of the reigning sovereign was a central expression of cultural cohesion, and the symbolism of the Crown as a pivotal image of imperial unity was further reinforced with Queen Victoria's elevation to Empress of India in 1876.[30]

The empire's expansion was accompanied by increasing discussion of nationhood and of the civilising mission Britain had been called upon to perform. The growth of the German navy, the unfavourable reaction to the Boer Wars, and the threat of a future European war prompted a moral reassessment of Britain's strength and helped to induce a ready identification between moral and martial might: pre-eminence in European and imperial affairs had to be defended at all costs.[31] British notions of manliness were underpinned by an understanding of effeminacy as a threat to the cultural self-perception that validated British international pre-eminence. As leaders of social opinion, physicians accepted the Victorian belief that privileged social status must be justified by superior morality. Social stability was founded on steady governance by the male head of the family, and medical writers saw it as their duty to preserve and maintain the authority of the *paterfamilias* as the cornerstone of British society.

Mentioning the 'unnatural crime' of same-sex practices in publications was seen as tantamount to corrupting public morality. A number of historians have recently highlighted how the attempt to prevent discussion of sodomy was a constant element in the British legal tradition. Many judges and legal officials believed that the public trials of sodomy that were reported in the press threatened to spread moral corruption and even to encourage the acts themselves.[32] Medical authorities avoided arguing against legal tenets that deemed male same-sex desires a vice, and thus Continental theories on sexual inversion were largely rejected. Medical discussion on male same-sex practices was also limited and cautious compared to Continental sexology's abundant literature. This is in contrast with what happened in Italy, where rather than physicians, it was the Roman Catholic Church that had shrouded same-sex desires in silence. Italian physicians identified silence with ignorance – itself the very source of endemic evil, superstitions, errors, and obscurantism – and set out to correct it by discussing 'pathological' sexual behaviours.

Silence on sexual matters did not mean that the Roman Catholic Church did not express views on the subject of sexuality. Theological treatises, moral casuistry, and manuals for confessors traditionally engaged with sexual vices, and at the turn of the century, these writings tackled the problems of sexual perversion and same-sex desires just as medical writings did.[33] Yet, for the Church the real problem lay in the public discussion of such a subject. The Roman Catholic Church was against the popularisation of sexual knowledge, and as late as the beginning of the twentieth century it continued to condemn books on sexual education that aimed to present medical knowledge to non-scientific audiences. In 1941, well after the decline of positivism, a priest and influential psychologist called Agostino Gemelli published a book called *La vita sessuale* [*Sexual Life*], where he argued that stating that the sexual instinct was a biological necessity that had to be fulfilled was a means to justify prostitution. He also criticised the numerous volumes on sexual knowledge published by positivist physicians at the turn of the century.[34] Although the Roman Catholic Church and Italian medical practitioners committed to secularising the country held firmly antithetical views about the popularisation of sexual knowledge, they did close ranks on a few occasions. For instance, in 1891, Catholic politicians called for a more severe application of legislation against pornography, a measure scientists did not hesitate to support.[35] However different their approaches to non-reproductive sexual behaviour were, the Church and much medical literature shared a negative view of homosexuality.

In explaining why, unlike Protestant countries like Britain, traditionally Catholic countries like Italy did not legally punish male same-sex practices, scholars have assigned great importance to the influence of the Roman Catholic Church in the sphere of morality and sexual conduct. Some studies have suggested that Catholic countries were more tolerant because the Church exerted a powerful influence on private moral issues, which succeeded in suppressing deviant sexualities. These studies highlight that because in Protestant countries religion did not interfere as much in private affairs, the sanctioning of sexual behaviour became the state's prerogative. According to these studies, the tolerant Mediterranean legislators sought to cloak deviant sexualities in silence; their intention was to prevent the disorder being named.[36] The historian Annamaria Buttafuoco has noted that in Italy even incest – which was frequent, and not restricted to the poorer classes – was not punished except in cases of scandal or *'querela di parte'*.[37] Buttafuoco has suggested legislators were not inclined to regulate family life, and so put forward laws that rather than altering the mechanics of domestic life, aimed at stopping sexual disorders from spreading beyond the family unit.[38] In the last decades of the nineteenth century, Church, medicine and *post-Risorgimento* sexual politics interacted in complex ways, which have been consistently overlooked by historians. The next chapters in this book show that the Catholic Church's influence on sexual matters was countered by the physicians' contributions to sexual debates. Medical opinions were couched in a considerable body of research on same-sex desires; both the existence of this body of research and the fact that scientists were actively involved in political life complicate historiographical accounts that overstate the passive role of the government in regulating sexual deviances.

Mostly characterised by strong anticlericalism, Italian physicians who wrote about sexual perversion disqualified the Church by remarking the high incidence of sexual perversion amongst priests. Positivist physicians such as Lombroso espoused the view that by imposing clerical celibacy, the Roman Catholic Church fostered homosexual behaviour. Because medical writers of the period argued that sexual activity was a natural and essential part of human instinct, they conceived of celibacy as an unnatural cause of psychosexual pathologies, which in the long term could lead to sexual crimes such as rape and even pederasty. There are frequent remarks about the incidence of pederasty amongst Catholic priests in medical literature. Lombroso, for example, believed that sexual crimes including rape and same-sex acts were more frequent amongst priests than other social groups.[39]

In nineteenth-century Britain, some groups also took exception to the sexual mores of Catholics, which were typically associated with decadent, feminine, and homosexual behaviours. In Britain, it was religious authorities rather than medical writers who accused the Anglo-Catholic Church of promoting effeminacy and homosexual behaviour. Evangelicals deplored Catholic sacerdotalism and ritualism as essentially un-English and unmanly. Since the 1830s the Oxford Movement, also known as Tractarianism, promoted a return to early Catholic doctrines from within the Anglican Church; in the 1860s the parish priest Charles Kingsley responded by accusing its members of being effeminate, homosexual, and un-English.[40] Kingsley's argument was in line with the view of Italian physicians: he abhorred the idea of celibacy as contrary to nature and further believed that celibacy was a sin against God.

Victorians held 'manliness' in high regard, which further explained their contempt for celibacy. Men were defined by their ability to work, to procreate within marriage, and by their courage, heartiness, and physical vitality. While it was less consciously articulated in Roman Catholicism, the spiritual significance of the world of work and the domestic realm assumed a critical importance in nineteenth-century Protestant attitudes towards manliness. Male employment was viewed not merely as a necessity for economic survival, but as a means of proving that a man was independent, honest, and competent; work was thus the main way a man asserted his dignity. These elements became key attributes in the Victorian understanding of manly virtue.

By contrast, virtuous Victorian women were expected to avoid the workplace unless there was no other alternative to avoid destitution.[41] Historians have argued that the rise of evangelical Christianity in the late eighteenth century helped transform attitudes to female sexuality by encouraging an ideology of female 'passionlessness'. Prior to the rise of evangelicalism in Britain, women were thought to be as licentious as men, if not more: given their limited ability to control their passions, it was believed women were weaker than men. The revitalised churches demanded women prove their noble character by being morally restrained. Historians have also suggested that women themselves embraced this link between passionlessness and moral superiority as a means of enhancing their status, gaining some control over their lives, and ultimately, expanding their opportunities.[42] As Michael Mason has shown, such an ideology was not only very closely associated with the religious practices of the period – especially the rise of evangelicalism – but was also considered to be a progressive ideological position because it was evidence of having transcended bodily desire.[43]

Regularisation of sexual behaviours

In the wake of rapid urban and industrial growth in nineteenth-century Western Europe, alleged rising rates of alcoholism, insanity, crime, prostitution, and syphilis fostered widespread fears amongst physicians, intellectuals, and the political elite. After 1850, many countries' governments developed concerns over falling birth-rates and the health standards of the population. Social agitation and political turmoil, class antagonism, women's emancipation, economic and social changes, ruthless competition, and the mechanisation of labour were perceived as threatening the stability of society, and it was feared they would cause European nations to deteriorate. Degeneration – at once a social and individual pathology – emerged as the disorder of the age.

The concept of degeneration was originally developed by the French doctor Bénédict Augustin Morel in his *Traité des dégénérescences physiques, intellectuelles et morales de l'espèce humaine* [*Treatise on the Physical, Intellectual and Moral Degenerations of the Human Species*] (1857).[44] European psychiatry promptly adopted Morel's theory, so that by the 1880s degeneration had become the dominant framework for understanding mental disorders. Morel had argued that mental illness was essentially a deviation or 'degeneration' from an ideal type, and stressed that most afflictions were caused by underlying physiological conditions – a hereditary weakness of the nervous system was the factor most frequently cited.[45] European psychiatry soon came to conceive of degeneration as a progressive decline towards self-extinction: physical and mental qualities deteriorated from one generation to the next, culminating in a final stage of sterility.

The lack of empirical evidence for physical causes of mental pathologies led medical writers to stress the importance of degeneration, which in turn became an explanation for virtually every pathological phenomenon. At the turn of the century, degeneracy had become an umbrella concept with the power to account for inexplicable diseases, and in this period the use and misuse of degeneration began to arouse scepticism amongst professionals. In 1892, the eminent British psychiatrist Daniel Hack Tuke acknowledged that his contemporaries mistakenly used degeneracy theory to explain every kind of deviation. Tuke criticised this practice on the grounds that a diagnosis of degeneration was implicitly and unnecessarily pessimistic, because treatment would often prove beneficial, and patients recovered from their ailments.[46] In 1896, the Italian criminal anthropologist Pasquale Penta, who had previously embraced degeneration theory enthusiastically, admitted that it was inadequate to explain the origin of psychiatric diseases. In essence, he said, degeneracy explained nothing.[47]

In the second half of the nineteenth century, politics and literature had increasingly relied on scientific language – especially on medical and biological metaphors – to explain how society worked, which in turn gave rise to the image of society as an organic body. With its suggestion that each human being was steadily and unavoidably decaying, degeneracy seemed to articulate the anxieties of society at large.[48] As Daniel Pick has pointed out in his study of this 'European disorder', degeneration took varying forms in different national cultures.[49] Thus, on the Italian Peninsula, the notion of degeneration was deployed to explain the dangerousness of the south, which was perceived as a threat in the delicate context of political unification. Bandits, endemic poverty, and constant unrest frightened the ruling class, which sought to distance itself from a southern world of delinquency and backwardness.[50] A host of factors explains why degeneration was not such a pressing concern in Britain as it was on the Continent: on the one hand, there was no founding text of the theory of degeneration comparable to Lombroso's authoritative *L'uomo criminale* (1876). Further, those who espoused a classical liberal conception of the individual, and therefore accepted the notion of free will, found it difficult to entertain the notion of a distinguishable degenerate type. Lastly, the prospect of extinction fostered by degeneration theory was unattractive to many who identified Britain with Imperial power. The image of the degenerate, however, articulated British fears of social pathology and the criminal subculture of the cities.[51]

No other aspect of human experience was as closely tied to the concept of degeneration as sexuality.[52] The argument ran that non-reproductive sexual practices contributed to the moral, physiological, and mental deterioration that afflicted the European population at the turn of the century. Especially in urban areas, pathologies were perceived to be characterised by physical and moral corruption, as well as an uncontrolled emotional state. In Europe, sexual perversion came to be regarded as belonging to a wider group of hereditary pathologies, which also included urban poverty, criminality, insanity, and abnormal gender behaviour. With the spread of Malthusian and Darwinian observations, sexuality in both Italy and Britain became increasingly regulated by the medical and legal professions and, as such, was turned into a scientific issue. New social conventions, sanitary prescriptions, policies of sex education, and practices of self-control were all put forward as a means of controlling disorderly sexual behaviours. These medical implementations, such as a periodical genital examination of prostitutes, were further reinforced by laws that affected conjugal life, relationships between the sexes, and families, such as the matrimonial acts that regulated divorce.

Limiting the lives of women

Amongst other things, the Italian unification meant that new homogeneous laws came into force on a national scale. The Italian Civil Code of 1865, which was based on the Napoleonic Code, established the principle of equitable inheritance. This meant that sons and daughters had more or less the same rights to their parents' property, and for women this meant that their proprietorship was neither occasional nor trifling. The Civil Code of 1865 not only ignored class distinctions amongst women, but also declared that the sexes were equal before the law. Moreover, minors of both sexes had the same legal position. Women and men came of age at twenty-one and were thereafter equally free to live wherever they chose. They could engage in trade, own and manage property, and make wills and bequests. Women could exercise these rights freely until they got married, but needed their husband's approval thereafter.[53] In practice, however, the new Italian Civil Code reinforced women's dependence on men: women were still not allowed to practise liberal professions, and they had no free access to secondary schools. Not only were they denied the right to vote in both national and local elections, but they were debarred from any public office or any other position where they had to represent the authority of the civil state. Women could not become lawyers or judges; they were forbidden from acting as arbitrators or notaries, and until 1877 they could not even be witnesses in legal actions.[54]

The supposed civil equality of the sexes declined even further after a woman married because she had to assume her husband's name and place of residence. Marriage conferred mutual rights and duties, and although both parties were equally obliged to each other, women had fewer rights than men. Married women could own, inherit, and bequeath property independently, but acts beyond the simple administration of their own property required their husbands' consent under provisions called 'marital authorisation'.[55] A woman who owned property could not even give it to her own children without her husband's consent. This meant, for example, that a woman could not act independently of her husband to provide a dowry for her daughter. With the property rights of mothers thus curtailed by marital authorisation, fathers added economic force to their legal power to determine whom their daughters were to marry. The 1889 Zanardelli Penal Code established that an adulterous woman could be punished with three to thirty months in prison, while a man could be punished only if he brought his mistress to the house he shared with his wife.[56]

The lives of Italian women in different regions of the country were affected in dissimilar ways by the Civil and Penal Codes. Women in the northern provinces lost rights they had enjoyed under Austrian domination: Lombard and Venetian women had been free from marital authorisation and, along with Tuscan women, had been entitled to vote in local elections.[57] These rights were abolished with the new codification, and in ensuing decades northern Italian women struggled not only to advance the rights of women in general, but also to regain lost ground.[58]

In Britain, the predominantly liberal approach to governance emphasised minimal state intervention in both the economy and family life. In the name of a broader national interest, the government tended to seek some balance between individual freedoms and public surveillance.[59] The (male) professional classes were fiercely protective of the autonomy of the individual, although this liberal approach applied more to men than to women. In the context of her battle against the Contagious Diseases Acts, the well-known feminist Josephine Butler denounced the 'government tyranny' that went against woman's 'absolute sovereignty over her own person'.[60]

In the mid-nineteenth century, a series of legal changes gave British women new rights and several significant pieces of legislation dealing with aspects of sexual conduct were passed. These were partly the result of attempts to tidy up and consolidate the inherited mass of criminal legislation, but also reflected the increasingly secularised regularisation of moral conduct, formerly under the sway of the Church. For example, in 1837 civil registration of marriage was introduced, and in 1857 the Matrimonial Causes Act made divorce more widely accessible. However, a husband could divorce his wife on charges of adultery, whereas a woman had to prove that her husband had been adulterous, and that he had committed an additional 'matrimonial offence' such as cruelty, desertion, bigamy, or incest.[61] Just as adultery was punished differently depending on whether it was a woman or a man that was being charged, women and men had unequal access to divorce. In both Italy and Britain, these legal differences illustrated contemporary ideas about the different sexual natures of men and women: adultery by a woman was so horrendous that a single act justified her husband's divorcing her. Furthermore, a man's right to divorce an adulterous woman had been traditionally established on the grounds that she might bring a spurious child into the family. In a man, however, the offence was so trivial that even when persistent, it was not regarded as sufficient reason for a woman to terminate her marriage unless it was aggravated by some other offence.

Despite the double standard, in Britain the Matrimonial Causes Act did represent a blow to the ecclesiastical doctrine of the indissolubility of marriage, and it did grant women some rights in cases of intolerable unions. Moreover, this Act gave divorced and legally separated women the status of *'femme sole'*, which enabled them to own property and sign contracts. Under common law's doctrine of 'coverture', the husband's identity and legal sphere subsumed the woman's upon marriage. Like Italian women, British women were unable to enter into legal contracts and had no right to own property.[62]

Limiting the bodies of women

Women could pose a problem for public order in nineteenth-century Italian and British cultures. Prostitution was widespread and visible in nineteenth-century cities; contemporary commentators saw the prostitute as the ultimately immodest woman and the embodiment of depraved sexuality.[63] Between 1860 and 1870, British, Italian, French, German, and a number of other European governments passed a series of laws that aimed to control women selling themselves. Scholars have attributed the growth of prostitution to both the expanding urban population and the new social possibilities that were made available for women at the time.[64]

The adoption of medical measures such as the compulsory internment in the lock hospitals to combat the spread of venereal diseases furthered control over women's bodies. Although in the nineteenth century venereal diseases were a considerable problem, science had only a limited understanding of the topic: the relationship of syphilis to its late manifestation of debility, paralysis, and insanity had not been established yet; there was no easy way to distinguish syphilis from gonorrhoea; and treatments were largely ineffectual, often severely debilitating, and sometimes even punitive (cauterisation and application of caustic substances).[65] It was believed that venereal diseases were caused by vice and promiscuity, and thus largely attributed to prostitutes. Men were defined as victims of venereal diseases rather than as instrumental to their circulation. Although it only had direct effects on the lives of women, the regularisation of prostitution also reflected a new enthusiasm for state intervention in the lives of the poor on medical and sanitary grounds.[66] The regularisation of prostitution shows that both in Italy and Britain the state combined legal sanctions of sexual practices with medical control of the human body and sexual behaviours.

Amongst other things, the unification of Italy meant that a homogeneous regulation with regards to prostitution was introduced. Most men, at least those in the administrative ranks of the national government, thought that it was neither possible nor desirable to ban prostitution altogether. Prostitution was believed to buttress the family by providing a sexual safety valve for single men who might otherwise try to seduce 'honest' middle-class women. It was in order to satisfy male sexual demands, yet prevent venereal disease, public immorality, and disorder, that officials decided to legalise, yet closely regulate, prostitution. Prime Minister Camillo di Cavour was a prominent champion of this cause.[67]

In February 1860, the *Decreto Cavour* [Cavour Act] introduced the regularisation of prostitution. The guidelines of this Act were clearly inspired by the Belgian experience, and aimed at concentrating prostitution in closed houses and at limiting, as much as possible, the freedom of movement of the women who were registered as prostitutes. The law also aimed to stop the apparent increase in the number of cases of syphilis. Prostitutes had to undergo a medical examination every second week, and medical treatment in a venereal hospital – a *sifilicomio* – was mandatory in case of infection. Until 1888, prostitutes had to report their movements from one city to another and even had to carry a special passport to identify their occupation.[68]

It was legal for a woman to sell sexual services both in Britain and in Italy, but in Britain there was no system of licensing or approving brothels. Most of the prostitutes acted individually rather than being organised in a profession, which posed a number of difficulties in the efforts to regulate their activities, but did not make it altogether impossible. In 1864, 1866, and 1869 the government passed the Contagious Diseases Acts, whereby any unaccompanied woman in garrison towns and sea-ports could be required to submit to a gynaecological examination to ascertain whether she was a carrier of a transmittable disease – in which case she would be incarcerated in lock hospitals until she was 'clean'.[69] To a certain extent, the regulation of prostitution was not homogeneous in Britain and this shows the extent to which Britain, like Italy, was characterised by a degree of national diversity. These Contagious Diseases Acts that sanctioned the coercive imprisonment and medical examination of any woman suspected of being a prostitute had never been applied in Scotland. However, even before these Acts were passed, some Scottish cities had developed similar strategies in an attempt to reduce the level of prostitution and contain venereal diseases. In fact under local Burgh Police Acts, in association with the lock hospitals and reformatory asylums, the police operated a policy of legal repression and medico-moral regulation for suspected prostitutes.[70]

British feminists and social reformers reacted strongly against the Contagious Diseases Acts arguing that the Acts served to legitimise men's urges and to restrict women's civil liberties; they also protested at the injustice of examining the prostitute but not the male client. Feminists objected to the assumption that prostitution was necessary because of the particular nature of male sexuality, on the grounds that the male sexual urge was a social rather than a biologically determined phenomenon. Led by the charismatic Josephine Butler, women attacked the Acts while claiming the right to promote wider moral and social reforms, and rallying for a single rather than a double moral standard.[71]

Under Butler's leadership, the Ladies' National Association for the Repeal of the Contagious Diseases Acts established a powerful national movement. Butler's campaign owed much of its popularity to its careful manipulation of Victorian ideals of femininity, for example by taking up the American Temperance Society's model of vigils in which white-clad genteel women would stand outside brothels to raise awareness of the private damage wrought by such establishments. While Butler used the symbolic capital of feminine purity, she encouraged women to reflect and debate about prostitution and the sexual exploitation of women. The considerable impact of these strategies on women and men alike far exceeded that of the sober suffragette campaign.[72] Nonetheless, although Butler's campaign challenged masculine privilege, it continued to equate femininity with purity, which echoed the medical idea of women as passive and responsive to male sexuality. Importantly, the campaign for the repeal of the Contagious Diseases Acts gave some British women the experience of discussing taboo topics, and the ability to engage with medical debates. Eventually, in 1886, the Contagious Diseases Acts were annulled, which arguably shows that feminists had a greater impact in Britain than in Italy, where women's movements had failed to influence the government.

In Italy, the feminist movement did not respond in any comparable way to the regularisation of prostitution. Italian feminism was linked to the *Risorgimento* movement, and had been involved in the struggle for Italian unification.[73] The Civil Code in general, and family law in particular, inscribed gender inequalities, which helped remodel women's *Risorgimento* nationalist fervour into a feminist commitment. Although feminist consciousness in Italy as in Britain was inspired by the egalitarian ideals of the Enlightenment, on the Peninsula the struggle for unification had disrupted the traditional social order and encouraged a deep revaluation of traditional attitudes and gender roles. Like the men involved in the *Risorgimento*, their commitment to the revolution meant female revolutionaries had suffered financial deprivation, exile, social ostracism, familial discord, and other personal consequences. Women who had helped the cause of unification had experienced men's life. Women involved in the *Risorgimento* had also been exposed to Giuseppe Mazzini's combination of protofeminism, moralism, and revolutionary nationalism.[74] It should not come as a surprise that feminists were attracted to Mazzini's principles, for his *Doveri dell'uomo* [*Duties of Man*], published in 1860, advocated equality between the sexes. Mazzini explained that men's ostensible superiority was due to their better educational opportunities and to the legal oppression they exercised over women.[75]

Although individual female emancipationists had struggled since unification to attain equal rights for women, they did not establish formal organisations until the 1880s. Only in 1881 was Anna Maria Mozzoni able to found the *Lega promotrice degli interessi femminili* [League to Promote Female Interests] in Milan, whose example was promptly imitated in several other cities. In 1864 Mozzoni had published *La donna e i suoi rapporti sociali* [*Woman and Her Social Relationships*], which sparked the feminist critique of Italian family law, but it took her close to fifteen years to launch an organised movement. These feminist leagues called for better working conditions for women, especially in the new textile factories, and restated older demands for civil rights. After 1890, the trend towards organisation accelerated, with bourgeois women demanding the vote in increasingly emphatic tones, while socialist feminists initiated a campaign for protective legislation for working mothers.[76] More radical feminists campaigned for both legal and social reform in the pages of the leading Italian feminist periodical *La Donna* [*Woman*], published regularly between 1868 and 1892 in Venice by Gualberta Adelaide Beccari. More than other women's journals, *La Donna* insisted on the solidarity between women regardless of class. Their maternal role was articulated as a shared fate and therefore a unifying existential and political element. *La Donna* reflected the concerns of educated women, and the middle class was its target audience. Major Italian feminist groups included the *Unione femminile* [Female Union] founded in 1899, the Italian branch of the International Council of Women, affiliated in 1900, and the more radical *Comitato nazionale pro suffragio femminile* [National Committee for Female Suffrage], founded in 1904.[77]

percentage of men could not vote until 1913, women had fewer grounds on which to blame sexism for their disenfranchisement. More importantly, the conservative attitudes of many Italian feminists led them to focus primarily on education, civil rights, and welfare projects. Most feminists embraced the *Risorgimento* myth of the '*madre italiana*' [Italian mother], who exercised an indirect power in society through motherhood.[78] This is to say that they neither challenged the scientific definition of womanhood based on maternity, nor questioned current sexual morality. Italian feminists were very careful not to mention sexuality in their protests. Nor were there clubs such as the English Men and Women's Club, which challenged the sexual order.[79] The relative inability of Italian feminism to challenge traditional sexual morality is proven also by the fact that in Italy, the New Woman was more of a rhetorical construction imported from abroad than a representation of local phenomena, at any rate until World War One.[80]

This does not mean that Italian – like the British – middle-class men were not alarmed by women's demands and changing social roles: almost all works in biology, sociology, psychology, and criminal anthropology written in this period are characterised by a degree of anti-feminism, and in many of them women's biological difference signalled their inferiority more or less explicitly.[81] Numerous studies of female delinquents by positivist criminologists suggest that, especially after the 1890s, women's participation in the industrial workforce generated anxiety over the 'modernisation' of gender roles. In Italy, such anxiety also seemed to be related to the opposite reasoning: if women were not as successful as men in the modernisation process, they would become symbols of Italy's political failure to develop beyond its pre-capitalist economy. However real these anti-feminist concerns were, to assert that physicians and especially criminal anthropologists were responding to feminism would not only be to overestimate the role of feminism in Italian society at the end of the nineteenth century, but to forget that Italian feminists were not radical. Scientists and many feminists agreed on the importance of motherhood, and thus upon the existence of cardinal natural differences between men and women.[82]

Limiting male same-sex desires

Before the unification of Italy, there were different standards for the punishment of male same-sex practices. In the Papal States, for example, male same-sex acts was punished with life imprisonment;[83] in the Lombardo-Veneto region (current Trentino-Alto Adige, Friuli-Venezia-Giulia, Istria, and Fiume), which was under Austrian domination, sodomy was punished with custodial sentences from six months to one year. In the kingdom of Sardinia, under the Savoy monarchy, male same-sex acts could land a man in prison for up to ten years.[84] The rest of Italy adopted the Napoleonic Code, whose silence on the issue of homosexuality meant that there was no legal framework for the repression of same-sex acts, which in practice meant they were not criminalised.

In 1861, the criminal code of the kingdom of Savoy was extended to the rest of the country. In southern Italy, however, the laws against homosexual acts were not enforced so that the tacitly lenient Napoleonic Code continued to be valid, thus creating a 'double standard'. The government acknowledged that in Mediterranean culture it was considered normal for young boys to engage in same-sex practices, thereby conceding a cultural difference between the country's northern and southern regions. The government authorities were aware that punishing same-sex acts in southern Italy would entail a complete transformation of the indigenous culture.[85] With the promulgation of the 1889 Penal Code (also known as the Zanardelli Penal Code), private homosexual behaviour between consenting adults stopped being a punishable offence, except in cases that involved violence, *'querela di parte'*, or 'public scandal'. Homosexuality was thus decriminalised in the whole of Italy and tolerated as long one did not 'cause too much of a disturbance'.

In the 1887 parliamentary discussion that had led to the decriminalisation of male same-sex practices, the minister for justice, Giuseppe Zanardelli, explained that in dealing with 'acts against nature', ignorance of the 'vice' was more useful than its advertisement through the law. This line of thought is informed by the doctrines of Giovanni Carmignani,[86] an expert in penal law who argued that the best way to confront the 'vice' of same-sex practices was to deny their existence.[87] Political caution moved Italian legislators to believe that unveiling sexual practices that had been covered by silence was against the public interest, the effectiveness of the legal system, and especially against traditional customs.[88] This political attitude can be interpreted as being in line with the tradition whereby Church and state occupy two distinct areas of intervention, and the regulation of sexual morality belongs to the former.

Nonetheless, unlike the British government, Italian political authorities never tried to stop public discussion of same-sex desires. The 1857 Obscene Publications Acts had a direct influence on contemporary British sexological discussions because it limited publications on sexual matters and established what had the power to 'deprave and corrupt'. Under this law, in 1888, Émile Zola's publisher, Henry Vizetelly, was sentenced to three months in jail for selling an expurgated translation of *La Terre*.[89] The Italian government, by contrast, did not discourage any kinds of publications on sexual perversion. The very fact that Zanardelli and his parliamentary colleagues were openly discussing such a topic is quite remarkable when compared to the British, who discouraged public and print discussions on the subject.[90] Further, given Catholic animosity towards the new Italian government that was trying to reduce the cultural influence of the Church, the broader social role of medicine cannot be underestimated. Symonds, who had closely followed the Italian reforms that had eventually led to the decriminalisation of male same-sex practices, suggested that the new legislation had come about due to the influence of 'Lombroso's school'.[91]

Men involved in acts of sodomy in Britain had been punished at least since the reign of Henry VIII, when the 1533 Buggery Act made such behaviour punishable with death.[92] During the eighteenth century, common law had made it possible to prosecute a number of relatively new offences grouped under the label of 'unnatural crimes'. This term covered sodomy, bestiality, and any same-sex act or invitation to the act, usually described as indecent assault or 'assault with intent to commit sodomy'. These definitions of offences affected both public and private expressions of homosexual desire.[93] In the nineteenth century the death penalty was seldom enforced as punishment for sodomy; in 1861 the law was modified so that sodomy was penalised with imprisonment for a period of ten years, and less for those conspiring to commit sodomy. It was not until the 1885 Law Amendment Act, put forward by Henry Labouchère, that there was any definite consensus on the legal status of the sodomite.[94] This Act dealt with sodomy, 'indecent acts', and bestiality, and raised the age of consent from twelve to sixteen; its approval meant that the ill-defined crime of 'indecent acts', which could include mutual masturbation and fellatio, would be punishable in both public and private instances. It was the 'private' clause, and indeed the notion of indecent acts, that set the English law apart from the Napoleonic Code adopted on the Continent by countries such as Italy, and against which many homosexual rights activists fought.[95]

Because the Labouchère Act introduced a new and very broad category of homosexual misdemeanour, prosecution was made easier and conviction more likely. This was the measure that inaugurated the era of persecution against homosexuals by blackmailers and moral purity brigades such as the National Vigilance Association. Nevertheless, it continued to be more dangerous to go against the prescriptions of class than those of the law. While heterosexual relationships across class boundaries were considered a minor threat to society's gender hierarchy, inter-class homosexual relationships had the potential to be thoroughly subversive. Hence aristocratic and bourgeois homosexuals with working-class consorts were more at risk than those practising in upper-class preserves such as public schools and universities.[96]

Despite the effort of the British government to limit public discussion of same-sex desires, newspapers occasionally reported scandalous cases of sex between men. One of the most famous homosexual scandals in mid-Victorian times was that of Boulton and Park, who had been arrested outside London's Strand theatre in 1870 for wearing female clothes. They had been charged with 'public indecency' and buggery, but it was impossible to prove that they were doing anything more than masquerading. Britain was further rocked by the discovery in 1889 of a male brothel in Cleveland Street (London), allegedly attended by aristocrats, and a few years later by Oscar Wilde's prosecution for 'gross indecency' in 1895. These two scandals, and especially the Wilde trials, played a crucial part in creating a public image of the homosexual as decadent, artistic, upper class, and male. The Wilde trials caused public attitudes towards homosexuals to become harsher and less tolerant. According to some historians, prior to the Wilde proceedings same-sex passions were met with a measure of pity, whereas homosexuals came to be seen as a threat after the trials.[97] The Wilde trials had further effects: they caused the public to associate art and homoeroticism, and to see effeminacy as an index of homosexuality. People with close same-sex relationships grew concerned about any behaviour that might suggest impropriety.[98]

Italian newspapers also created scandals around same-sex practices between men, but the protagonists of such stories were mostly foreigners.[99] Perhaps the most sensational case was that of the German Friedrich Alfred

Krupp, who was the wealthiest armaments manufacturer and trader of the time. In 1902 German newspapers accused him of having engaged in orgies with young men on Capri,[100] and an international scandal ensued that eventually led to his suicide. Krupp had spent his holidays on Capri for years and was well accepted by the local community because the island derived economic benefits from his luxurious lifestyle. It was only after German journalists broke their silence that their Italian colleagues followed suit, and the Krupp scandal became an opportunity for clerical and anticlerical intellectuals to fight over the sexual immorality of modern times.[101]

In both Italy and Britain, the literate middle classes became aware of love relationships between men in the context of journalistic discussions about scandals. Public disgrace and malevolent rumours about men who indulged in same-sex practices had appeared in both Italian and British newspapers since at least the 1870s, but at the turn of the century these scandals began to engulf famous figures too. However, neither in Italy nor in Britain were there any female cases comparable to that of Oscar Wilde. The press would occasionally feature cases of women passing for men and living with other women, or of female murderers who were moved to kill over their same-sex lovers, but these stories did not create a public image of the lesbian comparable to the image of male homosexuals that resulted from the newspapers' coverage of the Oscar Wilde trials.[102] This does not mean that there was no interest in the subject and, as the rest of this book will show, it was medical writers rather than the press who took it upon themselves to elaborate sophisticated figures of female same-sex desires.

Part II

Following a pattern of increasing specialisation, in the nineteenth century different medical sciences such as gynaecology, venereology, forensic medicine, and psychiatry explored the human manifestations of the sexual instinct in its 'normal' and 'abnormal' forms. Physicians – psychiatrists in particular – began to investigate systematically the sexual instinct, separating it from the reproductive function. In the course of the nineteenth century, physicians interested in sexual matters produced several classifications for pathologies of the sexual instinct with same-sex desires amongst them. Hereditary influences, milieu, marriage, age, and venereal excess were carefully analysed because it was believed they were predisposing causes of diseases. Extreme emotions and abnormal sexual desires were considered symptoms of insanity, and as such were material for scientific study.

The specific nature of women's 'normal' and pathological sexual desires was central to these debates. In the second half of the century, scientists agreed that maternity was the biological base of women; the maternal function thus strongly determined female sexual instinct. Biological 'facts' were used to demonstrate the intellectual inferiority of women: in anthropology, this gave rise to the notion that the smaller size of the female brain was synonymous with lesser intellectual ability. Female inferiority was 'scientifically' proven by two theories in particular: Spencerian theory, which claimed that in the organism – seen as a closed system of energy – the development of intelligence was inversely proportional to fecundity; and atavism theory, which claimed that women were on a lower rung of the evolutionary ladder than men.

The following two chapters draw attention to Italian and British medical debates about female same-sex desires. The figures of female same-sex desires in use on either side of the English Channel were very similar, but they were articulated in different medical languages. Italian sexological research quickly accepted and developed the concept of sexual inversion, and so female same-sex desires became a fashionable sexological subject. Conversely, British physicians were resistant to the Continental concept of sexual inversion, but this is not to say they did not discuss 'lesbian love'. The next chapters study the reasons that led to the rise of discrete medical discourses about female homosexuality in Italy and Britain.

3
Italy: The Fashionable Psychiatric Disorder of Sexual Inversion and other Medical Embodiments of Same-Sex Desires

In the second half of the nineteenth century, medical sexual knowledge was characteristically international, so that it was common for physicians to keep up-to-date with the latest developments in medical science from the United States and European countries. Despite such internationalism, each country formulated its own autonomous medical discourses. While it was not until the late 1890s that Ellis started to systematically study sexual inversion in Britain, the Italian case was completely different: Arrigo Tamassia had introduced the subject in the late 1870s, which triggered the publication of a considerable number of medical articles on the subject of sexual inversion between 1878 and 1890, plus numerous reviews of foreign and Italian works. From the mid-1880s onwards, the complex nosologies articulated by psychiatric treatises consistently referred to sexual inversion as a distinct mental disorder, so that same-sex desires were no longer interpreted as a symptom of other diseases, which had been the predominant interpretation in earlier decades. In the 1890s, this phenomenon was followed by an explosion of studies on sexual perversion, with numerous book-length scientific studies dedicated to the topic.[1] Pioneering historical research such as that of Giovanni Dall'Orto and Nerina Milletti has identified a number of medical works from this period that engage with same-sex desires. Roughly half of these medical studies addressed the question of female same-sex desires.

Following the trend established by their German colleagues, Italian psychiatrists collected case-histories of sexual inverts and debated the nature of their sexual inclinations, which made the psychiatric category of sexual inversion and its related assumptions known to practitioners in other medical fields. When addressing female same-sex desires, on the whole, these non-psychiatric writers tended to preserve old concepts of same-sex desires, which they modified slightly to fit the new psychiatric category of sexual inversion. This specifically Italian practice resulted in the emergence of several medical discourses on same-sex desires that did not necessarily share the same assumptions. Practitioners from different medical disciplines became increasingly interested in female same-sex desires, which also led medical writers to craft a number of stereotypes, amongst them the virago, the tribade-prostitute, and the *fiamma*. These figures had their roots in an earlier vocabulary used to describe sexual relations between women, and by the end of the nineteenth century, they conflated with one another, and with the newer medical category of sexual inversion. It is thus evident how, in the last decades of the nineteenth century, female homosexuality had become a major topic of discussion amongst Italian medical writers interested in sexual matters. Nowadays it is often assumed that late nineteenth-century medical writers scarcely looked at female same-sex desires, which has led many scholars to believe lesbianism was virtually unknown to medical science. Contrary to this widespread belief, Italian medical writers were well versed in the topic.[2]

The process that brought about such a significant amount of research on female same-sex desires, and on the broader area of sexual matters, was complicated by the wider historical circumstances in which Italian scientists found themselves. As has been mentioned in Chapter 2, after the unification of Italy, scientists were leaders in the process of secularisation: science, not religion, was to guide men and women in the moral sphere. Hence medical writers' interest in issues with deep moral implications such as birth control, sexual reproduction, and some areas of public health related to sexually transmitted diseases. Most physicians consciously strove to replace the Roman Catholic Church's influence over sexual behaviour.

The Italian case is further characterised by historical contingencies that affected internal history of psychiatry. While in Britain psychiatric institutions were established in the first half of the nineteenth century, it was only after the unification that Italian psychiatrists set up professional associations and psychiatric journals. The second half of the nineteenth century marks the beginning of a pioneering phase for Italian psychiatrists, who now aimed to make their research highly visible in international medical circles. Importantly, the official foundation of the Italian psychiatric association, *Società di freniatria italiana* [Society of Italian *Freniatria*] in the 1870s, coincided with the emergence of psychosexual pathologies as a subject of study.

This vigorous growth of psychiatry in the last three decades of the nineteenth century coincided with the rise of criminal anthropology, a discipline that in Italy had grown out of psychiatry and fostered debates about deviant sexuality. Criminal anthropologists interested in abnormality in women were drawn towards the study of prostitution, and would frequently engage with female same-sex desires too. Regardless of whether these discussions were held by criminal anthropologists, psychiatrists, or general physicians, these debates were still part of Italian medical writers' enthusiastic engagement with the study of sexual behaviours.

In this chapter I will show how studies of abnormal sexuality related to the specific historical circumstances that affected the work of Italian physicians and psychiatrists. I will start by outlining some of the most important steps that led to the establishment of an official psychiatry, which presents three distinct features: a professional independence from general medicine, a strong commitment to positivism, and an overt social role. I will move on to show how psychiatry theorised sexual inversion, and the medical issues raised by this 'new' category. Psychiatric manuals illustrate how sexual inversion was officially acknowledged as a mental disorder from the late 1880s, and show the conceptual developments that this new category went through. The first part of this chapter focuses on the psychiatric category of sexual inversion, and provides a detailed account of the first published clinical cases, both male and female. These early case-histories are significant because they shaped later scientific language and methodology (i.e. examination of the body, psychological investigation, and description of sexual inverts' lifestyles); the first examination of a female invert by Guglielmo Cantarano affords great interest in these respects. The last part of this chapter will focus on the various trajectories in medical science that produced such diverse medical formulations of female same-sex desires as the virago, the tribade-prostitute, and the *fiamma*.

The fact that doctors created and elaborated on these clearly defined figures, as well as on the concept of female sexual inversion, confirms that different ideas of female same-sex desires were highly popular amongst the Italian medical community. The Italian case illustrates the richness of medical representations, emphasising the extent to which the psychiatric concept of 'sexual inversion' coexisted with other medical ideas about female same-sex desires. Just as a curved mirror produces different images depending on the angle from which an object is viewed, the last decades of the nineteenth century produced a multiplicity of medical concepts that articulated female same-sex desires from numerous perspectives. Different medical disciplines looked at a series of diverse elements, combined them in different ways, and redefined endlessly the object under observation. Embodiments of female-same sex desires such as the tribade-prostitute had their roots in older medical works such as Parent-Duchâtelet's observations of love relationships amongst prostitutes, and earlier medical studies on hermaphroditism. The virago, the tribade-prostitute, and the *fiamma* have a few elements in common, such as an association of female same-sex desires with sexual excess and masculinity, or a perennial anxiety about women-only environments. Taken together, these medical representations of female same-sex desires complicate current historiographical accounts, which emphasise the conceptual ruptures within the history of medicine introduced by the mental disorder of sexual inversion.

Towards an Italian psychiatry

As early as the first half of the nineteenth century, Britain, France, Germany, and the United States had put in place consistent plans to reform asylums, but it was not until 1904 that Italy passed laws to protect the mentally ill. Because the economic relationship between local administrations and mental institutions was not regulated, asylums were often financed by the inmates' relatives. Mental institutions housed a highly diverse population, ranging from individuals who posed a threat to society and the mendicant poor, to the 'mentally defective' and the 'idiot'. This situation was permanently aggravated by the fact that a wide variety of scientific traditions with divergent definitions of insanity and, consequently, varied therapeutic approaches coexisted in Italy. This, in turn, meant that throughout the Peninsula there existed different kinds of institutions that cared for the insane.[3]

The first measure towards a more organised state of affairs was taken in 1852, when the Director of the Ospedale Maggiore of Milan, Andrea Verga, founded the first specialised psychiatric journal, *Appendice psichiatrica [Psychiatric Appendix]*, a bi-monthly supplement to the Milanese medical journal *Gazzetta medica italiana-Lombardia [Italian-Lombardy Medical Gazette]*.[4] Verga's long-term goal was to reach all the doctors who studied neurosis and diverse forms of madness and for the journal to be an instrument of professional cohesion; thus, from the first issue, the periodical campaigned for a new national model of psychiatry. Animated by a reformist spirit, the *Appendice psichiatrica* appealed to the patriotism of Italian alienists, and called for a professional alliance that would compensate the extremely precarious institutional situation in which they worked. Drawing attention to the most pressing institutional and political problems psychiatry faced, Verga's first move was to urge Italian governments to adopt a common law concerning the mentally ill. According to Verga, such a law would not only provide guidelines for the treatment of the mentally ill in asylums, but also define the social rights of the insane and their criminal accountability. Secondly, he stressed the need to set common policies for the construction and organisation of mental hospitals. Thirdly, he recommended developing a plan to regularly collect medical statistics and a handbook to train asylum staff. Finally, he advised that provisions should be made for insane criminals.[5]

Verga's efforts to raise Italian psychiatry above its almost chaotic circumstances were not restricted to the improvement of the institutional sphere, as strictly scientific matters were not beyond the scope of his suggested reforms. He urged the writing and adoption of an official treatise 'on the moral and the physical treatment of alienation' that listed all its possible causes and predisposing agents. Verga also believed that, in addition to the study of psychology and phrenology, psychiatrists should undertake periodical studies of the anatomical and physiological aspects of the nervous system to keep up-to-date with the most recent developments. In this same vein, he noted how crucial 'the diligent necroscopy of bodies' and the development of 'therapies for the mentally ill' were. New means for the treatment of the nervous system – such as 'animal magnetism, electricity, ether, chloroform, Indian hemp' – were also suggested as areas to be explored.[6] The scope and diversity of Verga's goals illustrate the inchoative state of Italian psychiatry, both at institutional and scientific levels, during this period.

It was not until 1864 that, in conjunction with Serafino Biffi and Cesare Castiglioni, Verga founded the specialised *Archivio italiano per le malattie nervose e più particolarmente per le alienazioni mentali* [*Italian Archives of Nervous Sicknesses and in Particular about Mental Alienations*].[7] This was a crucial development because it guaranteed a uniform circulation of the latest developments on the subject of mental disorders, but most importantly, it was instrumental in the setting up of a national professional network and a unified political plan for the future of the discipline. Verga and his partners were specialists based in Milan and its neighbouring areas, which was the home of a number of outstanding institutions for the treatment of mental illnesses, and some of the most respected medical researchers at the time. Out of the eighty alienists on the whole of the Peninsula, twenty worked in Milan's eight asylums.[8]

In 1873, the *Archivio italiano per le malattie nervose* called a meeting of Italian alienists at the eleventh Congress of Italian Scientists in Rome. The thirteen physicians who responded to the convocation became the founding members of the country's first psychiatric association – the *Società di freniatria italiana*. The term '*freniatria*' was chosen over the more common term 'psychology' to indicate a theoretical distance from the French model developed by the *Société médico-psychologique*, which welcomed interdisciplinary cooperation between psychology and philosophy.[9] Italian psychiatrists, by contrast, conceived mental illness as a strictly constitutional phenomenon, with no metaphysical implications. The association's programme envisaged its members would be strongly committed to positivist experimental methods in anatomical and physiological research.[10] This emphasis on experimental research aimed to give scientific credibility to the new profession, and to meld several traditional approaches to mental illness into a single coherent methodology.[11] The association rapidly became a major forum for the discussion of theoretical and institutional problems. A year after its inception, the *Società di freniatria italiana* hosted its first congress in the city of Imola: 88 members participated and the *Archivio italiano per le malattie nervose* was established as the *Società*'s official journal.[12]

In later years, psychiatrists had to wage two significant battles. On the one hand they fought to achieve social and cultural power in the new state, and were active promoters of legislative reform. This was the particular goal pursued by Verga and the rest of the so-called 'Milan school'. On the other hand, psychiatrists were committed to the advancement of science; establishing their own scientific credibility was the first step towards achieving professional autonomy from the older and more powerful community of general practitioners. This was the main concern of the 'Reggio Emilia school', led

by Carlo Livi, who was the Director of the asylum of San Lazzaro.[13] The
Reggio Emilia school believed that Italian psychiatry lagged behind France
and Germany, which were producing leading-edge research in the areas of
neuropathology, anatomy, pathology, and the clinical physiology of the nervous system. The Reggio Emilia school also regarded experimental methods,
evolutionary theories, and physiological and neurological studies as key
areas to be developed by positivist psychiatry.[14] In the autumn of 1874,
Enrico Morselli and Augusto Tamburini, Livi's assistants, decided to found
a new journal primarily concerned with experimental research on the links
between mental pathology and the nervous system.[15] Thus, in 1875, the psychiatric institute of Reggio Emilia began publishing the *Rivista sperimentale
di freniatria e medicina legale in relazione con l'antropologia e le scienze giuridiche
e sociali* [*Experimental Journal of Phreniatry and Legal Medicine Relating to
Anthropology and Legal and Social Sciences*].[16]

The Reggio Emilia school's approach to mental disorders was hugely influential on subsequent psychiatric studies of abnormal sexual behaviour. This
approach is synthesised in Livi's programmatic article, 'The experimental
method in *freniatria* and legal medicine', published in the *Rivista sperimentale di freniatria* in 1875. In it, Livi asserted that psychiatry should study
the lunatic patient by enquiring into the causes and symptoms of mental
illness, the somatic diseases that accompanied mental disorders, and their
possible remedies. He also outlined the tools of the experimental method
and explained that only 'experimentation', 'clinical observation', and the
'microscope' had the power to advance the scientific status of psychiatry.[17]
If psychiatry wanted to become a science of pure observation, it had to
embrace the experimental method. The bodies of the mentally ill held the
key to behavioural disturbances, and thus had to be examined and measured
with great precision.

Livi's article addressed both his intended methodological programme and
the relevance of the study of mental disorders. In this vein, he proposed
that the study of mental pathology would be a source of insight not only
into the behaviour of insane individuals secluded in asylums, but also and
mainly into that of normal people:

> The inner darkness of the human spirit is better discovered in the
> lunatic than in the healthy person. Factors such as civilisation, educa-
> tion, prejudices, and social conventions have removed certain structures
> and shapes from the primitive or natural man; madness restores this
> darkness to him. Often madness gives us the chance to anatomise, sep-
> arate into its intimate parts and study, this wonderful synthesis of the
> human intellect. [...] madness is a disease that dissolves the natural
> ties of the process of thinking and of the will: like a surgical knife it
> separates, analyses the human spirit; but as a microscope magnifies, it
> exaggerates too.[18]

In Livi's conception there were no set boundaries between mental illness and sanity: madness was an exaggeration of certain normal human characteristics, to the extent that observing how deranged people behaved could lead to a greater understanding of the very essence of human nature. Advancing the claim that psychiatry might reveal aspects of the normal human condition was itself an affirmation of the intrinsic value of psychiatric research. Such emphasis on continuity between normal and pathological states became crucial, as this notion would shape the Italian psychiatric community's approach to the study of sexual behaviour, which was starting to be systematised in the nascent discipline of sexology.

Livi also indicated how psychiatric research was to assist legislators and magistrates in the creation and delivery of just laws.[19] Until then, crime had been studied as an abstract entity. The offender was only punished, instead of being 'scrupulously' studied 'not in the moment of the crime, but in all his antecedent life[;] not in his moral being only, but in its organic complexity, in his physical imperfections, in his morbid hereditary germs, in the sinister influences of age, sex, temperament, illness, discomforts, misery and corrupt physical and moral atmosphere' in which he had always lived.[20] Thus, psychiatrists were encouraged to step beyond the boundaries of their own discipline, and venture into the field of human and social sciences.[21] This would put psychiatry on par with the new anthropological disciplines and, further, such a scientific contribution to the work of legislators and magistrates would be a step towards psychiatry's self-appointed role as society's guide.

The *Rivista sperimentale di freniatria* went on to become one of the most respected journals in the scientific community, and its welcoming interdisciplinary research with anthropology, forensic medicine, and legal studies reveals an important aspect of late nineteenth-century Italian psychiatry. Such interest in interdisciplinary research is confirmed by the *Rivista*'s collaboration with Lombroso, who at the time was practising as an alienist at Pavia. During the late 1870s, the 'Reggio Emilia school' had great expectations of Lombroso's research programme, which tended to establish a nexus between psychiatry and the natural sciences in general, and anthropology in particular.[22]

Lombroso subsequently founded a very influential journal devoted to the promotion of criminal anthropological studies, the *Archivio di psichiatria, scienze penali ed antropologia criminale per servire allo studio dell'uomo alienato e delinquente* [*Archive of Psychiatry, Penal Sciences and Criminal Anthropology to Serve the Study of the Mentally Ill and Offender*], which appeared for the first time in 1880.[23] Lombroso's *Archivio di psichiatria* would become one of most important medical journals that fostered the study of sexual psychopathologies until well into the twentieth century.

This movement towards the study of criminal anthropology also reflects one of the most salient features of Italian psychiatry: the manifest link between professional practice and public engagement that had been articulated since at least Livi's programmatic article. This political dimension of psychiatry was a specifically Italian characteristic, with roots in the *Risorgimento* movement. As the psychiatrist and historian of psychiatry Ferruccio Giacanelli has highlighted, in the years that led to its establishment as an autonomous discipline, there was a decisively 'militant' side to Italian psychiatry. This is confirmed by the fact that psychiatrists such as Livi and Lombroso fought in the *Risorgimento* wars; that Verga, Clomiro Bonfigli, and Leonardo Bianchi were members of Parliament; that psychiatrists like Virgilio, Tamburini, and Bianchi were often involved in the drafting of legislation; and that they were a constant presence in the administrative public sphere.[24] Psychiatrists who had participated in the *Risorgimento* promoted themselves as agents of the government in the building and organisation of a national network of asylums and, further, as instruments for the management of social disorders.[25] Physicians considered it their duty to put an end to the nation's great curses: poverty, illiteracy, alcoholism, and endemic illness. Psychiatrists were committed to giving 'scientific' solutions to these problems through ambitious programmes of social hygiene, as well as through the sanitary and moral education of the population. This reflected a well-established Italian tradition that linked scientific, and above all medical, practice with progressive political thinking and activism. At the core of this tradition is the notion that science was a crucial instrument of civil progress and emancipation, an idea derived from the Enlightenment, as well as a legacy of the Italian *Risorgimento*.[26]

As explained in the previous chapter, in the *post-Risorgimento* years it was a matter of some urgency for the new state to find a suitable replacement for the Roman Catholic Church's control over the moral sphere, and medical practitioners did much to secularise Italian culture. It was believed that the superstition and ignorance spread by religious authority needed to be cleared away, and that the authority of the Church in the sphere of sexual behaviour also had to be eradicated. A considerable number of Italian physicians actively promoted a new sexual morality based on the rational principles of public hygiene. While historians such as Giacanelli have acknowledged the contribution of psychiatrists in constructing a culturally progressive and unified Italy, less is known about their deep commitment to reforming the sexual morality of the new state. Still, in 1894, Lombroso commented that the morality of Italians from the north was different from those in the south, and concluded that this discrepancy had a defining effect on the broader issue of Italian cultural division:

> Even in evil, indeed, Italy is not bound together. [...] It is indeed clear that because of the sexual precocity of some regions, not only must someone who rapes a 12 year old girl in a region and someone who does it in another region not be punished in the same way, but also the age of consent is different in southern regions and especially on the islands.[27]

Sexuality was not only a legitimate field of enquiry for physicians: it was an instrument to survey the living conditions of Italians with a view to improving them so as to effectively unify the new nation. Same-sex desires were one of the areas that had to be looked into if Italians' living standards were to be raised, and such a high-minded aspiration had to be pursued with a scientific spirit.

'Inversion of sexual instinct': towards a new and fashionable psychiatric pathology

Forensic medicine

Nineteenth-century Italian medico-forensic treatises conventionally featured a section called '*Venere forense*' [Forensic Venus], which typically analysed the legal implications of rape, hermaphroditism, abortion, infanticide, and pederasty. A marriage could only be dissolved on the basis of hermaphroditism, in cases of impotency, or permanent infertility. Such cases also required medico-legal observation and judgement: physicians had to ascertain the sex of the person according to physiological and anatomical standards. Unlike ancient French laws, Italian laws did not give the hermaphrodite the right to choose his or her sex after puberty, which in turn gave the physician power over the civil and political rights of an individual, which were different depending on whether one was female or male.[28]

Drawing on the authority of Paolo Zacchia's *Quaestionum medico-legalium* (1621–35), since at least the seventeenth century, forensic doctors strove to identify signs in the rectal area that could prove unconditionally when there had been male same-sex penetration.[29] As explained in the previous chapter, with the Zanardelli Penal Code (1889), male same-sex acts were decriminalised, but pederasty between men was still legally punishable in case of violence, '*querela di parte*', or public scandal. Consequently, forensic medical treatises continued to deal with male same-sex acts throughout the nineteenth century. In this kind of text, doctors typically defined pederasty as an 'act against nature', followed by a list of signs for diagnosing whether sexual intercourse had occurred between two men, and whether the pederast being charged was 'active' or 'passive'. A book published in 1874 by the physician Secondo Laura is representative of this handling of the topic:

> Pederasty, either a miserable sickness that seriously affects even the good, or a vice that befouls vulgar souls, leading often to worse crimes, is the unnatural use of sex through anal intercourse between man and man or man and woman. *Sodomy* is sexual congress with animals.[30]

'*Tribadism*', or the 'sexual intercourse between woman and woman', mirrored the phenomenon of pederasty, but the difference being that female same-sex acts were not discussed in forensic medical treatises so frequently as male same-sex acts.[31] While forensic medical accounts of pederasty typically conveyed a fixed notion of sexual intercourse between males, sex between women was less identified with penetration. On occasions forensic medicine recorded sexual penetration by women mainly in the context of rape, in which an adult woman would typically use her fingers or other objects to break a girl's hymen; however, this kind of sexual assault between women was not interpreted as an act of 'tribadism'.[32]

Sexual inversion

In Italy, the medical category of sexual inversion emerged from the work of psychiatrists addressing issues at the intersection between forensic medicine and psychiatry, with the problem of criminal responsibility raising the question of whether it was possible to restrain sexual impulses. It was Arrigo Tamassia who, in an 1878 article published by the *Rivista sperimentale di freniatria*, introduced the term *'inversione dell'istinto sessuale'* [inversion of the sexual instinct].[33] Tamassia was a prominent Professor of Forensic Medicine at Pavia University who expressed a keen interest in British psychiatric research on the subject of moral insanity, and he considered Henry Maudsley a 'classic' in the field.[34] After concluding his medical studies in 1873, Tamassia had specialised in forensic matters, visiting some of the best psychiatric institutes in Berlin, Vienna, and Paris. His Italian translation of Maudsley's *Responsibility of Mental Disease* (1874) was published in 1875 and marks the beginning of Tamassia's study of moral insanity, one of whose subclasses was sexual perversion.[35] Upon returning to Italy in 1876, Tamassia succeeded Lombroso as Chair of Legal Medicine at Pavia University (see Chapter 5).[36]

At the heart of Tamassia's study of *'inversione dell'istinto sessuale'* was Westphal's 1869 diagnosis of *'Conträre Sexualempfindung'*. Indeed, Tamassia's article was a response to his German colleague's work. Tamassia thought that an individual's awareness of its 'individuality' developed according to the 'character of [its] own sex', and that most passions, tendencies, and human ideas were directly or indirectly caused by the sexual instinct. He explained that while physiologists recognised the sexual instinct only as an instrument to reproduce the species, psychiatrists and anthropologists conceived of the sexual instinct as one of the most powerful elements of psychological life, which had a decisive influence on every aspect of the individual's personality. To support his claim that there existed a causal relation between sexual instinct and psychological activities, Tamassia referred to puberty, when the recently appeared sexual instinct turned into mental activity, and to some forms of insanity that were caused by a 'perversion of the sexual instinct'.[37]

Building on previous studies of sexual perversion by Henri Lègrand du Saulle, Ambroise Tardieu, Casper, Lombroso, Westphal, and Krafft-Ebing, Tamassia argued that the anomalies of the sexual instinct could be divided into four groups: absolute lack of sexual instinct, exaggeration of the sexual instinct, manifestation of the sexual instinct earlier or later than 'phys-iological time', and 'perversion' of the sexual instinct – characterised by 'non-physiological' means to satisfy such an instinct.[38] This division was tacitly taken from an article Krafft-Ebing had published the year before (1877), in the *Archiv für Psychiatrie und Nervenkrankheiten*, in which the Austro-German psychiatrist had defined perversion of the sexual instinct as a degeneration of the nervous system.[39]

Tamassia explained that following Westphal's study on sexual inversion, scientists had gathered a small but sufficient sample of approximately twenty cases which, according to the Italian forensic doctor, delineated a clear enough picture. What singled out all cases of sexual inversion was lack of sexual attraction – maybe even disgust – towards the opposite sex and a premature manifestation of the sexual instinct with no visible physical variations of the genital organs.[40]

Tamassia himself considered the term 'inversion of sexual instinct' too vague, because it encompassed two different ideas. On the one hand, the sexual invert acknowledged being a member of one sex, but displayed all the psychological attributes of the opposite. The invert's whole way of thinking was shaped by 'this dualism between the sense of his own personality and the materiality of the organism'.[41] On the other hand, the sexual invert 'pre-ferred' people of the same sex as sexual partners.[42] This is to say that, in a startlingly insightful fashion for early psychological writings, Tamassia noted that the term 'inversion of the sexual instinct' conflated two distinct issues: on the one hand there was psychological gender inversion, or the sensation that one's mind is the opposite gender of one's physical sex; on the other, sexual inverts preferred satisfying their sexual instinct with an individual of the same sex.

In his attempt to overcome such confusion, first of all Tamassia rejected Westphal's adoption of a division between 'neuropathic' and 'psychopathic' states when interpreting sexual inversion.[43] As Tamassia explained, Westphal had indicated that the 'congenital perversion of the sexual instinct'[44] was accompanied by complex psychical and nervous conditions like the heredi-tary influence of a state of depression; however Westphal's understanding of 'true form of insanity'[45] did not include sexual inversion.[46] Instead, Tamassia considered the inversion of sexual instinct a fully-fledged disease, which ran counter to Westphal's conception of it as a partial alteration of the nervous system that could be the cause of other morbid symptoms, but that was not itself a genuine malady.[47]

Tamassia believed inverts lacked the basic element of mental normalcy, the 'unity of the self', or 'the general sense of [their] own existence'. Tamassia clarified what he meant, noting that even if the invert acknowledged he belonged to a specific sex, in his mind he refused the 'characteristics' associated with it, thus denying the truth. In conclusion, what defined sexual inverts was an erroneous perception of the 'self'.

More than same-sex desire itself, what made sexual inversion an authentic mental pathology was the psychological gender inversion, because this destabilised the individual's personality. It was this aspect – the feeling that one's mind was of the opposite gender to one's physical sex – that Tamassia emphasised as an unequivocal sign of disease, and it was this aspect that was at the centre of his analysis. Tamassia opined that *a priori* this perversion testified to:

> a limitation of the intellect, even without drawing upon cases already studied. [...] This is because the morbid impulse is so fatally accompanied by mental distress that it does not succeed in grasping the exact knowledge of its relationship with the species, and even if the latter is apprehended, in the invert the willpower is lacking.[48]

A healthy person's willpower was driven by intelligence, which made it possible to identify mistakes in the perception of reality; the sexual invert's discernment did not operate thus.[49] In most cases, Tamassia argued, sexual inversion could be traced back to human heredity: epilepsy, hysteria, general insanities, and other mental disorders and eccentricities always characterised the family history of sexual inverts.[50] It was not surprising, Tamassia concluded, that an all-round nosographic picture of sexual inversion included a diverse range of mental diseases or abnormal psychical states such as depression, irritability, and suicide attempts.[51] To summarise, sexual inversion was 'a deep psychopathic state' and – quoting Krafft-Ebing's 1877 work – a 'serious functional degeneration' as well.[52]

It follows from Tamassia's explanation of sexual inversion as a jointly 'neuropathic' and 'psychopathic' state that inverts could not make rational judgements. To demonstrate his theory, Tamassia presented the case-history of 'C.P.', a congenital invert from peasant stock, whose father was 'nervous'[53] and his mother 'half hysterical';[54] he had a 'half idiot'[55] brother, and an 'eccentric' one.[56] The case-history revolved around C.P.'s notoriously feminine general aspect, and every 'eccentric' aspect of C.P.'s life was taken to illustrate his mental disorder. Tamassia revealed that since the age of twelve, C.P. had preferred domestic work, at fifteen began to grow his hair long, and for a six-month period at the age of twenty, he had worn feminine attire. Having been imprisoned for the theft of a ring from a house where he was employed as a domestic servant, he was thirty-three years old and still interned when he first met Tamassia. C.P. had a normal physical conformation, and although he exhibited a degree of 'narrowness of mind',[57] he did not perceive any abnormality about his own 'perversion'.[58] Tamassia's article concluded with a discussion of sexual inverts' accountability for criminal matters: because sexual inversion should be considered a disease, inverts were not fully aware of their actions, not even criminal ones.[59] In this logic, C.P.'s sexual inversion was used to reduce his sentence for theft.[60]

The establishment of sexual inversion as a mental illness in psychiatric manuals

Tamassia's work paved the way for later studies on sexual inversion, and his term *inversione sessuale* [sexual inversion], rather than the German *conträre Sexualempfindung* [contrary sexual feeling], was first adopted by Charcot and Magnan in France in 1882, and subsequently in Britain and the United States.[61] Influenced by Tamassia's ground-breaking analysis, in the 1880s Italian physicians were moved to collect case-histories and scrutinise sexual inverts' lives, fantasies, and desires. Most of these case-histories were drawn from observations conducted in asylums and prisons, and consequently the subjects had often committed crimes or displayed anti-social behaviour. Physicians endeavoured to reconcile the two 'phenomena' Tamassia had described: individuals of one sex recognised themselves psychologically as members of the opposite sex, and sought sexual partners of their same physiological sex. These observations slowly drifted away from Tamassia's emphasis on the sensation that patients' minds were of the opposite gender to their physical sex, and equally addressed both the inverted psychological identification and same-sex longings. This first batch of Italian case-histories that followed Tamassia's diagnosis of sexual inversion as a mental pathology achieved remarkable notoriety in Europe.[62] But it was not only through case-histories that sexual inversion became a well-known subject for the Italian medical community; from the late 1880s it had become rapidly systematised in psychiatric manuals, and in 1902 it formally entered the official classification of mental disorders at the eleventh congress of the *Società di freniatria italiana*.[63]

While in the last two decades of the nineteenth century British psychiatrists seemed reluctant to engage with the concept of 'sexual inversion', their Italian colleagues routinely analysed sexual inversion as a symptom of innate characteristics in psychiatric manuals and textbooks. It was not until the twentieth century that British psychiatric treatises started assigning full chapters to the subject of sexual inversion; until then, many British psychiatrists were inclined to believe that same-sex acts and desires in general were a 'vice' – a position supported by the authority of Daniel Hack Tuke's *Dictionary of Psychological Medicine* (1892).[64] Instead, Italian psychiatric manuals used to train young medical practitioners in universities illustrate that the Italian psychiatric community had reached a consensus on the subject of sexual inversion, which was a mental disorder. These textbooks also show how, in Italian universities, there was no attempt to hide such topics from medical students, and young doctors were in fact encouraged to study sexual psychopathologies. Such encouragement would have been inconceivable in Britain, where the most important text on sexological matters, Ellis's *Sexual Inversion*, was not received favourably in psychiatric circles, and had even been banned from circulation amongst the general public. If a young British practitioner was to satisfy his curiosity, he had to rely on foreign sexological works. As Ellis himself had noted, in his own country it was a 'hard task for any student' to examine the 'problem of sex', even under the most favourable circumstances, because even scientific investigations of sexological issues were met with 'distrust' and 'opposition'.[65]

The situation was different in Italy, where the study of sexual psychopathologies entered the curriculum of conventional psychiatric training in the late 1880s. Enrico Morselli, Leonardo Bianchi, Eugenio Tanzi, and Sante De Sanctis published some of the most important Italian psychiatric manuals of the time, which makes it possible both to illustrate how sexual inversion became an established object of enquiry in psychiatry and to trace the conceptual shifts undergone by medical theories in their attempt to explain sexual inversion. In addition to displaying the ongoing nosologic developments on the subject of sexual inversion, these psychiatric manuals are interesting because they illustrate a progressive move away from explanations that emphasise the pathological aspects of same-sex desires, to accounts that emphasise its normalcy under certain circumstances. This shift runs parallel to the displacement between ideas of congenital inheritance and ideas of acquisition.

The first Italian psychiatric treatise to deal with sexual inversion in any systematic way was Morselli's *Manuale di semejotica delle malattie mentali* [*Manual of Semiotics of Mental Illness*] (1885–9).[66] Morselli discussed same-sex desires in the context of a more general theory of the reproductive instinct. He explained that the sexual function was a critical part of human health, because it was the origin of a vast array of feelings, thoughts, and tendencies, which in turn meant that if the sexual instinct was in any way anomalous, it could generate morbid manifestations across an ample spectrum of behaviours, practices, and feelings. In nature, Morselli explained, animals have two basic instincts: conservation of the individual and reproduction of the species. As in nature, the reproductive instinct is stronger than the self-preservation instinct; in humans, the sexual function has the greatest effect on psychological life.[67] Moreover, in man, the reproductive function triggered a wide range of psychological needs, some of which could lead to the appearance of sexual perversion. As a result of the process of civilisation, the brain had developed beyond the individual's natural physiological needs, and this gave rise to sexual perversion. According to Morselli, it was difficult to say exactly when the sexual instinct stopped being a natural manifestation and turned morbid, but it was certain that modern civilisation was consistently to blame for this deviance.[68] This attention to the influence of modern civilisation combined social and historical explanations of sexual perversion, and was related to current evolution theories.

Drawing on the German biologist and physician Ernst Haeckel's evolution theory and Lombroso's theory of atavism, Morselli put forward a phylogenetic explanation that accounted for both normal and deviant sexualities.[69] Haeckel had claimed that the biological development of an individual's organism (also called ontogeny), parallels and summarises its species' entire evolutionary development (or phylogeny). Ontogeny was interpreted as a purely biological unfolding of events whereby an organism gradually evolved from a simpler to a more complex level. Human development was characterised by an ontogenic process of sexual differentiation, which mimicked the phylogenic evolution from original primitive bisexuality to the civilised man's monosexuality. As part of the normal development of the human organism, the original bisexual disposition was replaced by monosexuality, which was reinforced by an increasing differentiation of male and female secondary sexual characteristics; the recessive sex left only a few abortive signs such as facial hair on females. Morselli explained that organisms had developed from being very simple but stable, to being much more complex or structured, and so more unstable than in their earlier stage.[70] Although the human organism was the most developed in the whole natural world, it could present the most aberrant perversions, which demoted the individual from its place as the most sophisticated product of evolution to a shameful reminder of a remote past. When regressing to primitive conditions, men returned to the primordial bisexual disposition, which also obscured the individual's secondary sexual characteristics. Morselli pointed out that although there had been 'unnatural intercourses' throughout human history, in contemporary times they were to be considered a result of atavism.[71] Sexual perversion was caused by obsessive morbid impulses, and represented 'stigmata of degeneration'. This framework did not prevent Morselli from suggesting that sexual perversion was caused also by modern conditions, which in turn means his argument included the environment as a causative agent of sexual perversion. He thought that especially in the higher social strata, sexual perversion was the result of vicious habits: civilisation stimulated the development of refined libidinal tendencies, and allowed them to be exercised for too long.[72] In Morselli's system, therefore, sexual perversion could be caused by either morbid heredity or a corrupted milieu, but only in those individuals who were already tainted by degeneration.

Morselli distinguished between sexual perversions based on their deviation from heterosexual intercourse: masturbation (*'Venere solitaria'* in men), *'clitoridismo'* (*'Venere solitaria'* or masturbation in women), pederasty (*'amor greco'*, Greek love), sodomy (*'rapporti preternaturali'*, relationships against nature between men, or *'mutui stupri'*, mutual rapes), tribadism or *'eufemismo'* (*'amplexus intra mulieres frictrices'*), bestiality (*'coitus cum bestia'*), sapphism (*'cunnilingus'*), *'fellare'* and *'irrumare'* (*'ore polluere'*).[73] Special attention was devoted to sexual inversion that manifested itself as 'psychic degeneration'; sexual inversion was included in the broader category of *'parafrenie'*, and was associated with brain lesions. In Morselli's nosology, *'parafrenie'* included criminal psychosis, *'mattoide'* insanity, and '*folie raisonnante*'.[74] Sexual inversion occurred only in individuals whose development had been halted at a primitive stage, a condition shared by congenital criminals, children, women, and the insane. The sexual instinct of the invert was more closely related to the instincts of the primitive world, which meant they were less advanced in the evolutionary process, and thus had to be kept under control.[75] 'Homosexuality' could also ensue after a breakdown in 'physiological power', quite often due to neurasthenia caused by frequent masturbation or a psychopathic state.[76]

Like most Italian sexologists between 1880 and 1890, Morselli adopted Krafft-Ebing's distinction between congenital and acquired sexual inversion.[77] This was significant because, while admitting it was impossible to treat congenital inverts, he stated it was possible to cure those who had allegedly acquired their inversion. On the one hand this theory accounted for any failures to steer individuals towards appropriate sexual behaviours, and on the other it validated the study of sexual inversion as a step towards containing the spread of the phenomenon. The picture would change closer to the turn of the century, as Italian psychiatrists who specialised in the study of sexual inversion began to move away from Morselli's explanation.

At the turn of the century, inversion was still considered a congenital phenomenon, a 'constitutional anomaly', an arrest of cerebral-psychic development, or a 'psychic degeneration', to quote some of the most common terms in psychiatric manuals. However, this period of Italian psychiatry is marked by a growing interest in embryological theories, the mechanisms of mental association and memory, and the influence of milieu in mental development. While Italian psychiatrists insisted that both congenital and acquired inversions were caused by a constitutional weakness or degeneration, they seemed to invest more energy in understanding acquired homosexuality. This new approach to sexual inversion privileged the study of the individual's social context and early life. This transition affected a further crucial shift in medical writing: the boundary between normal and abnormal moved, and same-sex desires were now considered normal phenomena, at least until puberty, and during adulthood in specific circumstances. Such a change in sexological writings is reflected in the works of a few of the most prominent psychiatrists of the time: Bianchi, Tanzi, and De Sanctis.

In 1904, when he published his *Trattato di psichiatria* [*Treatise of Psychiatry*], Bianchi was Professor of Neuropathology at the University of Naples, and Director of the Neapolitan asylum.[78] Bianchi's treatise was promptly translated into English in 1906 by James H. MacDonald, senior assistant-physician at the Govan District Asylum. Bianchi's work was relevant to British medicine not on account of its discussion of sexual inversion, or sexuality for that matter: Bianchi was known and respected by his British colleagues for his experimental research in the fields of histology and the localisation of brain functions, which had been published in the British journal *Brain*, and for his collaboration with the *Journal of Mental Science* since 1897.[79] A highly esteemed psychiatrist in both Britain and Italy, Bianchi was a very active member of the Italian medical community, where, amongst other things, his contributions to the sexological journal *Archivio delle psicopatie sessuali* had helped popularise sexological research (see Chapter 6).

Bianchi studied sexual inversion in his *Trattato di psichiatria* as part of the larger category of 'mental pathologies characterised by a lack of evolutionary psycho-cerebral development' that encompassed '*frenastenie*', '*parafrenie*', congenital moral insanity, epilepsy, hysteria, paranoia, fixed ideas, and neurasthenia.[80] He argued that in order to understand the normal and anomalous development of the sexual instinct, it was necessary to consider its phylogeny and ontogeny. From a phylogenic perspective, Bianchi thought that while in all higher animals the senses contributed to the satisfaction of the sexual instinct, humans were the only species for whom consummation was reliant on an 'aesthetic-intellectual factor'. This is to say that intellectual faculties interacted with the sexual instinct only in the case of humans. His ontogenic argument was organised around the fact that in the first months of life the foetus was sexually undifferentiated. This, according to Bianchi, was the source of the 'psychic hermaphroditism' present in all individuals, which was manifest until adolescence and remained 'latent' in adulthood. An evolutionary lapse of the sexual organs or the nervous system explained why male inverts looked like women, and vice versa. Sexual inverts with strong somatic traces of the opposite sex, or inborn homosexuals, were less common than 'occasional homosexuals', and should be considered untreatable. 'Occasional homosexuals' might have sexual relationships with both sexes and their physical appearances were coherent with their genital sex.[81] Bianchi believed that 'occasional homosexuality' was acquired, and as such it could be medically treated, or at least contained, by withdrawing the individual from the social and cultural conditions that led to sexual inversion.[82]

In the following year, 1905, Tanzi published *Trattato delle malattie mentali* [*Treatise of Mental Illness*].[83] As Professor of Psychiatry at the University of Florence, Tanzi was one of the country's most renowned psychiatrists of the day. Tanzi recognised that love might take different shapes and manifestations, but when love made reproduction impossible, it became a medical problem.[84] Sexual inversion belonged to the group of mental illnesses known as 'degenerative mental anomalies', which also included 'constitutional immorality', 'paranoia', and 'weakness of mind'.[85] That said, Tanzi pointed out that homoerotic feelings in environments like colleges and schools could be considered normal if they were temporary; it was only when such feelings became permanent that sexual inversion was considered a genuine mental disorder.[86]

There were several factors that contributed to the acquisition of sexual inversion, such as memory or associative intellectual functions. Tanzi believed that if in early memories pleasure was associated with a person of the same sex, each time the individual sought pleasure, it would have to reproduce the original equation, hence its inclination towards those of their own sex. A premature occurrence of the sexual instinct or life in certain social contexts such as schools or colleges would also increase the possibilities of developing inversion. Tanzi specified that even if homosexuality was acquired, it was acquired due to a 'constitutional anomaly', an organic weakness in the body.[87]

In later years, Tanzi wrote another manual, *Psichiatria forense* [*Forensic Psychiatry*] (1911), where he jettisoned all physical explanations of sexual inversion. Sexual inversion, he now posed, was purely 'psychological [in] origin'. He rejected the concept of congenital inversion and argued that same-sex desires were caused by erroneous mental associations.[88] He pointed out that most male inverts were virile and looked for other virile men, while most of the female inverts were feminine and desired women of a similar demeanour. According to Tanzi, this was evidence that homosexuality did not have a bodily origin, but an exclusively psychological one.[89] This shift in Tanzi's explanation of sexual inversion was related to his moving away from degeneration theory, which was bodily grounded, and towards Alfred Binet's work on fetishism, whose influence on Italian sexological works had been increasing steadily in recent years.[90]

Unlike Tanzi, other psychiatrists did not completely abandon organic aetiology, but even they were warming up to the possibility that psychological elements might explain sexual inversion. Thus, for example, in 1911, Sante De Sanctis combined degeneration and social causes to explain sexual inversion. He still maintained that the inversion of the sexual instinct was a 'psychic degeneration', as evidenced by the fact that sexual inverts had an undeveloped intelligence. Like all degenerates, sexual inverts were 'predestined' to be so – the origin of their sickness was hereditary. In De Sanctis's opinion, however, congenital inversion was a rare phenomenon. The cause of inversion was often to be found in the patients' milieu, especially if they lived in a school or a college. The imitative basis of human behaviour explained why sexual inversion was so widespread in such environments. According to De Sanctis, men experienced same-sex desires at puberty because there was a 'latent bisexuality' in all human organisms.[91] From Morselli to De Sanctis, Italian psychiatrists combined a biological and a social view of sexual inversion, which was in line with the Italian tradition of interpreting mental illness as a result of both social and organic causes.[92]

Bianchi's, Tanzi's, and De Sanctis's explanations of sexual inversion illustrate how, at the beginning of the twentieth century, the theory of degeneration had proved insufficient to account for deviant sexualities; theirs are a sample of the tentative approaches that superseded degeneration as an explanatory mechanism. At the turn of the century, Italian psychiatrists began to accept that all individuals were latently bisexual, at least until adolescence. This change also responded to conceptual shifts taking place in French and German psychiatric research, which influenced Italian theories of sexual inversion. Two theories began to gather momentum during the last decade of the nineteenth century: the 'embryological' and the 'association' theories. The former was originally put forward by the French physiologist and endocrinologist Marcel Eugène Émile Gley in the 1880s, and was subsequently popularised by the French physician Julien Chavalier and Krafft-Ebing in the 1890s. The 'association theory' was put forward by the French psychologist Binet at the end of the 1880s.[93] Italian scientists promptly embraced recent embryological and psychological studies in an attempt to overcome the failure of degeneration theory to give a comprehensive account of sexual phenomena.

Building on recent observations of the sexually undifferentiated embryo in the first weeks of gestation, in 1884, in the pages of the influential French *Revue Philosophique*, Gley advanced the hypothesis that the original anatomical bisexuality of the embryo left a physiological trace in human beings, and that these traces of bisexuality could be the starting point of sexual inversion. While this biogenetic hypothesis was further developed by a few American psychiatrists such as James C. Kiernan and G. Frank Lydston in the late 1880s, in Europe it initially passed largely unnoticed.[94] Only in 1893 did Chavalier base his explanation of sexual inversion on the bisexuality of the human foetus during the early period of gestation.[95] Finally, in 1895, Krafft-Ebing published 'Zur Erklärung der conträren Sexualempfindung' ['In Explanation of Sexual Inversion'], in which he subscribed to and elaborated on the biogenetic hypothesis of sexual inversion.[96]

The second theory that affected Italian psychiatric ideas on same-sex desires in the last decade of the nineteenth century came from Binet's research on sexual arousal in connection with non-living objects. Basing his observations on the phenomenon of 'fetishism' (a term he coined himself), Binet argued that prevailing medical accounts of sexual perversion in hereditary terms did not solve the more fundamental problem of how such aberrations had been acquired in the first place. While he did not completely reject degeneration as the fundamental cause of sexual perversion, he believed that heredity could not account for the specific ways in which fetishism and other sexual perversions developed in each patient.[97] Binet argued that sexual psychopathologies such as fetishism and sexual inversion were psychologically acquired by accidental exposure to events in early childhood. For instance, according to Binet, in individuals predisposed by heredity, when the first lively sexual excitement during childhood concurred with the sight or touch of a person of the same sex, it gave rise to a 'stable association', resulting in a predisposition for sexual inversion.[98] Binet explained that sexual inversion took place when there was an association between ideas and pleasurable feelings, without completely ruling out an organic link, because he observed sexual perversion only affected individuals who were already weakened by bad heredity. His aetiology of sexual perversion thus privileged psychological events in early childhood, and pushed degeneration theory into the background.

Until the beginning of the twentieth century, sexual inversion could still be classified as a 'psychic degeneration' caused by an 'arrest of cerebral development', as in Tanzi's *Trattato delle malattie mentali* or De Sanctis's *Trattato di medicina sociale* [*Treatise of Social Medicine*] (1911). Yet psychiatry's adoption of embryological and 'association' theories to explain sexual phenomena had helped to blur the boundary between normal and abnormal sexuality. Indeed, the argument suggests that a lapse in individual evolution caused the original bisexual nature of the embryo to reappear, which is a crucial development inasmuch as it implies that all humans had 'bisexual characteristics'.[99] As a result of these shifts, Italian psychiatrists at the turn of the century tended to hedge their bets and regard sexual inversion as both a mental disorder and a normal behaviour if it occurred in a single-sex environment before or during adolescence. This conceptual change echoed another shift in sexological debates: while in the 1880s most case-histories described prison or asylum inmates, towards the turn of the century medical writers were increasingly drawing attention to cases registered in 'normal' social settings. As I show later in this chapter, the active interest in school environments was part and parcel of this process, whereby in psychiatry, same-sex desires increasingly entered the domain of the normal.

Asylums as laboratories for tracking female homosexuality

Earlier in this chapter I have shown how male same-sex desires were discussed in a range of Italian psychiatric writings, from case-histories to psychiatric and forensic medicine manuals. Various monographs on sexual psychopathologies were published in the 1890s, such as Silvio Venturi's *Degenerazioni psico-sessuali nella vita degli individui e nella storia della società* [*Psycho-sexual Degenerations in the Life of Individuals and in the History of Society*] (1892), which covered the topic of sexual inversion at length. Within this rich medical literature, female same-sex desires occupied a prominent position. On the one hand, psychiatric writings treated female sexual inversion as the mirror of male inversion, so that where psychiatrists looked for feminine characteristics in male inverts, they looked for masculine features in female patients. On the other hand, psychiatrists and other kinds of medical practitioners acknowledged that female same-sex desires were not a mere mirror of the male phenomenon, but a condition with distinct features. The last two sections of this chapter will discuss the peculiar features of female same-sex desires as portrayed in a range of medical writings.

As the second section of this chapter has discussed, Tamassia should be credited with having introduced the study of sexual inversion to Italy in

1878, and having put forward the first Italian clinical observation of a male invert. The first case-history of female sexual inversion was published in 1883 in *La psichiatria, la neurologia e le scienze affini* [*Psychiatry, Neurology and Kindred Sciences*] by Guglielmo Cantarano, a Professor of Clinical Medicine and Neuropathology at the University of Naples.[100] This case-history became well known both in Italy and in other European countries such as Britain and Germany for being one of the first to focus on a woman.[101] At a time when very few case-histories of sexual inversion had been printed, publishing a case often amounted to fashioning a new subject within the international medical community. Indeed, new observations in an expanding field would guarantee recognition in the international medical community, because these cases (and the bold scientists who authored them) got a lot of exposure.

Case-histories of female sexual inverts are a relevant source of information in regard to the choice of subjects, the features physicians attached most importance to, and in general how female gender was theorised.[102] Cantarano's analysis of female inversion followed a pattern that would become the standard approach to case-histories of women in the 1880s, in that mental clinics served as the privileged location for observing female sexual inversion. Following the earlier study undertaken by Tamassia, Cantarano pointed out that:

> Inversion of the sexual instinct includes that form of psychical and instinctual anomaly, whereby the individual of a given sex feels the intellectual and instinctive characteristics of the opposite sex, and is driven to love persons of the same sex, showing rejection or indifference towards individuals of the opposite sex.[103]

According to Tamassia, the defining element in sexual inversion was a patently mistaken 'perception of the self', in itself a marker of serious psychopathology. Cantarano's theory contrasts with Tamassia's because the main point at issue was that the object of sexual choice conflicted with the individual's 'sexual biological organisation'. Against the grain of Tamassia's authority, Cantarano observed that sexual inverts could preserve the awareness of belonging to their biological sex.[104] Sexual inverts could appear to be almost normal, and it was observed that not everybody recognised they were 'crazy' people at first glance. The expert doctor, however, could look beyond this level and see inverts lacked the 'right harmony' which would enable their intellectual faculties to function properly. As Cantarano explained, a 'congenital' inversion might affect people who are:

> apparently sound of mind, and in whom only the scrupulous alienist finds the imbalance between the various intellective faculties, or the exaggerated predominance in the latter of feelings and instincts, or the perversion of some tendencies, or [...] a transformed morbid inheritance.[105]

Cantarano conceived of sexual inversion within the early nineteenth-century framework of partial insanity, also called *'folies raisonnantes'*. Early nineteenth-century alienists such as Jean Étienne Dominique Esquirol in France and James Cowles Prichard in Britain believed that some forms of insanity were characterised by a single pathological preoccupation in an otherwise sound mind, or uncontrollable passions, which otherwise did not stop the patient from reasoning and judging coherently. In other words, these forms of disorder expressed themselves as an obsession with certain ideas, which had no effect on the overall cognitive and reasoning faculties.[106] Cantarano suggested that sexual inverts were not prey to delusion because they understood the perceived contradiction between the fact that they had a biological sex but were attracted to those of their own sex. As he wrote:

> In them *the awareness of their own personality* [in relation to biological sex] *is retained*, and thus arises the tremendous dualism between their own acknowledged physical organisation and the opposite and conflicting sexual tendency. In this [...] category of patients, the inverted sexual feeling is usually congenital and immanent.[107]

That they were aware of this apparent conflict between biological sex and sexual tendency did not mean sexual inverts were mentally healthy; they were still unable to satisfy the sexual instinct in a normal physiological way. Cantarano believed that the sexual instinct was responsible for 'the search for [a mate of] the opposite sex and of the same species, with which it is possible to enter into genital intercourse for the benefit of reciprocal enjoyment'.[108] As sexual inverts broke the first rule of normalcy ('the search for the opposite sex'), it was necessary to study them in order to understand human nature itself, and thus satisfy psychiatry's ultimate purpose.

Cantarano's case-history epitomised the psychiatric anomaly of female sexual inversion in many aspects; first of all because of her unusually active life for a woman, which was a common element in many case-histories of female inverts. X, the young woman observed by Cantarano, was apparently a born sexual invert. She came from a relatively poor family and had been a little wild in her childhood years: X used to have tantrums and throw herself on the ground without any apparent reason and did not go to school, which was why she had never learned to read or write. Instead, she preferred to wander the streets begging, sometimes sleeping away from home. She had eventually run away from the family home and so was forced into a house run by a philanthropist, but she fled this house at a later date. She was then placed in an asylum for 'dangerous girls' where she became fond of Rosina, an ex-prostitute. They had sexual relations and X subsequently corrupted all the other 'weak' girls in the asylum. After escaping, she took to wearing male clothes and spending her time with prostitutes, which led to her being prosecuted for charges of disorderly conduct. When the police discovered she was a woman and still a virgin, she was returned to her father because she could not be listed on the prostitutes' register. In turn, her father sent her to a mental clinic, where she managed to corrupt other patients and number of nurses; Cantarano met her in this particular institution.[109]

Following the standard procedure for case-histories, Cantarano looked for a record of disorders in his patient's family and concluded that X's psychosexual pathology could in fact be traced back to her family's past, and her sexual inversion could be foreseen in her childhood given her 'degenerate' heredity.[110] Cantarano noted that X was the daughter of a second marriage, that her mother suffered from a thyroid condition, and that her brother was a wastrel with a propensity for insanity.[111] More than the physical conditions inherited from her family, it was the fact that X was quite devoid of modesty and disliked being a woman that caught Cantarano's attention:[112]

> She does not show any fondness for family, for domestic life, for women's attire or the tasks of her sex. She does not fancy being admired and courted by young men; she is not driven by a desire for marriage; nor does she show the discretion and modesty typical of young women. A vagabond existence, the choice of masculine jobs, the loathing of men and an attraction to the same sex, the boldness of a young and dissolute man, all give her a character that harmonises completely with that of a young man, of someone on whom the reins have been loosened.[113]

According to Cantarano, there was nothing exaggerated about X's sexual inversion: she displayed 'the entire and organised expression of the sexual feeling of a man in a woman's body'.[114] X was even able to control her behaviour to mask any attitude or action that might give away her sexual perversion.[115] Her facial physiognomy was not considered particularly feminine; her muscles were developed and her skeleton was not normal for a woman,[116] which does not necessarily mean X was not attractive – Cantarano in fact remarked that she was 'pretty' and that men flirted with her.[117] Her sexual organs, described in detail, were normal.

According to Cantarano, cases of female sexual inversion were less known because women were more skilled than men at hiding their 'perversion', and also because they were typically more prudish, and hence disinclined to discuss sexual matters.[118] He declared he knew only four case-histories of female sexual inverts: that of X, and those analysed by Westphal, H. Gock, and M. Wise.[119] At the time Cantarano was writing, case-histories of female sexual inversion circulating within the scientific community were rather scarce, which did not stop him from drawing the bold conclusion that love in male inverts was more pure and without carnal sexuality. Cantarano believed that male sexual inverts could be satisfied by the practice of onanism, while women were more sensual and were less skilful at controlling their will.[120] This last observation contradicted the findings of his own case-history, since he had pointed out that X's self-control was remarkable.

Cantarano had observed X, who was interned in an asylum in the hope that she would recover from her sexual inversion and her anti-social tendencies, which had led her to provoke some public disturbances. Therefore sexual inversion resulted in the patients' internment, but the asylum was also believed to be a cause of sexual inversion. Indeed, other psychiatrists at the time thought that the atmosphere of asylums, added to an unfavourable diagnosis such as insanity, could be conducive to sexual inversion. In the 1880s, asylums became laboratories for the observation of female inverts.[121] Italian sexologists sourced their cases from a variety of women-only environments: they focused on asylums and prisons at first, and in the 1890s turned to boarding schools, which were believed to be hotbeds for female inversion. While this rather voyeuristic fashion certainly led to an increase in the number of published psychiatric case-histories of female inverts, other types of medical writings continued to explore female-same sex desires, and eventually different embodiments of female same-sex desires overlapped with each other.

Figures of female same-sex desires

Since the beginning of the nineteenth century, Italian experts in jurisprudence had joined physicians in their efforts to delineate the specific characteristics of female deviancy. In the aftermath of the French Revolution, most of the states in the Italian Peninsula tried to control female sexuality by implementing laws regarding adultery, abortion, infanticide, and seduction. At the same time, following the ancient principle of women's weakness (*infirmitas sexus*), they also limited the punishment that could be inflicted upon the female body.

The French Revolution had prompted the rise of women's first claims to equality, which in turn urged jurists to justify civil and penal inequality between men and women. The sole weight of tradition proved insufficient to validate this inequality, and thus the jurists started to wonder about the specific emotional and behavioural characteristics of female physiology, particularly when it came to women's ability to reason and understand laws. The juridical paradigm of female inferiority was later linked to the scientific paradigm of the natural inferiority of women, thereby confirming the traditional argument about women's limited rationality. Women were different to men by nature, and therefore had to be made the object of scientific study. The unification of Italy in 1861 and the adoption of national laws that followed it generated a renewed interest in legal and medical debates surrounding women. As mentioned in the previous chapter, the Civil Code of 1865 not only ignored class distinctions amongst women, but declared all individuals to be equal before the law regardless of their sex. On the whole, however, women did not have access to many civil rights.

By the late nineteenth century, the increasing influence of criminal anthropology in Italy multiplied the number of Italian studies on the distinctiveness of female deviancy. The new medico-legal field paid particular attention to the specific ways women loved. According to nineteenth-century criminological theories, man was characterised by 'egoism' and woman by 'altruism'; this meant that in the sexual sphere, man looked for sexual pleasure, whereas woman sought to satisfy the need to reproduce.[122] Female deviancy was characterised by an abnormal sexuality, with prostitutes being considered the female offender *par excellence*.

All these elements contributed to a proliferation of medical discourses about the specific characteristics and deviations of female sexuality, including female same-sex desires. Within these debates it is possible to discern a few coherent and specific representations of same-sex desires other than sexual inversion: the virago, which synthesised ideas of excessive sexuality and masculinity; the 'tribade-prostitute', popularised by Italian criminal anthropology; and the *fiamma*, or homosexual relationships in 'normal' women-only environments, which featured mainly in psychologically oriented studies. These representations pre-dated the concept of sexual inversion, and point to existing medical research on psychological and physical attributes of those women who engaged in same-sex practices. The fact that in the last decades of the nineteenth century these various representations coexisted problematises current historiographic debates on same-sex desires, which insist the category of sexual inversion effected a full epistemological rupture.

Virago

Medical writers had traditionally paid attention to the specific traits of female nature, and noted that mental disease in women resulted from 'moral causes' such as the loss of a loved one, unrequited love, a mystical crisis, or other such causes. It was believed this susceptibility to mental disease was proof that a woman's intellect was dominated by sensitivity and sentiment.[123] A physician and Professor at the University of Pavia, Ferdinando Tonini, paid extensive attention to female same-sex acts in an 1862 book that partakes in the traditional enquiry into the 'natural' differences between women and men. His *Igiene e fisiologia del matrimonio* [*Hygiene and Physiology of Marriage*] argued that madness affected the unmarried more than the married. He explained nymphomania, satyriasis, erotomania, and onanism, and recorded stories of women who were dedicated to the *gioco lesbiano* [lesbian game].[124] Tonini's basic premise is that female same-sex desires are linked to an enlarged clitoris:

> An excessive development of the clitoris, if not itself a cause of sterility, might be so because it leads the individual to be more inclined to knowing other individuals of her sex than to be sensitive to male caresses. In some

women clitoral pleasure produces an irresistible need which continuously excites their imagination, and because they are lascivious in their erotic games, these tribades cultivate their mistresses' favours and are extremely jealous.[125]

Thus, female same-sex desires were characterised by excess manifested through abnormally large genitalia, overflowing sensuality, and overly intense passions. The enlarged clitoris allowed a woman to perform a man's role during sexual intercourse.[126]

According to Tonini, tribades usually lived in big cities, gathered in convents, colleges, and prisons, were not inclined to maternity and, most interestingly, looked like men.[127] Tonini subsequently described a tribade whose 'menstruation completely ceased; her large breasts disappeared, her skin lost the softness typical of the gentler sex, and resembled a man more and more every day. Thus, in less than one year she took on the physical and moral characteristics of a *virago*.'[128] Some of these ideas of female same-sex practices have old roots: long before the nineteenth century, popular medical books described the clitoris of lesbians as equivalent to a man's penis, while erotic books vividly described nuns having sex with each other in convents.[129] Yet, Tonini's description of the tribade recalls the typical female sexual invert in that the tribade had some 'moral' – that is psychological – characteristics that were not typically feminine. Indeed, as her body became masculine, she also acquired the psychological characteristics of men.

Medical writings would continue to feature descriptions of viragos along Tonini's terms, even after the inception of, and without reference to, the category of sexual inversion. Paolo Mantegazza's portrayal of female same-sex desires is one such example. Mantegazza was a Member of the Senate, Professor of Medical Pathology in Pavia, and Anthropology in Florence, but above all a sexual educator and hygienist.[130] In 1886 he published *Gli amori degli uomini: saggio di una etnologia dell'amore* [*Men's Loves: Essay on an Ethnology of Love*], which dealt with different forms of sexual desire.[131] This text, like many others he wrote, was harshly criticised by the Catholic Church and was promptly listed in the *Index Librorum Prohibitorum*, which did not stop Mantegazza's works from being extremely popular, both amongst physicians and the general public.[132] Mantegazza's views on sexuality were ambiguous: he accepted a measure of continuity between normal and pathological sexual phenomena. To this effect, he declared in his usually suggestive style:

It is impossible to erect the boundaries between love's physiology and its pathology. The highest rungs of eroticism may be the first steps on the ladder of perversion; and amid that hurricane of the senses, compounded of passion and imagination, in which a man and a woman who possess each other with desire are wrapped, it is only the sophist's casuistry which can distinguish that which is good from that which is evil. And even where this good and this evil are concerned, there is room for difference of opinion, according to how one considers the hygienic or the moral aspects of the problem.[133]

This declaration about the nullity of boundaries in love did not prevent Mantegazza from believing that sodomy and tribadism were 'sicknesses', and as such existed beyond the boundaries of normal eroticism.[134]

In order to explain both pederasty and tribadism, Mantegazza offered a well-known anatomical account.[135] He explained that the spinal nerve was linked to 'lustful desire': nerves distributed along the intestinal and rectal tract were closely connected to those that run down to the genital organs. As a result of an 'anatomic anomaly', the nerves responsible for sensual pleasure in the genitals were deflected to the rectum. This would also explain why some women enjoyed anal intercourse, or why lesbians enjoyed 'having the anus excited by a finger'.[136]

Mantegazza believed that women experienced sexual pleasure more easily than men because they had an organic relation to nature. He also denied that women's fate was to reproduce, and supported a certain degree of equality between women and men in the sexual sphere.[137] According to Mantegazza, masturbation was 'spontaneous and natural' in both sexes when for some reason men and women could not have sexual intercourse, and he noted that it tended to spring up at all times in all cultures. It was, nonetheless, a 'perversion' in the highly civilised countries in which economic, moral, and religious reasons hindered free socialisation between the sexes. Mantegazza reasoned that the modern practice of forcing men and women to live separately was an aberration in the first place, and so masturbation became a perversion because it was the result of unnatural causes. Colleges, schools, and monasteries were breeding places for such habits, and he referred to them as 'seminaries of masturbation'.[138] While in men mutual masturbation was merely an exchange of 'manual labour', in women it led to tribadism. Owing to the structure of female genitals and the wayward character of feminine imagination, 'lust is readily turned into a protean and special vice', which is to say that tribades were more lascivious than pederasts.[139] Relationships between women could be divided into 'lesbian love', when women practised oral sex; and 'tribadism', when a woman with 'an exceptionally long clitoris is thereby able to simulate the sexual orgasm [*amplesso*] with another woman'.[140]

Although at the time their work was well known in Italy and abroad, Mantegazza did not mention Tamassia's or Cantarano's studies.[141] In his 1886 treatise he did not use the term 'sexual inversion', even if he considered same-sex desires a sickness, nor did he look for effeminate physical characteristics in men or virile ones in women. Yet, like Tonini before him,

Mantegazza believed that tribades were characterised by an unusually active sexuality and a long clitoris, the implication of which is that tribades were able to perform the role of men during intercourse. In 1891, Mantegazza published *Fisiologia della donna* [*Woman's Physiology*] in which he explicitly associated female same-sex desires with masculinity by describing the tribade as a virile woman. In this work, the tribade was a sort of caricature of the female sexual invert Cantarano had drawn in the 1880s:

> The *virago* is well known to everybody: she has a moustache, hair on her breasts, narrow hips, dry muscles and a virile voice. [...] In love she hunts and does not wait, she never fancies a man, loves violent exercises, hunts, and sometimes she even loves war; she has gestures without grace and a heart without tenderness. She loves giving orders and curses the fate that did not allow her the supreme happiness to dress in boots and trousers.[142]

Interestingly, while there are no references to the masculine appearance of the tribade in *Gli amori degli uomini*, Mantegazza emphasised this aspect five years later in *Fisiologia della donna*, to the point that some descriptions of the virago come close to a parody. Given his remark that everybody 'knew' what a virago looked like, Mantegazza's *Fisiologia* is also an indication of the extent to which, at the beginning of the 1890s, the association between female same-sex desires and masculinity had become widespread. Even if Mantegazza disregarded recent sexological research and his approach to sexual matters derived from a tradition other than psychiatry, he nevertheless shared an enduring interest in female same-sex desires with Italian psychiatrists; he also shared their belief in the close relationship between excessive sexuality and women's masculine-like desires towards other women.

Tribade-prostitute

The idea that female same-sex desires were lustful also underpins the figure of the 'tribade-prostitute'. Decadent poets like Charles Baudelaire and writers like Théophile Gautier had drawn this literary cliché on the basis of medical studies conducted by the French physician Parent-Duchâtelet in the 1830s.[143] Since the beginning of the nineteenth century, prostitutes were increasingly seen as the main source of venereal contagion, which prompted a number of doctors such as Parent-Duchâtelet in France and William Acton in Britain to pay close attention to the different forms of prostitution observable in European cities. Late nineteenth-century Italian physicians were well acquainted with these studies, as evidenced by the fact that Parent-Duchâtelet's observation on the links between prostitution and tribadism was extensively quoted.[144] The two (frequently intertwined) types of medical research that stressed the allegedly widespread phenomenon of same-sex desires amongst prostitutes were criminal anthropology and medico-sociological-oriented studies.[145] In the last decade of the nineteenth century, criminal anthropology was Italy's leading scientific discipline and was involved in an extensive study of the extreme manifestations of female deviancy, which by definition meant sexual deviancy. The leading criminal anthropological journal, *Archivio di psichiatria*, paid close attention to prostitution because, as Lombroso himself had indicated, it was the most typical form of female crime.

While psychiatrists collected case-histories of female inverts in asylums, other medical practitioners conducted their observations in brothels and prisons, which led them to suggest there was a connection between these surroundings and the emergence of sexually deviant practices. In 1882, a surgeon practising in the penal institution of Messina called Carmelo Andronico remarked that tribadism was often common amongst prostitutes and prisoners:

> As in the case of female offenders, so too with prostitutes, the automatic feeling of love for the opposite sex is not enough, they form disgusting and revolting relationships with each other, and then they become excessively jealous; a similarly odd feeling of love is often displayed amongst the female offenders of this prison [Messina], where in some women tribadism is so bold that repressive means are used.[146]

It is not clear what means of repression Andronico used to contain tribadism, but in the same article he had mentioned that women's psychological disorders could be treated by removing the ovaries, and pointed out that psychiatrists cauterised insane women's clitorises in order to suppress onanism.[147] After all, Andronico was a surgeon and he may have resorted to drastic solutions.

During the 1890s, Italy saw a manifold increase in the number of studies published on the subject of prostitution, many of which favoured what we would now call a sociological approach to the phenomenon. Researchers were interested in the different kinds of prostitutes that existed, their relationships with their colleagues and customers, whether they had a partner, and what kind of life they led. Such studies frequently combined approaches and results from different disciplines so that, for instance, a single text would place Parent-Duchâtelet's tribade-prostitute and Krafft-Ebing's sexual invert side by side.[148] A representative example of this socio-cultural medical literature is a study on the criminal couple published by Scipio Sighele in the *Archivio di psichiatria* in 1892.[149] At the time Sighele was just twenty-four years old, at the beginning of his career as a legal and mass psychology expert, and already the author of a successful book, *La folla delinquente* [*The Mass Offender*] (1891). Sighele observed that tribadism was common in 'first- and second-class brothels', but did not occur in bottom-rank ones.

He began his observations on female same-sex desires by outlining the typical 'female urning' in terms reminiscent of Krafft-Ebing's description:

she wore her hair short, dressed fashionably, and pursued masculine sports and pastimes. Sighele took leave of Krafft-Ebing by pointing out that not all women who experienced 'love against nature' had 'abnormal' and intense feelings and desires. This did not mean that Sighele considered tribade-prostitutes normal individuals. Rather than classing female same-sex desires as an inborn mental condition, Sighele explained that a bad social milieu nourished such perverted passions.

The fact that some prostitutes sold their body both to women and men was evidence that many women were encouraged to have same-sex relationships because of their socio-cultural context. Such kinds of women did not fully meet the 'psychological requirements' Krafft-Ebing had set for the psychiatric model of sexual inversion. According to Sighele, they were not inborn inverts, but had acquired the habit of tribadism. While in many women this 'vice' was caused 'by nature', in the vast majority of cases, prostitutes became tribades because they were nauseated by their perverted male clients.[150] The absence of a *souteneur* also encouraged tribadism in luxurious brothels.[151] Even prostitutes, Sighele said, needed a stable source of affection: they turned to their colleagues in search of love because they felt no man could give them any.[152]

Sapphism could also be sparked by curiosity; the kept women of rich men sometimes tried tribadism, which is how the practice spread to lower social strata. Tribadism had originated in luxurious brothels and thereon to society by 'contagion'. The idea that tribadism operated as a sort of contagion was tacitly derived from Parent-Duchâtelet, who treated female same-sex desires amongst prostitutes as an incurable malady transmitted from one woman to another.[153] Sighele did not think prostitutes were the only population to engage in tribadism and stated that it was a well-established practice amongst married women, so much so that anybody living in a city knew at least one woman who regularly engaged in 'sapphism'.[154]

The lawyer G. B. Moraglia's data on the diffusion of same-sex practices were less alarmist than Sighele's: he had observed that only two out of fifty prostitutes practised tribadism.[155] According to yet another lawyer, Alfredo Niceforo, homosexuality was widespread amongst women, prostitution being the guaranteed means to acquire it, followed by masturbation at an early age or seduction by an older woman.[156] He claimed that a number of prostitutes had told him that almost all their colleagues were initiated into 'sapphic love', and that some men also paid to see two women having sex.[157] Although these medical writers diverged on the point of how widespread tribadism was, some of them like Niceforo and Sighele agreed that it was more common in high-class brothels than in those frequented by the lower classes.[158]

These accounts can still be found at the beginning of the twentieth century when the psychiatric model of sexual inversion was well established in the Italian medical community. For instance, in 1903 Ignazio Callari described the phenomenon of tribadism as a 'contagion' in an article published by the *Archivio di psichiatria*. Callari was a Sicilian physician who had conducted a study on three hundred prostitutes, twenty-two of whom said they practised tribadism. Callari had observed that many prostitutes whom men had abused for years became homosexuals. All the tribade-prostitutes he had encountered were between twenty-six and thirty-two years of age, and he reported it was a common habit amongst them to dress as men and imitate brothel customers. This acting was carried out in the evening when prostitutes gathered and entertained themselves with these imitations. According to Callari, tribadism was more frequent in brothels than in asylums and prisons, but he nonetheless believed that all these places to be a 'real school of tribadism' because one 'real tribade' was enough to taint the rest of the women.[159]

These socially oriented medical accounts of the tribade-prostitute tell us more about men's fantasies than about reality. Most of the authors did not explain how they had collected their data, and the descriptions seem more intent on arousing men's curiosity than on offering analytical information. Physicians turned a voyeuristic gaze on female homosexuality, and there is no doubt that their descriptions of the sexual practices and daily life of tribade-prostitutes afforded a degree of pleasure to the readers of highly regarded medical journals.

The appeal of the tribade prostitute was not only exploited by specialised and well-respected journals: at the beginning of the twentieth century, physicians disseminated sexual knowledge through an obscure genre of literature, a sort of erotic-scientific narrative about deviant sexuality. These were cheap books that announced themselves as scientific works, or even as records or diaries of supposedly anonymous women that exploited famous medical accounts of female same-sex desires.[160] An emblematic case was *Le miserie di Venere* [*The Trials of Venus*], published by a little-known physician, Francesco Stura, in 1904. Despite relying heavily on sexological knowledge, the book was not intended for a medical audience, but for the general public. Stura recounted that in Paris, tribades were so perverted that he knew of a brothel set up where female prostitutes satisfied the desires of upper-class women, and he said there existed at least 'four or five brothels' of this kind. Other such tales followed broadly similar lines.[161]

'Fiamma' [Flame]

In the last decade of the nineteenth century, scientists began to privilege schools over mental asylums or brothels as sources of empirical data on female homosexuality. At the turn of the century more bourgeois women entered schools and colleges, in particular teachers' colleges [*scuole normali*]; these became especially attractive for lower-middle-class women in search of the economic independence teaching could afford them.[162] This shift in the gaze of physicians thus coincided with the moment when women were starting to enter higher education; the importance of this turn is that by moving away from the observation of abnormal and pathological cases in asylums, they took the first step towards the study of same-sex desires in 'normal' circumstances, which implies a different perception, perhaps more threatening, of the phenomenon itself.

Medical writings seemed to be tapping into a long-standing interest in the sexual mores of women living together in places assigned to their education, where no men were allowed. The Catholic Church, for example, had paid attention to the evils fostered by the places devoted to women's education and close female friendships.[163] In 1760, Alfonso de Liquori, Bishop of S. Agata de' Goti, published *La vera sposa di Gesù Cristo* [*The True Bride of Jesus Christ*], a treatise on the education of nuns, which was expanded and re-edited several times in the following decades; later editions were unambiguous about the dangers of the 'wild loves' that could develop within the walls of God's house.[164] Certain 'special friendships' amongst nuns were more dangerous than close friendships with people outside the convent, because the former were difficult to prevent and represented an ever-present temptation. Hence the Bishop's recommendation that Mother Superiors should control 'poisoned friendships'.[165] In 1862, it was still acceptable for a pedagogic treatise for Catholic nuns to address the problem of 'dangerous' female friendships in convents in these terms: 'Just as pieces of wood catch fire from one another, so nuns with the fervent love of God *inflame* each other with holy ardour.'[166]

The terminology is certainly striking. In the late nineteenth century, Italian girls used the word *fiamma* to talk about a school friendship. In school slang, this term indicated both the loved person and the friendship itself, which was generally understood to involve a note of passion. Physicians used expressions like 'sweet feeling', 'cult', and 'adoration' to describe the kind of homoerotic feelings that could arise in girls-only schools. This was called '*amore fiammesco*' [flame-like love].[167]

Giovanni Obici, a psychologist, and Giovanni Marchesini, a moral philosopher, collaborated on an extensive psychological study of *fiamme*, which was published as *Le 'amicizie' di collegio* [*School 'Friendships'*] in 1898, and remains the most comprehensive survey of the subject.[168] The observations were conducted with the assistance of former pupils who had become teachers and focused chiefly on students of *scuole normali* aged twelve to twenty. There were both boarders and day pupils at these colleges; the boarders were easily inflamed to passion, but the sparks were provided by the day pupils. In the course of these relationships, the girls wrote and exchanged many letters, which in fact formed the chief material for Obici's and Marchesini's study. After reading over three hundred such letters carefully preserved by their recipients, the researchers concluded that the *fiamma* generally started when one girl admired and complimented the beauty and elegance of another.[169] According to Marchesini and Obici, the letters written in *fiamma* relationships were full of passion and 'physical excitement', and may be considered a form of 'intellectual onanism', which elicited feelings of remorse and shame in the writer, as if she had engaged in a 'physically dishonourable act'.[170]

They pointed out that the *fiamma* was an 'incomplete love', an imitation of a real love relationship between a man and a woman, and as such inferior to heterosexual love. *Fiamme* were not considered proper sexual inverts, but they had some elements in common with them: their behaviour was an example of 'deviation of purposes and means' in the sexual sphere because 'the goal cannot *directly* be that of procreation, and the means are inadequate to this goal'.[171] Unlike sexual inverts, however, the 'deviations' amongst *fiamme* were 'purely *formal*': they only had the appearance of a psychosexual pathology. *Fiamma* was a 'transitory' sexual phenomenon that manifested itself only during puberty, as a result of a specific milieu and its social models. Once out of college, these girls would almost certainly marry.[172] Marchesini and Obici underlined that the *fiamma* was a kind of 'homosexual love' felt by normal, not mentally disturbed, young girls, who got carried away by the typically 'intense' emotions of adolescence.[173] The *fiamma* might be thought of as an 'institution', an 'environmental necessity', and a custom caused by the women-only milieu.[174]

The authors recommended controlling these female friendships, especially when girls were 'highly strung', restless, temperamental, had 'sick' or pale faces, had a masculine attitude or a self-confident personality – all these types of women were considered prone to 'fall' into such relationships permanently. In a 'normal' environment, this 'perversion' was caused by unrelieved chastity, to which women were much more regularly subjected than men.[175] For this reason, Marchesini and Obici criticised religious education that shrouded sexual matters in 'mystery' and 'sin'. They advocated girls should be educated in sexual matters according to the principles of 'Scientific Pedagogy based on the laws of Physiology and Psychology', and recommended life at home rather than in a college.[176]

On the one hand, Marchesini and Obici underlined the pathological and degenerate nature of these relationships, and on the other, they emphasised how these relationships were a kind of substitute for the love of boys – a rehearsal towards heterosexual romantic love, a 'cerebral need', and thus a normal phenomenon which resulted from the specific conditions of women-only environments.[177] Morselli's introduction to this work contrasts sharply with this view: he highlighted these relationships had the same psychological characteristics as 'homosexuality', and were thus to be considered an 'aberration' of the sexual instinct.[178] Morselli also pointed out how moral qualities such as culture, intelligence, sweetness, and so on, instead of physical sexual attraction, played a key role in the rise of '*fiamma* love'.[179] In his comments on Marchesini and Obici, Morselli claimed that unlike hetero-sexual attraction, which depended on physical elements, the relationship between members of the same sex was based on an 'emotional' or 'mental association'. On the whole, Morselli's introduction marks a departure from the views he had espoused in the *Manuale* thirteen years before, so that now, while he clearly associated the *fiamma* phenomenon with sexual inversion, he seemed to tacitly endorse some of Binet's work, and therefore a psychological explanation for the cause of same-sex desires.[180]

Physicians believed boys had a naturally active sexuality that needed to be controlled and curbed, but was nonetheless an eminently normal phenomenon. The same, however, could not be said of female sexuality. Young girls, and in general all women who had an active sexuality, were thought to be deviant and classed as sexual inverts, prostitutes, or a combination of both (tribade-prostitutes); others were believed to be *fiamme*. During the 1890s there were a series of shifts in how sexological studies understood female same-sex desires, which resulted in their no longer being considered intrinsically diseased behaviour. There were two main developments that led Italian medical writers to affirm same-sex desires might be normal. On the one hand, from the late 1890s onwards, sexological studies began accepting that until adolescence bisexuality was a normal and universal phenomenon. Degeneration was increasingly judged to be insufficient when it came to explaining homosexuality and psychiatrists increasingly turned towards the childhood of sexual inverts for answers: this interest in early life was already present in the first case-histories sexologists had published, but as the century came to a close, this concern intensified as part of an increased awareness of psychological and social explanations of same-sex desires. The second theoretical development that led to a 'depathologisation' of sexual inversion in a certain milieus is related to sexology, which reformulated its object of study to encompass not only the poor classes, but mainstream society and middle-class women as well. Sexologists moved away from asylums and towards schools. Remarkably, at the turn of the century, female same sex-desires were no longer considered simply a cause or symptom of insanity or a sign of degeneration; they had become something worse. They were now a threat to physicians' daughters who were entering the higher education system.

The persistence of the figures of the virago, the tribade-prostitute, and the *fiamma*, with their origin in diverse medical fields and their independent discursive trajectories that range from the nineteenth to the early twentieth centuries, shows that the psychiatric model of sexual inversion did not fully replace other forms of understanding female same-sex desires. Different medical disciplines fostered opposing views on the origin of homosexual behaviour; by the end of the nineteenth century, these disparate anatomical, psychological, and socio-cultural medical observations had been condensed into different figures of sexual deviance that coexisted with one another, and frequently overlapped with sexual inversion too. These representations of the virago, the tribade-prostitute, and the *fiamma* sit awkwardly with current historiographical interpretations which suggest that, at the end of the nineteenth century, a 'new' psychiatric mode of reasoning obliterated other previous conceptions of same-sex desires. These constructions of female same-sex desires that synthesise older medical and cultural assumptions continued to be highly visible at the turn of the century in Italian medical writings, despite the advent of the 'newer' figure of the sexual invert.

Such variegated discourses on female same-sex desires also illustrate a distinctively Italian medical trend in the sense that the medical sources that engage with the topic are numerous and diverse. Current historiographical accounts err again in their suggestion that female same-sex desires were hardly visible if compared to male same-sex desires. At least in Italian medical literature, this is certainly not the case: debates on same-sex desires proliferated as part of broader studies of sexual perversion. While the Catholic Church promoted ignorance in sexual matters, psychiatrists – who considered ignorance a source of harm – unveiled a disorderly female sexuality. The conflict between Church and state – together with the more general process of secularisation – that ensued after the unification of Italy forced the state to intervene directly in the moral sphere. If the political authority of the secular kingdom was to be affirmed, it was critical to find a new set of moral principles that could replace religious influence. Science came to represent the new value-system and scientists became the agents of a new homogeneous Italian culture. They were often active intellectuals in the public sphere, as their presence in government, local administration, and public debates shows, and they were also instrumental in supplying a new sexual morality to replace religious ethics, which explains the boom in studies of sexuality. In particular, psychiatrists who dealt with sexual pathological phenomena could scarcely avoid coming across homosexuality. But why were they so interested in female sexual inversion?

Following the unification of the country, legal experts and politicians involved in the redrafting of the civil and penal codes were forced to examine the position of women. Medical writers supplied a scientific theory in support of the assumption of natural female inferiority, and in doing so they also explored female sexuality. While trying to disclose the biological roots of female behaviour, scientists switched with ease from descriptions to prescriptions about how healthy women should behave, and thus gave an apparently rational base to prescriptive rules. If the only difference between pathological and normal behaviour was of degree and not of substance, then abnormal sexuality could provide significant insights into normal sexuality. This medical assumption encouraged physicians, and above all psychiatrists, to unveil and study disorderly female sexuality. The masculinised, active styles of desire that were central to various definitions of female sexual deviancy did not fit with the typical nineteenth-century medical assumption that female sexuality was naturally passive and responsive to the inherently aggressive and active male sexuality. Medical descriptions agreed that female criminals, prostitutes, and lesbians were virile, which was evidence of a thwarted evolutionary process that did not culminate – as it normally should – in a full sexual differentiation. Female same-sex desires were perhaps the phenomenon that most failed to embody a passive sexuality. Female homosexuals longed for other women; they were not a passive receptacle for male input; their existence was somewhat equated to a state of regression because they failed to perpetuate the species and because their excessive sexuality threatened the moral order of society. Notwithstanding these perils, these physicians' overheated prose is testimony to how exciting, forbidden, and mysterious their object of study was to them.

Was such multiplicity of explanations for female same-sex desires an exclusively Italian phenomenon linked to the context of Italian unification or to the Italian medical tradition? The notions that underpin the Italian figures of the virago, the tribade-prostitute, and the *fiamma* are present in the British context, although formulated in different terms. This is highly significant because British physicians were, as the rest of this book will show, so resistant to the Continental category of sexual inversion. Medical communities operated under different historical and intellectual conditions, which explain the emergence of national-specific discursive directions and overtones in the debates about female same-sex desires.

4
Britain: Oblique Discourses Surrounding 'Lesbic Love'

British philosophical and scientific thinking had been enquiring into sexual matters since at least the end of the eighteenth century. In his *Essay on the Principle of Population* (1798), Thomas Robert Malthus rationalised and problematised procreative sexuality by linking it to economic and social issues. His main argument was that populations tend to multiply faster than the means required for their sustenance. The inevitable misery that would ensue could be avoided through sexual moderation: people should postpone marriage until they could support a family, and single people were expected to be strictly celibate. Although couples should not have many children, Malthus discouraged the use of artificial methods to limit reproduction because he considered them a 'vice'.[1] Charles Darwin's *The Descent of Man and Selection in Relation to Sex* (1871) reinforced the idea that sexual instincts were closely related to human developments. Paving the way for the study of sexual instincts as a 'natural' phenomenon, Darwin raised sexual impulse to the status of driving force in human evolution, and contended that man has two main instincts: self-preservation and gratification of the sexual instinct.[2] In some instances, members of the same species will fight each other to secure mates, rather than for food or living space.[3]

Late nineteenth-century European medical writers were greatly influenced by the theoretical contributions of British analyses of the importance of the sexual instinct such as Malthus's and Darwin's. Although these theories were effective starting points for rational and scientific examinations of the sexual instinct in the whole of Europe, in the last decades of the nineteenth century British and Continental lines of enquiry progressively diverged. While the Continent embraced medical studies of psychosexual pathologies, Britain was generally wary of this kind of analysis – initially, the study of sexual inversion was an almost exclusively Continental enterprise. Case-histories and theoretical studies of sexual inversion proliferated in Italian medical journals, but between 1869 and 1896 – before the publication of the English edition of Ellis's *Sexual Inversion* – only a small number of unsophisticated British psychiatric articles reported cases of either male or female sexual inverts. In this period, the psychiatric category of sexual inversion was not deemed worthy of independent chapters in British specialised treatises, whereas in Italy sexual inversion had been examined as an autonomous category in psychiatric manuals since the mid-1880s. Until the beginning of the twentieth century in Britain, same-sex desires would continue to be unsystematically described as symptoms of other mental disorders such as moral insanity and degeneration. Many British psychiatrists were inclined to believe that same-sex acts and desires were a 'vice' and were reluctant to take up Continental ideas surrounding the 'new' psychiatric pathology.[4] The publication of Ellis's *Sexual Inversion* in 1897 was a notable exception but, as I show in Chapter 7, the British medical community initially rejected this work. The general hesitation to take up of the concept of 'sexual inversion', however, did not mean that medical writers were silent on the subject of female same-sex desires.[5]

Female same-sex desires featured in other contexts such as medical debates on prostitution and nymphomania, gynaecological observations about the body, and the ever-present concern that girls at school might be engaging in sexual practices. Until the end the nineteenth and well into the twentieth centuries, British medical writers explained female same-sex desires mainly through women's anatomy and avoided sophisticated psychological explanations. In contrast to the overt and blatant Italian medical debates on female same-sex desires, in Britain these discourses were oblique and guarded. Generally speaking, British medical writers were cautious when referring to female homosexuality and always took great care to avoid exciting the reader's fantasies. As a result, it is often necessary to read between the lines of British medical texts because their writers resorted to euphemisms when writing about female same-sex desires. For example, instead of referring to sex between women, they could refer to the subject by reminding their readers of the 'most unbridled passions' amongst the prostitutes described by Parent-Duchâtelet. Only medical practitioners who kept abreast of Continental developments could readily understand such references to same-sex desires, leaving other readers oblivious to any allusion to 'unnatural practices'.

Rather than being concentrated around a boom in sexological studies, ideas about female same-sex desires were scattered throughout British medical writings in the nineteenth century. In order to trace the development of these various ideas, I have therefore included an earlier phase of writings that were printed as early as the 1840s. Some of the medical texts I draw on went through various editions in the course of the nineteenth century, which means that assumptions contained in some books first published in the 1840s were still popular in the last decades of the nineteenth century.

I will start by outlining some of the events that helped shape the structure of the British psychiatric profession and determine its role in society, so as to shed light on the contrasting backgrounds from which sexological discourses emerged in Italy and in Britain. Despite the fact that there existed a tradition dealing with abnormal sexual desires within British psychiatry, this did not evolve into a systematic study of sexual inversion, as happened on the Continent. In the second section of the chapter, I will show how deviant sexual behaviour was conceptualised in psychiatric taxonomies, and how same-sex acts were seen to be both symptom and effect of some mental illnesses characterised by perverted and unmanageable instincts – monomania and degenerative disorders amongst them. I will then focus on female same-sex desires, showing how these were incorporated into medical discourses about female sexual excess such as nymphomania and prostitution. My argument is that British physicians were more preoccupied with female sexual excess than with sexual practices between women. I will subsequently illustrate how British gynaecologists were particularly interested in identifying signs of same-sex practices in the female body, since they supposed genitalia and the reproductive organs could be marked by sexual perversion. The final section of the chapter examines physicians' warnings about the dangers of masturbation; amongst other things, it was thought that female 'self abuse' could lead to other kinds of sexual perversion, including female same-sex acts. Overall, this chapter shows that in Britain female same-sex desires continued to be tied to anatomical explanations well after the emergence of the psychiatric category of sexual inversion on the Continent. Bodily oriented explanations of female same-sex desires were so dominant in Britain that even the few psychiatrists who did address female sexual inversion associated this mental pathology either with dysfunctions of the reproductive system or with genital abnormalities.

British psychiatry in the nineteenth century

At the time Italian psychiatry was being established, British psychiatry had been an organised discipline for some time, with the result that it was subject to more regulations and government constraints than on the Italian Peninsula. While Italian psychiatrists and criminal anthropologists had championed theories that went against the grain of classic legal tenets such as that of individual responsibility, British psychiatrists were disinclined to argue against legal principles. In these terms Continental sexological research was unattractive to many British psychiatrists, as it was not possible to adopt its latest developments without disputing the doctrine of free will and British law against 'unnatural crimes'. A survey of the history of psychiatry in British society contributes to appreciate why British physicians were, generally speaking, so reluctant to engage with Continental modes of analysing psychosexual pathologies.

The 1808 County Asylum Act aimed to solve some of the difficulties workhouses and prisons were facing as a result of their variegated population, which placed insane people alongside criminals and paupers. This Act expanded the jurisdiction of the Justices of the Peace and appointed them to form committees in order to erect lunatic asylums out of towns, inspect them regularly, and levy a county rate for the purpose. In this way, the government financed the creation of a network of asylums.[6] In the following decades regular committees of inquiry were set up to inspect and improve the situation of madhouses. England and Scotland put forward independent bills for the reform and regulation of asylums in part as a response to the reports issued by these committees. The 1828 Madhouse Act, the 1845 Lunacy Act, and the 1890 Lunacy Act were some of the most important pieces of legislation relating to mental health passed during the nineteenth century. They stipulated, for example, compulsory alimentary provisions for asylums built at public expense, regular inspections by observers with no ties to the boards running the institutions, and state endorsement of the medical profession's authority in the sphere of diagnosis and treatment of mental illness.[7] The 1890 Lunacy Act prescribed everything in great detail: from admission processes to duration of treatments, and from the ban on intimate relationships between doctors and patients, to the prohibition on employing males for the custody of females.[8]

These laws, which aimed to set standards for the handling of the insane, led to the appointment of new medical superintendents. An increasingly well-defined group of doctors specialising in mental illness also began to emerge as a result of these new regulations. The British psychiatric profession, therefore, was regularised throughout the nineteenth century. Its continuous legal codification was accompanied by considerable conflict between doctors and the government, and between medical and legal experts, who had the power to determine whether the mentally ill could be held responsible for criminal acts in judicial settings. Throughout the nineteenth century, asylum doctors sought to consolidate their status as the sole professional group able and entitled to oversee the caring for the insane by contending that mental illness was a disorder of the brain which required medical intervention.[9] In November 1841, Samuel Hitch, Resident Physician at Gloucester General Lunatic Asylum, joined eleven other doctors in the launch of the Association of Medical Officers of Asylums and Hospitals for the Insane. This was a crucial event because it was the first official organisation that sought to represent the professional interests of British alienists. In 1865, this association was transformed into the Medico-Psychological Association, and by the end of the nineteenth century it had almost six hundred members organised into five divisions: three English, one Scottish, and one Irish.[10]

The British psychiatric profession was riven by conflicts between practitioners working in private asylums and those employed by public institutions, with two main journals representing their respective views. Forbes Winslow's *Journal of Psychological Medicine* mostly endorsed the agenda of private asylum doctors, and was founded in 1848.[11] The *Asylum Journal of Mental Science* was founded in 1853 by fifteen members of the Association of Medical Officers of Asylums and Hospitals, amongst them the eminent physician Bucknill, who assumed responsibility for the journal.[12] The *Asylum Journal of Mental Science*, published as the *Journal of Mental Science* since 1858, championed the cause of public asylum doctors, and had a more practical approach to madness than the *Journal of Psychological Medicine*. With the 1876 publication of *Mind* there emerged a more philosophical and psychological journal.[13] In 1878, the neurological journal *Brain* was founded by John Charles Bucknill, James Crichton-Browne, David Ferrier, and John Hughlings Jackson.[14] Along with these various journals, which show the expanding market for British psychiatry, in 1858, Bucknill and Hack Tuke published *A Manual of Psychological Medicine*, the first reputable textbook about insanity since James Cowles Prichard's 1835 *Treatise on Insanity*, which the *Manual* set out to replace.[15] The *Manual*'s second edition appeared in 1862, followed by two subsequent editions, with the last appearing in 1879. For almost a quarter of a century, Bucknill and Tuke's *Manual* summed up the knowledge on which alienists rested their claims to exclusive jurisdiction over the treatment of the insane.[16]

Unlike Italian psychiatrists, who had experienced a highly enthusiastic phase between the 1860s and the early 1890s, the outlook of British psychiatrists was bleak. The early optimism over possible cures for the insane through moral treatment faded when hereditary explanations of mental illness began gaining ground in the second half of the century. Hereditary explanations of madness appeared to confirm the widespread notion that mental disorders were incurable, the result of constitutional weakness and an unwillingness to cease depraved and unhealthy activities.[17] This argument was especially attractive because it justified therapeutic failures in asylums.

A further element contributed to nourish the psychiatric pessimism of the second half of the nineteenth century in Britain: in this period, British public opinion did not hold psychiatrists in high regard. Allegations of abuse following government inspections of asylums fostered the idea that psychiatrists were corrupt, that they exploited the insane, that they were unable to cure the mentally ill, and were therefore essentially incompetent. Broadly speaking, the role of psychiatrists was restricted to the routines of asylum management, which left few opportunities for theorisation based on clinical studies. It was also difficult to sustain professional organisations, and the strong role of the government in regulating the treatment of the insane made matters worse.[18] A very different atmosphere animated Italian psychiatry in the second half of the nineteenth century. After the Italian unification, the importance of psychiatrists' cultural and social functions was widely acknowledged, mainly because psychiatrists participated actively in the political life of the new kingdom, and also because of their contribution in fighting the cultural hegemony of the Catholic Church.

Criminal anthropology played a substantial role in the enthusiastic phase of Italian psychiatry of the second half of the nineteenth century. Led by Lombroso, criminal anthropology was a growing discipline followed by numerous practitioners, with professional journals and national and international conferences – all of which made criminal anthropological studies highly visible and well regarded. Although at the end of the nineteenth century Italian criminal anthropology was increasingly contested by the international scientific community, it had acquired international fame, which energised Italian medical practitioners and produced a huge amount of research on human deviancy. In Britain, criminal anthropology did not play such a crucial role. By the 1870s and 1880s, British psychiatrists had already abandoned earlier attempts to characterise 'criminals' in pathological terms. In the 1880s, leading figures such as Tuke and Maudsley were distancing themselves from their earlier claims about the 'morbid psychology' of criminals. While Continental criminologists were developing new categories such as 'born criminals' and 'the criminal type',[19] Maudsley openly repudiated the 'lamentable extravagances into which some disciples of the latest school of criminology have been betrayed', and pointed out that it was risky 'to play to the gallery and so to burlesque science'.[20] During this same period, British criminal courts did not hear psychiatric evidence that contradicted basic legal axioms about individual free will and responsibility. Thus hemmed in, psychiatrists gradually restricted themselves to the confines of their practice so as 'to minimise conflict between psychiatry and law'.[21] This development is in line with David Garland's point that British medicine tended not to isolate men as psychological 'types'. Instead, Garland suggests that British psychiatry was a therapeutically oriented practice based on a system of classification of mental disorders that discussed the condition separately from the patient who suffered from it. Within the classificatory schemes of morbid psychology, criminals were said to exhibit a variety of conditions including insanity, moral insanity, degeneracy, and feeblemindedness. But generally speaking the criminal was not conceived of as a distinct psychological type.[22] As I will discuss later in this chapter, Garland's observation about British psychiatrists' refusal to conceive the criminal as a psychological type seems to coincide with psychiatrists' widespread reluctance to think of sexual inversion as an inborn psychological condition.

Psychiatry and same-sex desires

Taxonomies of mental disorders

Throughout the nineteenth century, British psychiatrists were engaged in the development of taxonomies. These classifications of mental disorders illustrate the role played by sexual behaviours in the study of mental pathologies. David Skae's and Thomas Smith Clouston's nosologic work is an example of how psychiatry addressed the topic of same-sex desires within an established academic context. Both Skae and Clouston worked at the Royal Edinburgh Asylum; Skae had been the founder of the Edinburgh School of Psychiatry and Clouston had been his student. In 1873, Clouston published a set of lectures on insanity originally written by the late Skae.[23] When presenting his teacher's work, Clouston explained that until at least the 1860s, the only method of classification of mental disorders was that proposed by Philippe Pinel, and subsequently modified by Esquirol. This method was founded entirely on mental symptoms, and the forms of insanity were divided into mania, melancholia, monomania, and dementia.[24] Clouston himself followed this classification when teaching a course on mental diseases at the University of Edinburgh in the 1880s.[25] In contrast to this classic understanding of psychiatric nosology, Skae had developed a classification based on the bodily conditions that accompany insanity.

Skae first unveiled his classification at the 1863 Annual Meeting of the Association of Medical Officers of Asylums and Hospitals for the Insane. He had described his taxonomy as a 'natural history' of disease that linked forms of insanity to bodily functional disorders.[26] Skae distinguished the following conditions: 'Insanity with Epilepsy; *Insanity of Pubescence*; *Insanity of Masturbation*; *Satyriasis*; *Nymphomania*; *Hysterical Insanity*; *Amenorrheal* [sic]; *Post-Connubial Insanity*; *Puerperal Insanity*; *Insanity of Lactation*; *Insanity of Pregnancy*; *Climacteric Insanity*; *Ovarian Insanity*; Hypochrondriacal Insanity [sic]; Senile Insanity; Phthisical Insanity; Metastatic Insanity; Traumatic Insanity; Rheumatic Insanity; Podagrous Insanity; *Syphilitic Insanity*; Delirium tremens; Dipsomania; Insanity of Alcoholism; Malarious Insanity; Pellagrous Insanity; Post Febrile Insanity; Insanity of Oxaluria or Phosphaturia; Anaemic Insanity; Choreic Insanity; General Paralysis with Insanity; Insanity from Brain Disease; Hereditary Insanity of Adolescence; Idiopathic Insanity'.[27] As is evident from this classification, almost half of the various forms of insanity were related to the sexual apparatus or sexual behaviour.[28]

In his nosologic work, Skae addressed the issue of same-sex desires in the context of 'Insanity of Masturbation' and 'Hysterical Insanity'. He noted that 'Insanity of Masturbation' seemed to affect more men than women, and amongst other symptoms he listed the 'dislike of female society', a hint at same-sex desires.[29] Skae's reference to passions towards people of the same sex became clear later in the piece. In his account, satyriasis and nymphomania were diseases closely allied to the insanity of masturbation, but differed from it because 'the sexual [feeling] arises from *desire* towards the opposite sex, [...] the origin of the morbid and incontrollable passions is in the nervous centres, and not in the testes or ovaries or other sexual parts'.[30] This is to say that insanity of masturbation differed from satyriasis and nymphomania in its origin: testes or ovaries in the former, nervous centres in the latter. Further, while insanity of masturbation is connected to sexual desire for people of the same sex, satyriasis and nymphomania drew patients towards those of the opposite sex.

The most interesting condition was 'Hysterical Insanity', described as a form of insanity that affected mainly the female sex. According to Skae, during puberty girls' bodies underwent a series of changes that predisposed them to insanity or to hysteria, and made them prone to acquire the 'habit of masturbation'. These changes could easily terminate in 'Hysterical Insanity'. At first, this would manifest itself by great excitement, laughing, crying, incessant talking, restlessness with occasional sleeplessness, efforts to run out of the house, and screaming. In more serious cases, hysterical insanity led to the development of 'varied sexual and erotic symptoms', so that the patient would seem to be 'truly maniacal'.[31] Skae illustrated the characteristic traits of hysterical insanity by reproducing a letter written by a young girl to her father. According to Skae, the patient presented 'acutely maniacal symptoms' of hysterical insanity:

> Do you remember the verse 'There are, &c.' (12th verse, 19th chapter of Matthew), about Eunuchs? Then I beg to inform that according to Scripture and my conscience Jessy, your cook, is a man, and Janet, the mad devil, is a man; and Denham and Henry boys who can have children. Aunt Isabella is a man, and yourself also, both made of man; and I am a boy made of Dr. C. and Dr. Z. Mrs. T. is a man, made of men. They are ignorant on this subject here. [...] I have at times since I came here passed the shadow of death, and therefore am authorised to speak in opposition to all men and women, gentlemen and ladies, who oppose me. *I am, I can swear, as you want to know what sex I belong to, a mixture of nymph and half-man, half-woman, and a boy, and a dwarf, and a fairy.*[32]

The young lady must have written this while in the grip of severe delusions. Indeed, she noted that she had been forced to take certain drugs because of her sickness. Nevertheless, she was clearly confused about the issue of her sexual identification, especially as regards her own sex. She did not know whether she was a man or a woman. Did she feel like a man? Did she feel she had a man's nature? Did this girl desire other women? It is difficult to decide because Skae did not provide a full case-history of this girl. Yet, Skae's nosologic research is an example of how, in British psychiatric works, a number of forms of insanity could include same-sex desires and other forms of sexual perversion, which in their formulations problematised the established notion of gender. Individuals were supposed to conform to the appropriate feminine and masculine roles, which, if violated, could be considered a symptom of a mental disorder and, on occasion, explicitly associated with same-sex desires.

Skae also clarified that even though he had adopted a classification of mental disorders based on bodily symptoms, it was still necessary to divide mental disorders into mania, melancholia, monomania, and dementia to fully understand the various forms of insanity. He highlighted that his courses routinely addressed the different classes of mental disorders. In his discussion of mania, Skae drew attention to the fact that an alienist may encounter every degree of 'maniacal excitement – incessant talking and gesticulation, and destructiveness, and filth and nudifying [sic] – without any intellectual impairment'.[33] In this case, passions and emotions might be excited without the intellect being deranged. Skae explained that such cases belonged to what Pinel called '*reasoning madness*', and Prichard '*moral insanity*'.[34] According to both authors, 'the symptoms of kleptomania, pyromania, dipsomania, erotomania, satyriasis, and nymphomania are not generally accompanied by intellectual delusion, unless they are mere accidental symptoms of some such form of insanity as general paralysis or some other forms'.[35] Skae's clarification spelt out the extent to which mental disorders were conventionally linked to unrestrained emotions and sexual passions, and therefore were a matter for study in British psychiatry. It also shows that in Britain there was an established debate on the intellectual competence of those whose emotions and passions were perverted. It was within this tradition that Clouston recognised same-sex desires.

Though he had been a pupil of Skae, Clouston's approach to the classification of mental diseases followed a more conventional pattern. When he delivered clinical lectures to his students at the University of Edinburgh during the 1880s, his classification was still based, as mentioned above, on mental symptoms.[36] After asserting that all varieties of mental disease 'find their origin in and flow out of excesses, defects, and irregularities in the physiological functions of the brain', Clouston pointed out that the most important 'human instincts, appetites, and organic necessities' were: '1. Love of life, with efforts to prolong it. 2. Desire to reproduce the species. 3. Love of offspring'. Eating and drinking were respectively fourth and fifth items in his list.[37] Normalcy was defined by the presence of the instinct to reproduce. When such instinct was absent, individuals lacking any inhibition displayed 'perverted instincts, appetites and feelings'. Examples of these perverted instincts were sodomy, along with the 'uncontrollable impulse towards sexual intercourse', self-abuse, rape of children, nymphomania, and bestiality.[38] Clouston did not develop his analysis on sodomy, nor was tribadism added to the list of sexual perversions; he merely specified that there were cases in which 'there are irresistible impulses towards sodomy and incest'.[39] He also believed that people were not always accountable for acting on their sexual urges.[40] Consequently, it would seem Clouston believed that sodomy was a mental disorder rather than a crime, but as he did not develop the implications of these cursory observations any further, it is not possible to assert this conclusively.

The careful reader would have been able to discern some sketchy references to same-sex desires and other forms of sexual perversion amongst the abstruse debates about psychiatric taxonomies. Skae's and Clouston's respective classifications are two examples of this period's typically circuitous and unsystematic attempts to explain same-sex desires. There are, however, other taxonomies that focus strictly on sexual instinct. In 1867, an anonymous author published an article in the *Medical Times and Gazette* in which he grouped aberrations of the sexual instinct as follows:

> 1st, in tendencies to sensual gratification without any union of the sexes. [...] 2ndly, in union of the sexes without due provision for the ends for which such union was intended – viz., in casual and temporary union. [...] 3rdly, in marriages which are artificially barren. [...] 4thly, there are aberrations of a more innocent character, because often forced upon individuals by a worrying moral or political condition of society, amongst which *androgynism* [...] [or] the intrusion of either sex, voluntarily or not, into the province of the other.[41]

The first deviation of the sexual instinct was masturbation, the second was sexual intercourse that was not intended for reproduction, the third included secular free-lovers, and the last one, *'androgynism'*, was explained as an 'intrusion of one sex into the province of the other'. In the clarification of this term, the author associated 'androgynism' with:

> those dark crimes which, as the law says, are not to be named of Christian men, and which are shortly catalogued in the Epistle to the Romans, Casper's 'Medical Jurisprudence,' Gall and Spurzheim's 'Phrenology,' etc. They illustrate the dangers of unchaste thought, even without unchaste act.[42]

After this reference to same-sex practices, the author devoted considerable attention to those types of 'sexual aberrations' of the sexual instinct that most affected women, such as sexual incontinence, being unmarried, or being married without 'the fruits'. This last aberration was more 'unnatural' and 'detestable' than same-sex acts.[43] It had its 'root first in selfishness – the desire to enjoy life without trouble', but in America it was aided by another 'aberration': the 'women's rights system', which sought equality for the sexes. The women's rights movement caused 'too great stress on the intellectual, and too little development of the nutritive functions'.[44] Females who promoted women's rights – a characteristically American phenomenon – were thus typified by the author of the article: 'big foreheads, flat bosoms, skinny legs, sharp features, coarse voices, and a strong taste for criminal statistics'.[45] The author explained that the scarcity of women in American society was to blame for the development of such physical aberrations. The women's rights movement, therefore, ultimately caused women to become masculine. But he warned doctors on the other side of the Atlantic that a similar phenomenon was being experienced by British women for the diametrically opposed reason, namely want of a husband.[46]

Subsequently, the author of the *Medical Times and Gazette* article made explicit why he believed that there was a relationship between the women's movement and same-sex desires. As he went on to explain, there were two types of *androgynism*, which, as shown above, were associated with same-sex practices through the reference to 'those dark crimes not to be named of Christian men': when men performed traditionally 'feminine' responsibilities, such as housework or bookkeeping, or when women carried out male duties. Despite the author's declared support for female emancipation, by which he meant improvement of women's education and wages, he thought they

> make a mistake undertaking duties which are better fulfilled by men. It is an aberration of the sexual instinct in any girl to aim at occupations which are incompatible with the duties of maternity, and an equal aberration to smother those maidenly instincts which should lead her not to intrude into the occupations which custom has associated with the male sex. There is no intrinsic sin in riding a horse, or in wearing boots and breeches, but there is harm in violating those decent rules by which the conduct of either sex is regulated. We say it in all kindness, that for a girl to present herself at a public Medical examination is as great an aberration of the sexual instinct, as it would be if a young man were to leave the dissecting room and apprentice himself to Madame Elise or Mademoiselle Couturière.[47]

The 1867 *Medical Times and Gazette* article shows that, well before the publication of the first case-history of sexual inversion in 1881, British medical writings described same-sex desires in a language not altogether dissimilar from that of gender inversion.[48] The author also linked the women's rights movement and ideas of gender inversion but, so far as I have been able to tell, in this respect his article stands alone in the period ranging from the late 1860s to the 1880s, although such an association did indeed come more readily to physicians' minds in the 1890s in Britain.[49]

Perversion of emotions and instinct

As seen above, Skae's comments on mania touched upon the psychiatric issue of the intellectual capacities of individuals whose emotions and passions were perverted. These preoccupations were positioned alongside the debate about criminal responsibility, which was occupying lawyers and psychiatrists in both Britain and on the Continent in this period. Psychiatrists throughout the nineteenth century struggled to understand those people who displayed uncontrollable passions, but who were otherwise able to reason and judge coherently. In Britain, as on the Continent, psychiatrists were concerned with forms of insanity in which derangement was limited to a small number of ideas, emotions, and patterns of behaviour. These forms of insanity were first identified by Pinel as *'manie sans délire'* [mania without delirium], or 'reasoning madness' according to Skae's translation. Esquirol's monomania, Prichard's moral insanity, and finally Morel's degeneration were subsequently concerned with behaviour patterns characterised by impulsive, irrepressible acts and perverted emotions.[50] In Britain as in Italy, same-sex desires were regarded as expressions of an uncontrollable will, perverted emotions and instincts; they were first classed as one of the so-called 'partial insanities' such as monomania, and later recognised as a manifestation of degeneration.[51]

For instance, in Bucknill's and Tuke's *Manual of Psychological Medicine*, monomania was defined as a perversion of the emotions and instinct.[52] They listed 'Unnatural Crime' – a term commonly used to refer to same-sex acts in the nineteenth century – alongside various forms of monomania:

> Connected with the question of general Diagnosis of Insanity is that of Unnatural Crime, which in this country is so strange and uncommon as to lead to the supposition that persons guilty of it must be out of their minds. We have known instances in which persons who were simply criminals of this kind have escaped from the punishment which the law would have awarded them, by willingly availing themselves of the imputation of lunacy. In these cases the habit of unnatural crime had been acquired in foreign countries where it was common.[53]

On the one hand, Bucknill and Tuke ambiguously classified same-sex acts as monomania, from whence it follows they recognised them as a form of insanity. On the other hand, they said that under British law, same-sex acts were considered a crime, so that some men pleaded insanity in order to avoid punishment for charges of sodomy. This suggests that individuals on trial had themselves related same-sex acts to mental disorders, which illustrates that such an association was not the prerogative of psychiatrists.

As with the Continent, during the last decades of the nineteenth century British psychiatrists referred less to monomania and more often to degeneration to explain mental disorders characterised by loss of willpower and perversion of the feelings. Same-sex desires fell under the heading of those degenerative mental disorders which presented loss of willpower and perversion of the feelings. For instance, the work of leading psychiatrist George Savage linked degeneration to sexual deviancy; his work *Insanity and Allied Neuroses* (1882) explained the extent to which insanity was often a cause of sexual perversion.[54] He pointed out that insanity could appear during puberty as a result of the sexual changes undergone by the body, and that girls were particularly vulnerable given their reproductive function. Their reproductive system put women at a disadvantage in regards to men – in Savage's view, menstruation, parturition, lactation, and menopause could often cause mental derangement in later life.[55] Thus, Savage, like other British psychiatrists, conceived of women's anatomical characteristics and their special role in reproduction as sources of insanity. During puberty some children experienced early sexual excitement and became 'utterly unmanageable in consequence of the development of sexual desires [...], giving way most openly to masturbation; or, if boys, making attempts of every description to gratify their lusts upon children or women of any age'. These children's and boys' sexual desires were 'almost invariably immoral'.[56]

Moreover, it was thought that modern society's 'unnatural and arbitrary' way of promoting relationships between the sexes was a fertile ground for the multiplication of sexual perversions. According to Savage, the original function of the sexual instinct was to perpetuate the species, but men had long cultivated it as 'a special source of pleasure', with no connection to the reproductive function.[57] The modern way of life stimulated sexual indulgence and immoral sexual behaviours.[58] In the same way that modern life facilitated the appearance of 'depraved' sexual behaviour, so insanity in general was a vehicle of sexual perversion. Savage believed that 'unnatural offences' might easily occur in connection with insanity, which by definition was characterised by lack of self-control. The 'imbecile' or 'idiot', the offspring of 'nervous parents', of 'the moral imbecile', or of the 'senile dement', might be guilty of sexual perversion. Savage summed up his argument thus:

> unnatural or brutal offences of a sexual nature may result from undeveloped higher control; from loss of the same control; or from degeneration or disease, in which case they may follow simply from delusions. I have met with one case of a young single man given to masturbation who had complete sexual perversion, so that he lusted after men, not women.[59]

In brief, Savage argued that the phenomenon of men who desired other men was caused by degeneration and loss of willpower.

Like many Continental medical authors, the psychiatrist Charles Mercier related degeneration to same-sex desires.[60] In 1890, he explained there was an 'inversion of sex' when a person of one sex took on the attributes of the opposite sex, and such 'inversion' was associated with degenerative signs.[61] Mercier argued that, under certain circumstances, some individuals 'of one sex will assume many of the characters of the opposite sex, the reversion being to [their] ancestors of this sex'.[62] This was caused by the constraint of 'reversion' and 'latency'. According to Mercier, when an attribute existed in an individual and was subsequently absent in his or her offspring, but reappeared in the third or other subsequent generation, it was said to be *'latent'* in the generations in which it did not appear. The individual in whom it later appears is said to *'revert'*, so far as the attribute is concerned, to the ancestor in whom it had been present.[63]

In women, 'inversion of sex' was related to the reproductive system. Mercier explained that insanity was dependent on and connected to reproductive functions in so many ways, 'that to attempt to treat the former without reference to the latter would render the book a mockery and an imposture'.[64] Specifically, when going through menopause, some women grew a beard, saw their breasts shrink, acquired a deeper voice, and developed other masculine secondary characteristics. Women so affected 'revert[ed] as to habits and mental qualities to some remote feral or semiferal ancestor of man, and [might] in consequence exhibit such inability to adapt herself to civilized surroundings as constitutes actual insanity'.[65] It was not coincidental that in every asylum he visited, Mercier encountered 'a certain number of bearded and bass-voice women, whose insanity is usually of a very intractable type; and I have had under care at the same time, two men, whose hairless faces, large mammae and shrill voices betokened an assumption of the secondary characters of the other sex, and whose insanity was notably intractable'.[66] Mercier's observations of women characterised by male secondary sexual characteristics and masculine physical traits acquired in old age were not the first of their kind: similar cases of 'inversion of sex' observed in asylums had already been published on the Continent.[67] The fact that he interpreted some of these women's symptoms in a language reminiscent of sexual inversion arguably suggests he was familiar with Continental literature. Yet Mercier did not refer specifically to psychological gender inversion, nor did he analyse the psychological dimension of the phenomenon he called 'inversion of sex'. As discussed in Chapter 3, by the 1890s, Italian psychiatrists had embraced the investigation of the psychological characteristics of sexual inverts. Mercier's analysis, however, was still heavily reliant on physical and other external signs of 'inversion of sex', such as the growing of beards in women. Regardless of this difference, Mercier interpreted this phenomenon through the lens of degeneration, just as his Italian colleagues before him had done.[68]

Sexual inversion

As has been shown, British psychiatry approached same-sex desires in different ways. This included listing them in psychiatric taxonomies under the heading of mental disorders such as monomania; within the debate about criminal responsibility; and through degeneration theory. Occasionally, same-sex desires were explained through the framework of gender inversion, as with the case of the anonymous physician who had described 'androgynism' as the 'intrusion of one sex into the province of the other' in 1867. On the whole, however, British psychiatrists were reluctant to engage systematically with the study of 'sexual inversion', even after Westphal broached the subject in 1869. It was not a coincidence that the topic was introduced in Britain by Julius Krueg, a German physician who published an article in the neurological journal *Brain* in 1881.[69] Although Krueg's piece did not deal with any British cases, it is crucial because it was the first time the 'new' psychiatric category appeared in the pages of a British psychiatric periodical.[70]

Krueg's article summarised the most important cases of sexual inversion recorded by Westphal, Krafft-Ebing, Tamassia and others, and then went on to recount a case-history he had drawn from the life of Herr N.[71] Krueg's subject differed from most of the case-histories published in the 1880s because the subject of his study was neither a convicted criminal nor an inmate of a lunatic asylum, and yet his sexual activities were, as Krueg made clear, 'the result of a morbid impulse'.[72] Herr N.'s first sexual experience had taken place at school, where he and other boys masturbated together and formed close friendships. His inclinations and behaviours consistently pointed to gender inversion: when he was a child he enjoyed watching naked men, and as an adult he was involved in women's fashion.[73] Krueg stressed that in well-defined cases of sexual inversion, 'there has almost always been evidence of an inherited psychopathic, or, at all events, neuropathic constitution'.[74] Herr N. was not an exception: he came from a neuropathic family and had a 'hysterical' mother.[75] As Krueg suggested, it was frequent for patients diagnosed with sexual inversion to show signs of degeneration, which gave rise to the question of 'whether the perverted sexual instinct is not in itself, isolated'. The answer is for the negative: it may be 'a sufficient indication of degeneration. Biologically considered, there cannot be a doubt that an impulse so adverse to propagation must be a mark of degeneration.'[76] In other words, Krueg – like Tamassia in 1878 – was responding to Westphal, who had wondered whether sexual inversion should be considered a mental pathology in itself, or a symptom of other mental disorders. While Westphal had not provided a clear answer, both Krueg and Tamassia believed that sexual inversion was a mental pathology.[77]

Krueg also introduced a very short case of a woman whose attraction for another woman 'never got beyond the platonic stage'. This female patient, 'F', was a 25-year-old maidservant. Little was known of her previous history. 'She says she cares nothing for men, and other circumstances confirmed this'; instead, she was 'passionately devoted' to the ladies for whom she worked, and cried 'all day if she thinks one or other of them has looked black at her, going about moaning that Miss does not want her any more, &c. She is a tolerably big woman, with strong but pretty features, not at all unwomanly in appearance, and, as far as can be judged, normally developed.'[78] Krueg devoted only a few lines to this woman and stressed that it was difficult to obtain 'any definite information as to the frequency of this condition in women'. He supported this assertion by noting that, of the fourteen case-histories known to the international community until that date, twelve had to do with men and only two with women.[79] In Krueg's opinion, the situation of women was less known due to the fact that 'it is easier for women to escape detection than [it is for] men'.[80] This reasoning that pointed to women's skills in hiding their inclinations and assumed that their prudish attitude discouraged discussion of sexual matters was typically adopted by physicians until the late 1890s to explain why there were fewer case-histories of women than of men.[81]

Even counting Krueg's article in 1881, the publication of male case-studies outnumbered female case-histories in medical journals, a trend that would be confirmed in the ensuing years. In Italy, research into female sexual inversion flourished, but this was not the case in Britain, where occasionally psychiatrists remarked that they had encountered cases of female inversion, but that they were unable to supply detailed case-histories. Only four other short and unsystematic articles on the topic of sexual perversion involving same-sex desires were published in the six years that separate Krueg's study from Ellis's *Sexual Inversion*. In 1884 Savage put forward a case of a male homosexual in the *Journal of Mental Science*; in 1890 the Secretary of the Medico-Psychological Association, Dr Urquhart, delivered a paper featuring another case of male inversion; in 1893 William Sullivan recorded a case of acquired sexual perversion involving same-sex desires; and finally, in 1896, Savage and Mercier published a joint study of insanity, including two cases of inverts.[82] Savage's and Urquhart's publications (1884 and 1890, respectively) briefly mentioned a case of female inversion. On both occasions, female inversion was explained in terms of the reproductive system, as I illustrate later. Overall, the few British studies on sexual inversion published before 1897 manifested a faint interest in the female phenomenon, but given their unwillingness to explore the inner life of female inverts, their analyses were neither greatly developed nor systematic. But, as I will show, physicians did deal with female same-sex desires in the context of other medical discourses.[83]

When viewing Britain and Italy side by side, one cannot help wondering why British psychiatrists were by and large so reluctant to engage with the study of sexual inversion. In an article he wrote for Tuke's well-established *Dictionary of Psychological Medicine* (1892), Conolly Norman expressed the view that same-sex practices were a 'vice', and therefore not a mental disorder.[84] Many British psychiatrists were inclined to agree with him.[85] A further insight may be gleaned from Savage's and Mercier's 'Insanity of Conduct', a joint paper published in the *Journal of Mental Science* in 1896.[86] Although their article did not deal with sexual inversion exclusively, they discussed two cases of male same-sex desires and argued that sexual inversion was a 'disorder of [the] mind'. The recognition that sexual aberrations were expressions of mental illness did not, however, entail diminished responsibility in the case of 'criminal acts'. In reference to the Oscar Wilde trials, Savage and Mercier pointed out that 'the practices of sexual perversion are of themselves regarded by legal authorities as crimes and punishable as such, and not as evidences of insanity'.[87] There is no further discussion of sexual inversion in their paper. Yet the fact that they rejected legal unaccountability in cases of sexual inversion suggests that they did not want to challenge those legal authorities who regarded same-sex practices as a vice and, as such, subject to punishment. During this same period, their colleagues in Continental Europe argued that sexual inversion was a disease and not a crime, but British psychiatrists were not inclined to agree with them because it would set them on a collision course with the law.[88]

Female sexual excess: nymphomania and prostitution

Conventionally, physicians viewed nymphomania and prostitution as expressions of female sexual excess and as abnormal forms of female behaviour that merited close scrutiny. A strong sexual drive could lead a nymphomaniac to satisfy her desires through the practice of prostitution and, according to some physicians, prostitutes were often tribades.[89] While, generally speaking, sexual inversion in Britain was not an autonomous category within psychiatric nosology, physicians nonetheless engaged with the study of same-sex desires within their observations on various manifestations of an exaggerated female sexual appetite. The analysis of British medical writings about the sexual feelings of nymphomaniacs and prostitutes shows that physicians spoke about female same-sex desires more than historians routinely acknowledge, and that female same-sex desires were considered a 'temporary sexual aberration', rather than an innate psychological characteristic. Ultimately, these British medical discourses call into question the tenets of current historiography, showing the extent to which, in countries like Britain, medical writers did not employ a 'psychiatric style of reasoning', to use Davidson's words, when approaching female same-sex desires. Instead, British medical writers continued to adopt older models of thinking about same-sex desires. Female same-sex desires were continuously redefined in different medical disciplines, but such rearticulation, as this chapter shows, did not coincide with Continental developments. Local circumstances highly constrained the ways in which female same-sex desires were understood across medical fields.

Nymphomania

Nymphomania featured in a number of British medical analyses of female diseases, whether psychiatric or gynaecological. The sexual drive of the average woman was believed to be less powerful than the average man's, but when 'erotic feeling' was present, it was capable of exciting her nervous system and influencing her behaviour, which resulted in mental and physical disorders.[90] There was no consensus on the aetiology of this disease: some physicians believed nymphomania was caused by the uterus, others by the clitoris, others again by the ovaries, and a few of them by the cerebellum.[91] In their *Manual of Psychological Medicine* (1858), Bucknill and Tuke had analysed both erotomania and nymphomania as forms of insanity.[92] They defined erotomania as an 'error of understanding' in which 'the imagination alone is affected'. In principle, nymphomania was different from erotomania because each disease originated in a different way. In the nymphomaniac, 'the evil originates in organs of reproduction, the irritation of which reacts upon the brain', whereas erotomania originated through a disorder in the brain.[93] In practice, nymphomania and erotomania could coexist.[94]

A vivid illustration of the disease is offered in the appendix to Bucknill's and Tuke's *Manual*, which features a case of 'Acute Nymphomania, with suicidal Impulse'. The case-history was based on a thirty-year-old dressmaker referred to as 'J.M'. Three years before the alienists visited her, she had had a couple of maniac attacks, from which she appeared to have recovered. She had converted to Roman Catholicism and her first attack had to do with her change of faith. No cause was identified for the second attack, which had taken place when she went into a church and threw herself to the floor, making 'a scene' during the service. Although she had not displayed any other sign of insanity leading up to this event, she was found to be 'raving mad' upon being removed from the church. Afterwards she tried to kill herself and was admitted to an asylum, still covered in bruises and gravely disfigured due to her suicide attempts.[95] Once in the asylum, 'she was sensible and did not appear to be suffering from any delusion'. However, she made repeated efforts to beat her head against the wall and said she wanted to die. Consequently, she was never left alone, as she seized every opportunity to suffocate or strangle herself. In the course of five or six days the symptoms abated greatly and nine days after her admission her mental health appeared to be perfectly restored: she was in 'perfect right of mind', occupying herself with needlework.[96]

Five weeks after her admission, while she was menstruating, J.M. suffered a sudden relapse: 'she made a violent and indecent attack upon a woman whom she believed to be a man'. Again, she tried to injure herself in any possible way; by thrusting her hand down her throat, by beating her head against the wall or the floor, and by attempting to drown herself in the bath. 'Several nurses were with her night and day, and upon them she made constant attacks of indecent nature. She moaned and exclaimed, "Oh my God! Oh, blessed Jesus! Oh save me!" and evidently suffered great mental anguish.' Large doses of opium were administered without result. At the end of the fifth day, after the last attack of nymphomania, she died. The post-mortem examination revealed the hymen was 'perfect', while the 'uterus and its ligaments, and the ovaries, were greatly congested', and her brain was in 'tangibl[y] morbid conditions'. The rest of the organs were healthy.[97] This case is relevant to the purpose of this book because all the patient's sexual desires and initiatives are directed towards other women – the text does not mention men at all. Nonetheless, Bucknill and Tuke classified this woman's case under nymphomania rather than under 'unnatural crime' – also included in the *Manual* – or under other mental diseases.[98]

This case-history shows that in this period, British doctors would typically group female same-sex desires under the heading of nymphomania or erotomania.[99] This case does not hint at gender inversion, but it indicates the sexual excess of nymphomania could encompass sexual desires towards women.

In 1892, Gustave Bouchereau wrote the section headed 'Nymphomania' in Tuke's *Dictionary*. Bouchereau defined nymphomania as a 'morbid condition peculiar to the female sex, the most prominent character of which consists in an irresistible impulse to satisfy the sexual appetite'.[100] Although the 'morbid love' could be merely intellectual, in other cases it was accompanied by 'a violent, irresistible sexual appetite which must be satisfied, regardless of age or any other consideration'. Nymphomania could be caused by 'a disease of the genital apparatus: eruptions on the labia majora and minora, inflammation of the vagina, uterus, Fallopian tube, and organic affections of the Uterus and the commencement of the vagina'.[101] Bouchereau warned that nurses and servants to whom the care of children was confided 'should be kept under strict surveillance by the parents, because it is not uncommon that under the influence of hysteria or of a morbid disposition, they subject the children to manipulations which affect their health and compromise their existence'.[102] Finally, Bouchereau made it clear that nymphomania could also take the form of homoerotism:

> For many years a whole literature of romance and plays has been occupied in the description of Lesbic love, to great damage of young girls and neuropathic women; curiosity at first attracts and soon misleads them; the sensation experienced enslaved them, and then, aided by the use of morphia, ether and cocaine, nymphomania establishes itself. The word has spread from the unfortunates to the women of the theatres, and from thence has taken possession of unoccupied women of all classes of society with unsatisfied desires.[103]

Bucknill's and Tuke's *Manual of Psychological Medicine* and Tuke's *Dictionary of Psychological Medicine* were amongst the most important British psychiatric manuals of the day. Both of them illustrate that female same-sex desires were believed to be no more dangerous than excessive sexuality which, I would suggest, was what these authors considered the proper affliction. So far as physicians were concerned, it made no difference if a nymphomaniac had sex with many men or with a woman. The real issue was the possibility that women might have a boundless and active sexuality because this contradicted the tenets of medical science, which asserted female sexuality was passive and responsive to male aggressive sexuality. British psychiatric manuals did tackle the issue of female same-sex desires, but conceived of them as mere symptoms of other mental disorders. Same-sex practices were caused by nymphomania, which was a proper mental disorder that originated as a bodily malady, generally in the genitals or reproductive organs, and then affected the brain.

This medical way of thinking about female same-sex desires continued to have currency at the turn of the century. Fred Smith edited the 1905 edition of Alfred Swaine Taylor's classic contribution to the field of forensic medicine, *The Principles and Practice of Medical Jurisprudence*. In this version Smith decided to expand Taylor's *Principles* and to explain 'tribadism', which previous editions had ignored. Tribadism was defined as 'the gratification of the sexual desire of a woman by a woman' and was listed under the heading 'unnatural offences'.[104] It was clarified that neither tribadism nor masturbation were 'indictable offence[s]' unless done publicly, unlike 'sodomy', 'bestiality', and 'indecent exposure', which were punished by British law in all circumstances.[105] The fact that there is little chance of 'medical evidence establishing proof of the act' of tribadism was also remarked upon. Yet, the editor of the manual decided to publish a letter written in 1897 to Dr Stevenson by another unidentified physician. The letter said:

> DEAR DR. STEVENSON—Is tribadism a criminal offence? and [sic] if so would the commission of it by a married woman enable her husband to obtain a divorce supposing it could be proved? These problems were submitted to me by a medical friend, and I find so very little on this unsavoury subject that I venture to see if your experience can help me.
>
> My friend sent the married lady in question abroad a short time ago to travel with a nurse on account of great sexual excitability and suspected masturbation, and it seems that whilst abroad she has been guilty of tribadism with a lady friend. This having come to the knowledge of the husband, he has been to his mother-in-law and told her that her daughter has been guilty of a criminal offence, and that he is going to obtain a divorce.
>
> I apprehend that the proof would have to be furnished entirely by other than medical evidence. Have you any experience on this subject?
>
> My own impression is that the problem is purely a legal one.[106]

Smith, on behalf of Taylor, reiterated beyond any doubt that tribadism was not a criminal act. He went on to note that '[t]he act commonly arises in women with nymphomania, and its alleged occurrence would suggest an inquiry into her mental state'.[107] Therefore, Smith agreed with Dr Stevenson in his interpretation of sex between women as a simple 'act', a sexual practice common in women suffering from nymphomania, but without associating such practices to innate characteristics or to the psychiatric category of sexual inversion. Smith also suggested that practitioners should enquire into the mental condition of women engaging in same-sex acts – this view of nymphomania as a psychiatric disorder was well accepted as it had been articulated in Bucknill's and Tuke's *Manual of Psychological Medicine* and

Tuke's *Dictionary of Psychological Medicine*. Bearing in mind the extensive body of medical literature on homosexuality to be found on the Continent and the recent publication of Ellis's *Sexual Inversion* in Britain, Smith's comments on tribadism betray the reluctance of established British medical and legal practitioners to embrace the concept of 'sexual inversion'.[108] Smith did not even mention the psychiatric disorder of sexual inversion, preferring instead more traditional interpretations of female same-sex desires.

Prostitution

By 1840, Parent-Duchâtelet's work on prostitution had already reached its second English edition.[109] According to the historian Judith Walkowitz, British researchers were unable to duplicate Parent-Duchâtelet's meticulous research because they did not have access to the kinds of official records that had furnished the French hygienist with most of his statistical information.[110] Over the next four decades, however, British writers from medical and non-medical backgrounds displayed a considerable familiarity with the Frenchman's research on prostitution, and occasionally tried to apply his catalogue of sexual customs and habits to the Victorian world. The study of female prostitution attracted British men from the world of organised religion, laymen influenced by evangelical doctrine, but mostly doctors, such as Acton.

British physicians who drew on Parent-Duchâtelet in their analysis of prostitution were generally keen to deny the existence of same-sex practices in their country; they claimed that tribadism was a foreign vice, or at any rate a rare phenomenon in Britain. Yet some medical writers suspected that female same-sex acts amongst prostitutes were not being recorded, not so much because of a lack of information, but because of British physicians' reluctance to survey the issue. In 1837, *The Lancet* reviewed Parent-Duchâtelet's work in five lengthy articles, which address love relationships between prostitutes amongst other aspects of his studies.[111] It was reported that a quarter of all prostitutes had a 'depraved and unnatural taste', which led them to choose women as lovers. These women were called 'tribades' and this sort of relationship was thought to be very common in prisons where the prostitutes were detained. *The Lancet* reported that Parisian police forced every woman to sleep in a single bed in order to avoid this 'horrible vice'.[112] The author of the article stressed that British physicians had not investigated this aspect of prostitutes' lives yet. In 1839, a physician practising in London called Michael Ryan agreed with *The Lancet* and noted that the Secretary of the Society for the Prevention of Juvenile Prostitution – a man named Talbot – had no information on the incidence of tribadism amongst British prostitutes. Although Talbot judged it rare, Ryan insisted that 'secret depraved pleasures of lust' did exist, but were not studied in Britain.[113]

George Drysdale joined a number of mid-nineteenth-century medical writers who complained about the lack of analysis and discussion of sexual problems in British medical journals and books. He and other neo-Malthusian doctors formed a small group of radical free thinkers who advocated that non-procreative sex was healthy and that contraceptive methods should be used to limit reproduction.[114] In 1854, Drysdale published *Elements of Social Science*, a radical and greatly successful textbook, which was reprinted and re-edited several times between 1854 and 1914.[115] In his book, Drysdale drew attention to the condition of women. Hoping that one day they would develop their own intellectual skills and contribute to culture at large, he expressed the view that men and women had 'different thoughts, feelings, and modes of judgement', and that men ought not to impose their own moral and physical codes on women.[116] According to Drysdale, given the influence of religious beliefs on women's lives, physical virtues were seldom thought to belong to her province: 'strength, vigour, courage, and activity, are not considered feminine virtues, but, if possible, rather detract from woman's peculiar charms in the eye of spiritualism'.[117] Hence the physical character of women was, as a general rule, degraded to the utmost degree: 'poor, weak, nervous, delicate beings who can scarcely walk half-a-mile, whose muscles are unstrung, and whose nerves are full of weakness and irritability'.[118] To make women's condition worse, at 'boarding school and in other places of instruction, bodily strength and physical courage and activity are not regarded as female excellences, but rather looked upon as unfeminine; and gentleness, quietness, and timidity are cherished'.[119] Current women's education thus brought about weakness of mind and body, a state of affairs the reformist Drysdale believed could be improved upon if the education system was bettered.

Drysdale's arguments seem to highlight essential differences between the sexes. But this is unsettled by his view that 'in the higher animals, including man, there is great evidence to show that each individual is really hermaphrodite'. In the embryo there was no difference between a penis and a clitoris, as these organs become differentiated only in later stages of development. Thus, on the basis of embryological evidence, 'the difference of sex is rather apparent than essential, and all of us are truly hermaphrodite beings'.[120] On the one hand, Drysdale believed that women and men were essentially different, although current women's weakness was caused by a detrimental education system that did not allow them to develop their possibilities. On the other hand, like many physicians of his time, he was fascinated by recent embryological studies and was inclined to think that men and women shared a fundamentally hermaphroditic nature. Although Drysdale himself did not resolve such a contradiction, other late nineteenth-century physicians did so by using evolutionary theories and reflecting on sexual dimorphism as a result of evolution.

Drysdale also analysed female and male same-sex relations. He criticised the social customs that separated the sexes and disallowed intimacy. As a result, instead of a 'proper amount of natural sexual intercourse', 'morbid tastes' took the place of the healthy ones and led to 'unnatural indulgences'.[121] He explained that the incidence of sodomy in contemporary society was the result of a prurient and stringent morality, which prohibited all sexual activities outside marriage, thus generating a range of sexual problems that included unnatural vices.[122] Although Drysdale did not approve of 'unnatural' sexuality, he argued that a society that had ultimately triggered sex between men through an excess of moral rectitude should critically re-examine its attitude towards 'vice', if only to understand and eradicate it more effectively.[123]

While 'sodomy' was prevalent in prisons, same-sex acts between women were 'much more frequently seen in the female venereal hospitals, especially in the one annexed to the prostitute prison'.[124] Drysdale went on to describe same-sex love and practices amongst prostitutes, drawing extensively on Parent-Duchâtelet's study. After emphasising his point about the frequency of 'mutual loves between prostitutes' in certain women-only environments, he stressed the highly passionate nature of such relationships:

> In this singular connection, two prostitutes enter into sexual relations with each other, with all the ardour, impetuosity, and tenderness of passion, that the most intense normal sexual love could inspire. They devote themselves to each other, and practice together all devices of unnatural voluptuousness. They feel for each other the conflicting sexual passions, now burning with jealousy, now melting with tenderness; they are distracted at separation, and follow each other every where. If the one be committed to prison, the other gets herself also arrested, and they seek to leave it together. They are much more jealous of desertion by their female lover than by a male one; and if one has proved false, her companion will seek revenge in every way.[125]

Drysdale was not the first medical writer to insist that women having sexual relations with each other were particularly jealous and passionate; this idea was widespread in Italy, both in studies on the tribade-prostitute and in other medical writings such as Tonini's, which contained elements that anticipated the concept of sexual inversion. This suggests that Italian and British physicians were drawing on common sources such as Parent-Duchâtelet, who had underlined that love between prostitutes often took the form of a fierce devotion with a strong component of jealousy.[126] Even writings that did not deal with the figure of the tribade-prostitute – Tonini's amongst them – remarked on the passionate love of these women, which reveals that cultural ideas about relationships of this kind circulated easily across national borders.

According to Drysdale, it was no surprise that 'this class of unnatural lovers' [tribades] was common especially amongst prostitutes. On the one hand, young prostitutes were confined together in prisons, often for several months. On the other hand, older prostitutes whose sexual experiences with the other sex 'have been so painful and degrading' came to abhor all men, and only take pleasure in these 'unnatural relations'.[127] Here Drysdale was extending Parent-Duchâtelet's observations about tribadism in prostitutes. The French physician had noted that prostitutes often became tribades after working in the sex industry for a period of six to ten years, and that prostitutes ended up 'loathing all men'. Yet Parent-Duchâtelet did not explicitly say that prostitutes disliked all men because they had been degraded by them.[128] Drysdale's *Elements* achieved remarkable success in Italy after being translated and published for the first time in 1874 under the title *Elementi di Scienza sociale ossia religione fisica, sessuale e naturale* [*Elements of Social Science, Physical, Sexual and Natural Religion*], which was subsequently reprinted four times in seven years. Hence some physicians – notably Lombroso – adopted Drysdale's argument that prostitutes engaged in homosexual relationships as a result of the abuses male clients inflicted on them.[129] While Lombroso did not directly acknowledge Drysdale in his work on the tribade-prostitute, given Drysdale's wide circulation in Italian medical circles, it is reasonable to infer he had read his observations about love between prostitutes.

Parent-Duchâtelet's and Drysdale's works helped cement the link between prostitution and tribadism in Britain. Ellis spoke openly about the figure of the tribade-prostitute in both *The Criminal* (1890) and *Sexual Inversion* (1897). Reporting Parent-Duchâtelet's, Albert Moll's, and Lombroso's research on the phenomenon of same-sex desires amongst prostitutes, Ellis informed the reader that a number of medical researchers agreed that one quarter of prostitutes were 'homosexuals'.[130] But, unlike in Italy, the tribade-prostitute did not become a widespread stereotype in the British medical community in the second half of the nineteenth century. British physicians may well have been aware that Parent-Duchâtelet and Drysdale drew a relationship between prostitution and tribadism, but, in contrast with Ellis, were unwilling to speak openly about it, in the same way that they were generally reluctant to speak openly about female same-sex relationships.[131] Instead, they resorted to circumlocutions and coded references that betrayed their familiarity with medical literature on female same-sex desires without discussing the topic any further.

For instance, in 1874, an Irish Professor of Midwifery called Fleetwood Churchill published *On Diseases of Women*, where he engaged with the work of Parent-Duchâtelet who, as Churchill said, was greatly respected in Britain.[132] In his discussion of women's enlarged clitorises, Churchill spoke about those 'females of the most unbridled passions' who made up a distinct group amongst prostitutes;[133] he described them as being particularly 'lascivious' and having a 'lustful sexual appetite'.[134] It is possible but unlikely that Churchill was discerning between prostitutes with more or less passionate temperaments. His reference to Parent-Duchâtelet research was probably hinting at tribade-prostitutes, whose passions he did not name as such. Colleagues with the same medical background would understand Churchill's allusions to female same-sex desires and the literature on the topic, unlike less knowledgeable readers, who would not.

Gynaecology: sexual excess in the female body

In Britain a medical group that advanced hypotheses to explain female same-sex desires, and whose explanations of the phenomenon were adopted by other medical specialties, were gynaecologists. British gynaecological writings are interesting in this context not only because they have been neglected by modern historians interested in same-sex desires, but also because, as far as I have been able to tell, Italian gynaecologists did not show an equivalent interest in same-sex desires.

Much in the same way as forensic doctors had traditionally looked for signs of male same-sex acts in the anus, gynaecologists believed that the female genitalia and the reproductive apparatus could reveal whether same-sex practices had taken place.[135] Gynaecologists argued that diseased ovaries or disordered menstruation could lead to injuries to the nervous system and brain, which in turn could cause mental illness. Contrary to psychiatry, gynaecology transferred the focus from the brain to the genitals and the reproductive system.[136] Redness, soreness, or itching of the genitals, along with an enlarged clitoris or labia, were believed to be the main indicators of female same-sex desires. British physicians observed that enlarged female genitalia could sometimes resemble a small penis, a characteristic that had been associated with tribadism since the early modern period.[137] Although Parent-Duchâtelet's study on French prostitutes refuted the widely held belief that female sexual excess would result in hypertrophy of the genitals, British gynaecologists throughout the nineteenth century continued to draw attention to the size of the clitoris and other parts of the female body as evidence of sexual unrestraint. The irregular sexual desire of women who indulged in 'lesbian love' was revealed not only by an enlargement of the clitoris, but by an atrophic uterus, irregular physiology, or typically masculine secondary sexual characteristics.

In 1877, the gynaecologist Heywood Smith, a member of the Royal College of Physicians who at the time worked at the Hospital for Women and at the British Lying-in Hospital in London, stated that hypertrophy of the clitoris could be congenital or the result of masturbation. Amongst the symptoms of a disproportionately sized clitoris, he listed 'inordinate sexual desire', which included 'craving for sexual intercourse' and nymphomania. To cure the body of the supposed discomfort caused by an enlarged clitoris, Smith suggested cold applications on the female genitalia, physiological rest, and finally, clitoridectomy.[138] Even though Smith did not openly mention same-sex longings when discussing the excessive size of the clitoris in relation to nymphomania, some of his colleagues may have taken his vague reference to 'inordinate sexual desire' to mean same-sex desires. This is so not only because nineteenth-century medical writings commonly held that one of the most usual outcomes of nymphomania was same-sex desires, but because European medicine had traditionally associated a hypertrophied clitoris with tribadism.[139]

The size of the clitoris, and in general the anatomical structure of female genitalia and the reproductive system, was also crucial in determining the individual biological sex in cases of hermaphroditism. When dealing with the malformation of the female genitalia in *Diseases of Women and Surgery* (1889), the gynaecologist Lawson Tait pointed out that sometimes it was difficult to determine the sex to which an individual belonged.[140] 'There are many historic cases of males having been married as women, and of women who have been placed in the positions of men.'[141] For clinical purposes, cases were divided into individuals whose underdeveloped male sexual organs made them appear female, and 'those in which an excessive development makes the female organs resemble those of a male'.[142] While the first class was more common, Tait suggested that in doubtful cases 'it is a good rule to assume that it is a male child unless the contrary can be shown, for in this way lamentable mistakes can be avoided'.[143] The underlying assumption of why patients should be considered male by default was that a woman brought up amongst males could do 'little harm' to herself and other people, while a man brought up as a woman could be 'raped to death'.[144] He then recounted some cases of women passing as men and vice versa. He recounted the story of a woman he met in a male prison who had lived all her life as a man, and who 'had never entertained any partiality for either sex – facts probably due to an infantile condition of the internal organs as marked as that of the external'.[145]

Tait had frequently observed that girls' external genitalia were deformed, and that one of the most common forms of genital deformity was an 'abnormal development of the clitoris'.[146] He reported having seen a clitoris so large as to resemble an 'infantile penis', which 'according to the statement of the patient, [was capable] of distinct erection during sexual excitement'.[147] However, he promptly warned the reader that 'the stories we read of women having this organ so large as to be capable of having, and desirous of, connection with other women' should be recorded as cases of men mistakenly registered as women.[148] Despite Tait's belief that these individuals with a big clitoris were actually men passing as women, he informed the reader that medical practitioners read and knew of 'stories' and cases about women having a big clitoris, and having or seeking to have sexual intercourse with other women.

Tait was not the only medical practitioner prepared to discuss the relationship between the female body, and in particular the genitals, and same-sex desires. Indeed, the interest in the topic was such that it went beyond the boundaries of British gynaecology. On 8 October 1890, Robert Barnes, a consulting physician at the St George Hospital in London, gave a paper at the British Gynaecological Society 'On the Correlations of the Sexual Functions and Mental Disorders of Women'.[149] The session was well attended, with fifty-one fellows of the Society and twenty-three visitors. The psychiatrists Savage, Tuke, and Mercier were in the audience. As has been shown, both Savage and Mercier had engaged with the topic of Barnes's paper in their treatises. In *Insanity and Allied Neurosis* (1882), Savage established a link between insanity affecting girls and their reproductive function; in *Sanity and Insanity* (1890), Mercier associated 'inversion of sex' with women's reproductive system. Barnes began by praising the advances in gynaecology, anatomy, and the physiology of the nervous system that had in recent years allowed the study, 'with scientific precision', of the correlation between sexual functions and nervous phenomena in women.[150] Scientific knowledge had established a close association between nervous disorders and disease, or disordered functions of the sexual organs. The paper itself was not markedly different from many other gynaecological studies on the influence of menstrual and ovarian diseases on nervous illnesses that were common at the time.[151] The discussion that followed Barnes's paper, however, ended up focusing on same-sex desires.

In the course of this debate, Savage raised the question of whether masturbation was a symptom or a cause of disease.[152] He then went on to discuss the question of 'sexual perversion', observing that it was a 'marked symptom in a certain number of cases'. He reported an example of a woman who had 'an infantile uterus about the size of a nut, although the vagina was of normal capacity', and who 'had always manifested a strong passion for another woman'. Savage remarked that with regard to physical characteristics, 'degenerated' women tended to resemble men. Therefore, he concluded, a woman with degenerative signs such as 'unnatural' growth of hair on the face and other masculine features was particularly prone to sexual perversion.[153]

His comments at the British Gynaecological Society were based on his own previous research. In 1878, he had published a 'Case of Malformation of Genitalia in Insanity' in the *Journal of Mental Science*.[154] On this occasion the case centred on Elizabeth B., a 48-year-old single woman. Until 19 she had been 'decidedly pretty', but at that age Elizabeth had her first period, and started to develop oddly. She had no pubic hair and no breasts; her hands were 'man-like', and she became very tall and broad, assuming a masculine appearance.[155] Before Savage treated her in Bethlehem asylum in London, she had been at a private asylum, and 'on one occasion during the night, [she] tried to take indecent liberties with a female night nurse, and she got into bed with another woman, and said she must have sexual intercourse with her, and tried to effect this in the masculine way'.[156] A post-mortem examination of her genitalia revealed that although the external labia were large, the clitoris was not visible. The vagina was 'very large, thin-walled and baggy'.[157] In this case Savage concluded that 'the amorous desires of the patient towards females shows [sic] the complete change in the mind and affections, produced by purely bodily conditions'.[158]

In 1884, Savage returned to the topic of female same-sex desires in the *Journal of Mental Science*. He published the first British case-history of male sexual inversion, which briefly considered female sexual inversion towards the end. Savage reported that while working at Bethlehem asylum, he had met a female patient who showed 'powerful lust towards those of her own sex', whose 'infantile uterus was discovered' upon carrying out an autopsy.[159] In his 1884 article, Savage reiterated his 1878 argument whereby the female body was responsible for same-sex desires, but this time associating the latter with a diseased reproductive system. As Savage's observations illustrate, psychiatrists showed interest in gynaecological studies because they believed, like gynaecologists, that there existed a close relationship between reproductive apparatus and a woman's brain. Psychiatrists were interested in gynaecology for practical reasons too: they had to be familiar with its practice because they often had to perform gynaecological examinations when admitting a patient to an asylum. This was a routine procedure that aimed to assess the general health of patients. It was also common for psychiatrists searching for explanations for mental disorders involving abnormal sexual behaviours to conduct post-mortem examinations of the reproductive apparatus.

Barnes's paper 'On the Correlations of the Sexual Functions and Mental Disorders of Women' at the British Gynaecological Society must have sparked an animated debate, because the session had to be adjourned to 22 October 1890. On this second occasion, the gynaecologists Heywood Smith and Percy Smith spoke again about masturbation in women. Dr Hugh Fenton referred to Savage's previous comments and alluded to the 'conditions of eunuchs' in relation to same-sex desires.[160] Barnes, for his part, made reference to some points raised by Savage, noting that women displaying masculine attributes illustrated one of the difficulties with which gynaecologists had to contend. On the subject of Savage's question, he pointed out that female same-sex desires, 'sometimes associated with degeneration or imperfect development of the generative organs[,] had been known more or less for years, before eunuchs or spayed women were known'.[161]

As Barnes implied, the removal of sexual organs was supposed to be one of the causes of same-sex desires. In the nineteenth century, physical explanations of female same-sex desires were supported by observations following the removal of ovaries, and gynaecologists believed that women could acquire sexual perversions after such operations. In most gynaecological reasoning, the underlying assumption was that normal and healthy ovaries, or a generally healthy reproductive apparatus, necessarily gave rise to a normal (reproductive) sexual desire, whereas a defective reproductive system triggered sexual perversion. For instance, in 1894, the physician C. H. Routh published a study of ovarian diseases in the *British Gynaecological Journal*. He stressed that eunuchs were 'the vilest creatures of the human race, cowards and deceitful because they are weak; envious and spiteful because they are unfortunate'; they seemed to resemble females in their bone development, lack of courage and muscular force, and were more sensitive and affected by nervous diseases and 'low spirits' than normal men. Castrated women frequently grew hair on the face, and were affected by insanity (generally in the form of melancholia); they experienced disorders of the senses, 'perverseness of the sexual function', and finally the loss of 'sexual feeling'. In some women the passion exceeded all bounds, resulting in nymphomania.[162] In the typically cautious British manner, Routh refrained from openly discussing same-sex acts, but when discussing the associated masculine physical characteristics of castrated women, he associated them with the 'perverseness of the sexual function', which strongly suggests that he had female same-sex desires in mind.[163]

Throughout the nineteenth century, British physicians in general, and gynaecologists in particular, associated female same-sex desires with masculine secondary sexual characteristics and paid attention to the bodily signs of female same-sex acts, such as a hypertrophied clitoris or an abnormal anatomy of the ovaries and uterus. The kind of bodily explanations of female same-sex desires promoted by gynaecologists were popular in other medical circles, and even a few British psychiatrists embraced anatomically based arguments when discussing the topic.

Gynaecologists and psychiatrists shared the assumption that female same-sex acts were signalled in women's bodies. Reluctant to take up Continental ideas about sexual inversion, British psychiatrists clung to 'old' anatomical explanations of female same-sex desires, showing the extent to which bodily explanations were firmly grounded in medical discourses at large until at least the end of the nineteenth century. While Italian, German, and French case-histories of sexual inversion consistently agreed patients had no abnormal genitalia, on the other side of the English Channel psychiatrists kept encountering abnormalities in the anatomy of women who displayed same-sex proclivities.

Young females do not sleep in the same bed: masturbation and adolescence

Historians have convincingly shown that anxieties about the perils of masturbation for British boys and young men pervade nineteenth-century medical literature.[164] Lesley Hall has noted the surprising fact that for all the fears expressed about the various sicknesses engendered by masturbation, there were no equivalent concerns about adolescent homoerotic experimentation leading to permanent 'inversion' or homosexuality.[165] Sean Brady has stressed that British doctors avoided discussing or examining sexuality between men, noting that British medical treatises that dealt with the threats of masturbation did not link 'self-abuse' to 'inversion'.[166] Conversely, the association between 'self-abuse' and female same-sex acts was relatively common, although British physicians did not agree on whether women masturbated as much as men did. Medical writers highlighted the perils of girls sleeping in the same bed or in an overcrowded room, warning that the 'vice' was often learned in boarding schools.[167] They thought woman-only environments, especially where young girls gathered, fostered sexual feelings that were difficult to contain. It was a short step for the reader to assume that women could easily go from masturbating together in the same bed, to pleasing themselves mutually and then actually having sex with each other. Physicians hinted at female same-sex acts within this framework of anxieties.

In these studies physicians seemed preoccupied not only with women-only environments, but with female adolescence. For psychiatrists, adolescence could be the cause of specific forms of madness, such as the 'insanity of pubescence', which was studied by the psychiatrist Skae, as has been mentioned earlier. It was well established in British medical practice that puberty was a dangerous period for both boys and girls because their nerves and imaginations were still very impressionable and all forms of stimuli could prematurely excite the genital centres. Amongst other things, this could cause an addiction to masturbation.[168] During puberty, the development of the sexual organs triggered a generalised excitability of body and brain, which could easily lead to various diseases. Furthermore, medical writers stressed the role female education played during adolescence in moderating sexual feeling. For instance, too much reading might augment girls' sexual appetite and thus lead to sexual perversion. Most physicians agreed the impressions received during puberty had a great influence on female development and that the young girls' surroundings were crucial for an appropriate sexual development.[169]

The problem of physical changes occurring during puberty was related to the issue of the balance of the nervous system. One of the major concerns of nineteenth-century psychiatry was that the demands of civilisation exhausted the nervous system. Childhood mental development was central to such debates. The well-known psychiatrist, Crichton-Browne, linked the problem of exhaustion in childhood to the excessive demands of overstretched Victorian education. Crichton-Browne was also concerned with childhood mental disorders, and argued that boys and girls who indulged in prolonged, intense reveries might develop mental depression and feelings of lethargy.[170]

The specific focus of these anxieties on girls' sexual awakening and their education in boarding schools bore a distinct resemblance to Italian medical literature's treatment of the *fiamma*. Nonetheless, as had been the case in discussions about love between prostitutes, British physicians were more cautious and indirect than Italians in their approach to same-sex desires in women-only environments. Moreover, while Italian physicians had debated the *fiamma* phenomenon mainly in the 1890s, their British colleagues disseminated encoded information on the topic throughout the century, but without ever going so far as to formulate a structured discourse on the phenomenon or a psychological description of the young girls who engaged in same-sex acts.

Historians have pointed out that in Britain the public debate over the right of women to access higher education reached its peak in the 1870s and 1880s, but from at least the mid-nineteenth century physicians displayed a growing concern about the appropriateness of women's education.[171] Within this context, medical manuals had advised those responsible for supervising girls during puberty 'not to allow the young to sleep in the same bed with the old, nor even with those advanced in age or debilitated, nor with too many – not more than three – in the same sleeping apartment, which ought to be large and well aired'.[172] Boarding schools frequently overlooked these considerations, thus, physicians thought, contributing to the spread of diseases.[173] In his discussion of hysteria, the London physician James Copland warned the reader that whenever females lived in close proximity during puberty, and especially where several used the same sleeping apartment, and were subjected to:

> a luxurious and over-refined mode of education, some will manifest a precocious development of both mind and body; but in proportion to precocity will tone and energy be deficient, and susceptibility and sensibility increased. [...] There can be no question, although the subject has been but rarely approached by British medical writers, that indulgences in solitary vices and sexual excitement, is not an infrequent cause of this, as well as of other disorders.[174]

Inasmuch as lifestyle, environment and education could affect a woman's health, it was believed these factors had an influence on physiological development. The surgeon Thomas Laycock was following Copland when in 1840 he alerted other physicians to the perils of female association, which could result in masturbation:

> Young females of the same age, and influenced by the same novel feelings toward the opposite sex, cannot associate together in public school without serious risk of exciting the passions, and of being led to indulge in practices injurious to both body and mind.[175]

It was no specific medical discipline's appointed task to wage a public war against women's schooling: they all agreed that girls learned the habit of 'self-abuse' from each other. In the second half of the century, both psychiatrists and gynaecologists joined the fray. For instance, in 1882, the psychiatrist Savage, who had previously written on sexual inversion and same-sex desires in women, criticised, as mentioned, the social conventions that led to the segregation of the sexes, pointing out that this custom fostered unhealthy sexual behaviours such as masturbation, which in a high proportion of cases was 'an educated vice. It is taught by one to another.'[176] The gynaecologist Tait, who had written about women passing as men, believed that 'self-abuse' was common in boys but rarer in girls. He also believed the practice originated in different ways, so that while boys discovered it by themselves, girls learned it from other girls. In girls, masturbation was 'the result of direct contamination'.[177] According to Tait, 'the most pernicious effects are met with when the contamination reaches a congregation of young women, as in a girls' school'.[178] Girls learned how to masturbate from other girls at schools, as well as from servants:

> The method of practice of the vice is usually by the finger, but devices of a still more mischievous character have come under my notice. In young children, masturbation is often associated with defective mental development, and it should always be a ground for placing them under special care. [. . .] In every instance where I have found a member of congregated children to be affected, the contagion has been traced to a servant.[179]

Tait believed girls' 'self-abuse' to be so dangerous a problem that 'clytoridectomy [sic] might be beneficial' in incorrigible cases. While he reported having performed the operation himself on one occasion, he did not inform other practitioners whether his method had been effective.[180]

Since his earliest works, the surgeon and gynaecologist H. Macnaughton Jones had consistently condemned not only female boarding schools, overcrowded sleeping apartments, heated rooms, ill-ventilated sitting rooms and bedrooms, prolonged sedentary employment, too much stooping or standing, excessive study and long school hours, but also want of suitable outdoor exercise or amusement, excessively violent exercise, and masculine attire. These bad habits resulted in uterine troubles later in life.[181] In later works Jones explained that masturbation caused a vast range of nervous and mental perturbations, from 'immoral tendency' and nymphomania, to all sorts of sexual perversion and 'unnatural tendencies'.[182]

It was not only male doctors who believed that there was a link between female same-sex desires and women-only environments. As late as 1915, the feminist and campaigner for women's rights to reproductive control, Stella Browne, gave a paper at the British Society for the Study of Sex Psychology, where she openly associated female sexual inversion with the separation of the sexes and with female education. She believed that much of the 'unhealthiness of sexual conditions' at that time was due to the habit of segregating the sexes during childhood and in later life, so that as they grew, they became each other's 'alien enemies'.[183] Even though British physicians had approached the topic circuitously and there was no systematic analysis comparable to Marchesini and Obici's in regards to the *fiamma* phenomenon, Stella Browne's remarks about female homosexuality reveal that in Britain, ideas coupling female same-sex desires with the school environment were as deep-rooted as they were in Italy.

Physicians in both countries established a connection between schools and female same-sex desires by proposing that same-sex practices spread through a sort of 'contagion' in women-only milieus. Unlike their Italian colleagues, British physicians never embarked on a psychological study of the phenomenon they were describing. In both countries there appeared other representations of female same-sex desires that had some aspects in common. The tribade-prostitute derived from common medical sources the British and the Italians drew upon, such as Parent-Duchâtelet. Likewise, the stereotype of masculine women was associated with same-sex acts in both countries. The similarities between British and Italian medical discourses were nonetheless betrayed by a different rhetoric and diverging underlying assumptions of the figures of female same-sex desires. Overall British medical literature on female same-sex desires was characterised by discretion and circumlocution, while Italian physicians indulged in titillating images of sex between women. It would be tempting to see this choice in language as the direct result of the allegedly typical Victorian prudishness in dealing with sexual matters. Yet this would merely reproduce Lytton Strachey's portrayal of the Victorians, which has been complicated and challenged by the accounts of innumerable historians of sexuality, which have shown another side to this period.[184]

Nonetheless, there is no doubt that British psychiatrists come out as prudish when compared to their Italian counterparts and one cannot help wondering why. British physicians had very practical reasons to be circumspect; above all, they had to avoid titillating readers because the Obscene Publications Act (1857) had extended the government's power to control and prosecute the writers and publishers of writings considered obscene.[185] In principle, physicians were allowed to write about sexual matters because they were experts and professional scientists. In practice, however, they had to use cautious language to describe sexual matters to avoid repression, legal backlash, and ostracism. The government also exerted control over other aspects of medical practice by conducting regular inspections of psychiatric asylums, and approving regulations to protect the insane and define the limits of psychiatric practice. On the Italian Peninsula there did not exist any equivalent institutional constraints, and while this lack of rules had the disadvantage of leaving Italian psychiatrists in a chaotic situation when it came to the standardisation and regularisation of their practice, it left them free to pursue research that could be judged immoral by some, or that contradicted legal tenets. Indeed, long before male same-sex acts had been decriminalised in the whole of the Italian territory, psychiatrists advocated that inverts should not be held legally responsible for their acts.

In the specialised medical literature produced in Britain, information about female same-sex desires was presented piecemeal. British medical writers, however, dealt with female-sex desires to a greater extent than historians have suggested, although they were guarded and oblique when approaching the topic. This distinct way of articulating ideas concerning female same-sex desires did not prevent such information from being disseminated in the same medical contexts as it had in Italy. Both in Britain and Italy, same-sex desires were part of psychiatric nosologies, and were discussed in studies about the influence of sexual instinct and passions on mental disorders, within debates on moral insanity, and degeneration theory. Yet, when engaging with British medical sources, historians need to interpret an almost encoded language.

The second important point where the two medical communities diverged was the meaning of masculine features associated with female same-sex desires. For Italian physicians, masculinity was taken to mean not only physical features but also personality traits; for their British counterparts, masculinity remained a predominantly physical characteristic. Anatomical explanations of same-sex desires underpinned figures such as the virago, which had currency in Italian medical discourses. Yet by the end of the nineteenth century, there was a strong tendency to explain same-sex desires in terms of psychology, which was not restricted to sexual inversion, and encompassed other categories such as the virago and *fiamma*. The popularity of bodily explanations of female same-sex desires even outside the discipline of gynaecology set Britain apart from Italy. British gynaecologists explained that anomalous genital or reproductive organs could be evidence of same-sex acts. The gynaecological representation of masculine women, one might argue, was not so far from Continental psychiatrists' accounts of inverts, which, while framing homosexuality in terms of gender inversion, paid attention to masculine secondary sexual characteristics of female inverts. Yet British gynaecologists discussed tribades' physical characteristics but did not go any further: there is no record of the personality of those women experiencing same-sex desires. Remarkably, British psychiatrists also endorsed gynaecological explanations of same-sex desires; in adopting anatomical accounts of same-sex desires, British psychiatrists evidenced their refusal to think about homosexual behaviours in terms of psychology. This was closely related to their reluctance to embrace the recently introduced medical category of 'sexual inversion'. Where Italian psychiatrists wasted no time and rushed to publish case-histories of sexual inverts that would secure them privilege and recognition from their European colleagues, the British held back.

This did not mean that British psychiatrists were not aware of Continental studies on sexual inversion. Psychiatrists such as Savage and Mercier were familiar with recent research on sexual psychopathologies, and it is not coincidental that they were closely related, if not to criminal anthropology, at least to the study of criminality.[186] In Italy, as in Germany and France, criminal anthropologists such as Lombroso had played a critical role in promoting the study of sexual psychopathologies. But in Britain medical circles seemed ever so slightly wary of criminal anthropology. National overtones of ideas about same-sex desires were inevitably influenced by various disciplines that had different weights within their medical communities. The history of the transformation of the concept of same-sex desires is also, at least partially, the struggle of different medical disciplines dealing with sexual behaviours to secure symbolic capital within the medical profession. While psychiatry was an important discipline in the field, it was not the only one.

Part III
Case-Studies

Part II has traced ideas and figures of female same-sex desires in the broad context of Italian and British medical research; Part III presents these ideas as they appear in the works of individual physicians who engaged with sexological research. Focusing on single medical writers enables a close study of how concepts and texts were constructed. Medical writers' adoption and elaboration of ideas about same-sex desires were not without complications. As Part III shows, sexological research could serve a number of different professional and political purposes. The pursuit of sexual knowledge itself was a conservative or radical endeavour depending on who the practitioner involved was, and on his professional and cultural context. Part III focuses on individual medical writers, who are examined as case-studies that illustrate the complex implications of writing about female same-sex desires in the late nineteenth and the early twentieth centuries.

I have selected two physicians from each country: Cesare Lombroso and Pasquale Penta for Italy, and Havelock Ellis and William Blair-Bell for Britain. The analysis of their work within their national contexts further shows the developments of broader medical discourses about female same-sex desires in each country. Lombroso and Penta were two physicians who had specialised in criminal anthropology; they represent mainstream sexological research in Italy. Ellis is generally considered to have been a psychologist of sex; his approach to sexual inversion is a remarkable exception in the British medical community. Bell was a gynaecologist and typifies the British scientific community's stance on female same-sex desires. For each country I have chosen a representative of the 'conservative', and of the 'progressive' position on homosexuality. Lombroso's and Bell's views might be interpreted as traditional, whereas Penta's and Ellis's radical views on sexuality went against the grain of their respective scientific and cultural contexts, and ultimately led to their professional ostracism. Above all, the four case-studies prove that none of these medical writers were monolithic in their approach to sexual matters.

5
Cesare Lombroso and Italian Criminal Anthropology

Cesare Lombroso is best remembered as the founder of modern criminology and author of 'odd' theories of the 'born criminal' that strike modern sensibility as both ridiculous and horrific. However mocked, at the time of their inception, Lombroso's descriptions of the physiognomic characteristics of criminals – their heads were meant to be asymmetrical, their upper lips thin, their ears large and protruding, their bushy eyebrows met over the nose, their eyes were deep-set, and even their toes were pointy – were neither unprecedented nor unique. By relying on constitutional explanations of deviancy, Lombroso, like contemporary British psychiatrists such as Maudsley, conflated crime and disease as part of the same phenomenon. Educated by positivism's ideals, Lombroso followed the trend of Italian psychiatry of the last three decades of the nineteenth century and believed that there existed a continuity between phenomena like madness and normal physiological states, so that the passions of the insane person were considered an exaggeration of tendencies present in healthy people. Studying pathological behaviours was thus a way to gain a better understanding of the nature of 'normal' men.[1] Nevertheless, Lombroso's endless catalogues of deviancy appear to stand at odds with this assumption because his extensive lists of physiognomic markers made the abnormal visually distinguishable and separate from the 'normal'. It was this last aspect of Lombroso's research that made his theories very popular: the international medical community was the first to be seduced by it, and the popular imagination of places like Italy, Europe, and North and Latin America soon followed.

Sexual abnormality was one of the many forms of deviancy Lombroso investigated in over thirty uninterrupted years of research. Ever since the 1870s Lombroso had been interested in sexual deviancy – an area he would explore in depth in the following two decades. By the 1890s, he was well established in the international medical community as a pioneer of the emerging medical field of sexology, which was increasingly drawn to the study of sexual perversion, and in particular to sexual inversion. Over the course of his career, Lombroso articulated different explanations of same-sex desires, with the result that his work on the topic is particularly obscured by contradictions and shifting positions – as is his broader work on criminality. But although historians from Italy and elsewhere have written extensively on Lombroso's theories of criminal anthropology, they have consistently overlooked his research on sexual perversion.

This chapter contributes to filling this gap and explores how Lombroso's concepts of same-sex desires shifted in the course of his career, paying particular attention to how older ideas of pederasty and tribadism were intertwined and confused with the 'new' psychiatric category of sexual inversion. Some aspects of Lombroso's concept of 'pederasty' anticipated the psychiatric idea of sexual inversion, for instance by indicating that all pederasts had certain specific psychological characteristics. It is possible to argue that even if Lombroso applied the term 'sexual inversion' in a psychiatric context, and 'pederasty' in a legal context, his work confirms that pederasty was more than a juridical term referring exclusively to same-sex practices. Further, Lombroso interpreted the concept of sexual inversion as a continuation of older medical ideas of monomania. In linking the psychiatric category of sexual inversion to the older medical debate of the *'folies raisonnantes'* and partial insanity, Lombroso did not recognise any conceptual innovations surrounding the debate about sexual inversion. This was not due to his lack of knowledge about the growing field of sexology; on the contrary, he was well abreast of the latest research on sexual perversion. Moreover, Lombroso's emphasis on the constitutional elements in the emergence of sexual inversion and his obsession with bodily markers of human deviance prevented him from developing any substantial psychological analysis of sexual inversion, which other psychiatrists of the time were pioneering.

The analysis of Lombroso's ideas on male same-sex desires will be followed by a brief section exploring his role in the popularisation of sexology in Italy. Sections three and four closely examine his work on female same-sex desires. Throughout his career, Lombroso argued that sexuality was the key to distinguishing between appropriate and deviant female roles. In *La donna delinquente, la prostituta e la donna normale* [*The Criminal Woman, the Prostitute and the Normal Woman*] (1893) Lombroso claimed that the prostitute was the equivalent of the male 'born criminal'; in his characterisation of female deviancy, women's active sexuality played a critical role. Thus, by definition, the sexuality of female deviants was irregular, but played only a secondary role in defining male criminals. Given that female deviancy was informed by sexual abnormality, Lombroso and other authors published by the *Archivio di psichiatria* were bound to engage with female same-sex desires in the course of their investigations. Indeed, their commitment to shedding light on existing forms of deviation compelled them to study one of the main examples of female active sexual behaviour. In the long term the enthusiasm both Lombroso and his followers displayed for studying human deviancy fostered a proliferation of documents relating to female same-sex desires that was unparalleled in Great Britain.

Ultimately, Lombroso's studies on tribadism exemplify how new explanations of female same-sex desires incorporated and continued to mobilise older medical and cultural assumptions: the figure of the tribade-prostitute, the connection between female same-sex desires and sexual excess, and the socially oriented explanations of love between women are all examples of this discursive practice. This is not to say that Lombroso's writings did not also borrow from the new and sophisticated German concept of sexual inversion. Notwithstanding this, his explanations of female same-sex desires were still strongly linked to anatomy: the clitoris and the reproductive system still played a central role in his analysis of sexual inversion. Lombroso's attempts to treat female sexual deviancy through surgical practices such as the cauterisation of the clitoris or removal of the ovaries are incontestable evidence of the persistence of anatomical conceptions.

The Criminal Man, pederasty, and sexual inversion

The Criminal Man

Lombroso was born in Verona in 1835 to a Jewish family; he studied medicine in Pavia, Padua, and Vienna, and completed his degree in 1858. From the beginning of his medical training, Lombroso was influenced by German materialism, French and Italian positivism, and was also interested in English evolutionism. Although he read Darwin, Lombroso's conception of evolution – like that of many Italian scientists of his time – was a synthesis of philosophical, biological, and anthropological theories drawn from authors such as Herbert Spencer, Jacob Moleschott, Ernst Haeckel, Bénédict Augustin Morel, and Pierre Paul Broca.[2] When the Italian unification wars broke out in 1859, he volunteered as a doctor in the revolutionary forces, and in 1862 was stationed in Calabria. Many young intellectuals of the period identified science with the ideals of the *Risorgimento*, and Lombroso was one of them. He firmly believed empirical research would help create a new political order based on rational principles that would eradicate the obscurantism that the old regimes had spread throughout the Peninsula. During his posting in Calabria as an army doctor, Lombroso had begun conducting anthropometrical research on some three thousand soldiers with a view to investigating racial variations in Italy.[3] Measuring, classifying, and differentiating the human body remained a central obsession throughout his career. In the logic of biological determinism, the body was a legible text and through the study of its surface and interior it was possible to discern its meaning; Lombroso's anatomical measurements were an attempt at reading the text of the body. After concluding his military service, Lombroso began climbing the academic hierarchy at the University of Pavia. In 1863, he was appointed lecturer in Clinical Mental Disorders and Anthropology; in 1866, when he was thirty-one years old, he became '*Professore straordinario*' [Professor] of Clinical Mental Pathology; in 1874 he obtained the Chair of

Legal Medicine, along with that of Psychiatry. At the same time, between 1863 and 1872, Lombroso was head of the psychiatric department of the hospitals in Pavia, Pesaro, and Reggio Emilia, which granted him access to a wealth of case-histories he would elaborate on in the course of his research. In 1876 he moved to Turin to take up his position as Professor of Legal Medicine and Public Hygiene, and he went on to become Chair of Psychiatry and Criminal Anthropology, in 1896 and 1905 respectively.[4]

His publications attest to the ample scope and variety of Lombroso's interests, with one element remaining stable: his fascination with deviant behaviours, and with madness in particular. Lombroso's life-long study of the brain and insanity had begun while pursuing his medical training under the supervision of Camillo Golgi, who was awarded the 1906 Nobel Prize in Medicine. An unwavering faith in biological determinism had driven Lombroso to question the degree of individual responsibility in phenomena such as madness and crime, and reject the still dominant juridical doctrines developed by Cesare Beccaria in the eighteenth century. Works of classical penology like Beccaria's relied on the principle that the individual could exercise free will and advocated that punishment be proportional to the crime; this notion mirrored the Catholic doctrine of free will whereby the individual is responsible for its own sins. Lombroso's theories were opposed to this view: he argued that crime was caused by biological, psychological, and social factors, and advocated that punishment be proportional to the dangerousness of the criminal for society, rather than to the crime itself. For instance, even if they perpetrated the same offence, a recidivistic criminal was to be punished more harshly than an offender who committed a crime out of a temporary passion.[5]

His interest in biological determinism led Lombroso to study Franz Joseph Gall's phrenological works and several physiognomic and anthropometric surveys of deviant subjects. In a self-celebratory account of how he first formulated the concept of atavism, Lombroso explains that upon examining the skull of the brigand Vilella in 1870, he recognised traces of a primitive ancestry in the man's cranium, which he believed were the source of contemporary deviant phenomena.[6] While posing that deviance and criminality were caused by a person's 'nature' rather than by that person's autonomous will to engage in reprehensible acts, Lombroso argued that deviant behaviour was the expression of atavism, that is: the reappearance in a modern context of elements typical of earlier stages of the development of the human species.[7] Lombroso's use of the concept of atavism to explain deviance implied that if the criminal belonged to a more primitive and underdeveloped stage of human development, the normal civilised man potentially had in himself the germs of atavism. Lombroso supported this idea by adopting Haeckel's evolution theory: if the biological development of an individual's organism (ontogeny) parallels and summarises its species' entire evolutionary development (phylogeny), then every individual organism carries traces of criminality that for the most part remain latent, because criminality is a typical feature of the first stages of human evolution.[8] Deviance thus became a latent threat; it was science's task to identify the signs of atavism and tainted individuals as a means to protect society as a whole.

Criminal anthropology thus emerges as an attempt to respond to such threats, and one of the major contributions to the field was Lombroso's *L'uomo delinquente*, first published in 1876, and subsequently expanded in four editions that appeared in 1878, 1884, 1889, and 1896–7 during Lombroso's lifetime. He laid the foundations of his deterministic explanation of deviancy by accumulating a wide range of data on criminals' bodies: he measured their skulls, feet, height, weight, and strength, and recorded the shape of their noses, ears, and foreheads, as well as any other physical anomaly. Accumulating statistics on age, marital status, sex, profession, diet, and class of criminals, Lombroso concluded that social conditions were to blame for a fraction of criminal behaviour, and that the overwhelming majority of the offenders were constitutionally so. All criminals had not only some physical characteristics in common, such as a distinctive facial asymmetry, but they also shared a few psychological characteristics. Lombroso portrayed the male criminal as a vain, vindictive, lazy, and bloodthirsty person who delighted in taking part in orgies.

The notion that criminality was essentially congenital had underpinned Lombroso's work since the first edition of *L'uomo delinquente*, but he only started using the notorious term 'born criminal' in the third edition (1884).[9] Neither the concept of atavism nor the notion of the constitutional and hereditary origin of criminality was unprecedented. Lombroso's formulation of atavism was certainly influenced by various versions of evolution theory, including the Darwinian, but as the historian Renzo Villa has explained, Lombroso's theory of atavism owed more to pre-Darwinian comparative anatomy, medicine, and linguistics, than to Darwin's own work. In particular there is evidence that indicates that Lombroso first encountered the notion of atavism in the course of his studies of botany.[10] While a Darwinian influence on Lombroso's theory of atavism is open to debate, it is certain that British scientific thought supplied the view of criminality as inherently constitutional. In the 1860s and 1870s, medical officers working in the British prison system had conceptualised the criminal as belonging to a relatively homogeneous group with distinctive physical and mental traits, which was precisely the kind of assumption underlying Lombroso's later conception of a 'born criminal' type.[11] In the early 1870s, the *Journal of Mental Science* began publishing occasional articles by prison physicians and surgeons who engaged in the description of the physical and psychological characteristics of criminals. For instance, in 1870, a physician at Scotland's Perth prison called James Bruce Thomson published a significant article based on degeneration theory in which he argued that many criminals were born as such.[12] Lombroso drew on Thomson's work and acknowledged his debt to the British physician in subsequent editions of *L'uomo delinquente*.[13] In the same years, the *Journal of Mental Science* published a series of articles by David Nicolson, a physician at Portsmouth prison in England, on the abnormal psychology of criminals.[14] In 1872, Maudsley, whose work was read widely in Italian medical circles, spoke of 'instinctive criminals' in his *Responsibility of Mental Disease*, where he remarked that the 'criminal class constitutes a degenerate or morbid variety of mankind, marked by peculiarly low physical and mental characteristics'.[15] Maudsley maintained that crime was often hereditary and spoke about 'the true thief' as 'born not made'.[16]

It is thus clear that Lombroso embraced the medical categories used by British doctors in the early 1870s; British medical communities, on the other hand, would soon reject criminological generalisations and take distance from Lombroso.[17] But as Neil Davie has noted, this was not an instant reaction: Britain did not develop a strong anti-Lombrosian rhetoric until the 1890s.[18] As late as 1887, Tuke spoke highly of Lombroso's theories in the *Journal of Mental Science*, reporting the 'astonishing resemblance between the criminals of different European races', the 'dull, cold and fixed' expression of assassins, and the 'restless, oblique and wandering' look of the thief described by Lombroso.[19] Even after the British medical community had dismissed his work in the last decade of the nineteenth century, there was a marginal group of physicians who continued to read and be influenced by Lombroso's criminal anthropology; most of them were working in the emerging field of sexology. Ellis and Clouston, for example, had referred to Lombroso as a 'man of genius' as late as 1894.[20] References to his ideas in literary works such as Bram Stoker's *Dracula* and Robert Louis Stevenson's *Dr. Jekyll and Mr. Hyde* suggest that Lombroso's theories had also impregnated British popular imagery.[21] It is probable Oscar Wilde had read Lombroso carefully, because when he wrote to the Home Secretary to request a pardon after being convicted for 'gross indecency', he quoted Max Simon Nordau's and Lombroso's works to substantiate his claim that his sexual preference was not a vicious behaviour, but an inborn characteristic.[22]

Pederasty and sexual inversion

Often considered the founding text of modern criminology, *L'uomo delinquente* brought Lombroso international fame. Although sexual abnormality was not the defining element for the male offender as it was for the female, the criminal man had been associated with a certain degree of sexual deviancy since the first edition of this work. Lombroso consistently described the offender as a man with no control over his passions, who indulged in wine and orgies, and whose body did not present typically male secondary sexual characteristics and thus resembled a woman's.[23] The depraved atmosphere of urban modern society was conducive to 'sex crimes' because it stimulated sexual excitement and desires; Lombroso stressed that the rise in sexual offences and prostitution was directly related to the current state of civilisation in which an increasingly complex society inevitably brought about a refinement of sexual vices.[24]

Lombroso remarked that in ancient times, people were not immune to sexual oddities. In his brief discussion of the phenomenon of male same-sex desires, Lombroso explained that in ancient Rome and Greece, pederasty was a normal 'moral custom'. His argument is couched in terms of moral and cultural relativism: he suggested that certain behaviours should not be considered a crime if they were widespread in a given culture. This is why, when dealing with antiquity, he did not regard pederasty as a moral offence. In modern times, however, it was a sign of atavism because it represented a regression to ancient Roman and Greek times. The atavism of modern pederasts was evident in their 'aesthetic tastes', meaning that their artistic preferences coincided with those of the ancient Greeks.[25] At the end of the nineteenth century, the assumption that artistic inclinations often coexisted with effeminacy was widespread, and in his *L'uomo delinquente* Lombroso was just taking up such a preconception and associating effeminacy with modern pederasty. Despite the fact that Lombroso did not offer further comment on pederasty, in the 1876 edition of *L'uomo delinquente* he described pederasts as a group with a number of characteristics in common, amongst them their aesthetic preferences and an undeveloped psychology. Lombroso's engagement with same-sex practices was not restricted to establishing that pederasty was a sign of atavism, and he explored the subject further in his later work.

In the second edition of *L'uomo delinquente* (1878), Lombroso rehearsed the same arguments about pederasty without elaborating on the topic any further; he would depart from this sketchy analysis in an article that appeared before the third edition of *L'uomo delinquente* in 1884. This seminal essay was entitled '*L'amore nei pazzi*' ['Love in the Insane'] and was published by the *Archivio di psichiatria* in 1881. This was a widely influential piece that continued to be acknowledged for many years to come in important sexological texts such as Krafft-Ebing's *Psychopathia Sexualis*, as well as in other medical fields. Even medical writers only vaguely familiar with the study of psychosexual pathologies recognised the pioneering nature of Lombroso's 1881 article on love in the insane. In reviewing the most salient developments in Italian psychiatry during the 1880s for the British *Journal of Mental Science*, Raymond Joseph Gasquet praised the *Archivio di Psichiatria* and pointed out that Lombroso's 1881 article on the sexual passions of the insane was one of the 'the most striking' and 'remarkable' papers published by the Italian journal in the early 1880s.[26]

Lombroso's article tacitly followed Krafft-Ebing's earlier attempt to classify various forms of non-procreative sexual behaviour. In this important 1881 work Lombroso identified five main types of pathological love: '*necrofilomanie*' (broadly speaking, necrophilia); '*eroto-maniaci*' (a form of mystical love); '*amore zoologico*' (love for statues or animals); '*amore paradosso*' ('paradoxical love'; broadly speaking, fetishism and exhibitionism), and finally *'amore invertito'* ('inverted love').[27] The section dealing with *'amore invertito'* included eight case-histories of sexual inversion.[28] Lombroso reported two original case-histories of male inversion, and summarised six other cases previously presented by Westphal, Tamassia, Gock, and Krafft-Ebing. Three of the eight cases dealt with women; of those, one had been recorded by Westphal, and the other two by Krafft-Ebing.[29]

Lombroso's case-histories consisted of a list of physical stigmata which confirmed the subject was a degenerate, followed by a catalogue of illness or anomalies drawn from the subjects' family trees. Lombroso recorded a single psychological trait in his subjects: the tendency to display feminine behaviour, such as dressing up as women or preferring to work at home. He also noted that many sexual inverts presented anomalies such as sparse beards or narrow intelligence, which was taken as evidence of an impediment in cerebral development. According to Lombroso, sexual inversion was a 'transition point' for the 'species of pederasts' who had been drawn to the 'vice' of same-sex acts due to some inborn characteristic. Such pederasts, Lombroso wrote, had been noted by the German legal expert Casper.[30] In 1852 Casper had indeed pointed out that the vice of pederasty was in some cases 'hereditary', and appeared as 'a kind of mental hermaphroditism'.[31] Therefore, Lombroso's reference to Casper's work on pederasty suggests that the Italian psychiatrist was linking the new psychiatric category of sexual inversion to an older forensic medical understanding of same-sex practices.

'L'amore nei pazzi' was also concerned with the broader significance of deviant sexual behaviour in the insane. Lombroso shared the view of many psychiatrists of his time who insisted there was no clear demarcating line between sanity and insanity; he argued that 'insanity is also linked to physiological conditions', and that 'love in the insane reproduces the tendencies – but exaggerating such tendencies – of the healthy man'.[32] According to conventional psychiatric knowledge of the time, mental illness disclosed the deepest nature of man, so that by examining pathological manifestations of love it was possible to understand normal love. Lombroso embraced these ideas: just as it was not possible to draw a line to separate sanity from insanity, it was not plausible to separate between normal and pathological love in absolute terms. In the same way as necrophilia was thought to recall the violent love that our ancestors felt for each other in the dawn of time, and *'eroto-maniaco'* was an exaggeration of platonic love, so too:

> inverted love reminds us of Lesbian and Socratic horrors and it explains them. Perhaps sexual inversion goes farther as it is linked to the hermaphroditism that Darwin recognised in our oldest ancestors; and we can have a vague idea of that hermaphroditism in the first months of the foetal period, [...] and also in that analogy of sexes that I discovered in criminals.[33]

Lombroso associated the culture of ancient Greece (Socrates and Sappho) with sexual inversion. Earlier comments on pederasty in the first edition of *L'uomo delinquente* had already pointed in that direction; moreover, as with the case of pederasty, Lombroso explained sexual inversion through his theory of atavism. Interestingly enough, if the sexual invert represented a primitive type, it follows that all men are latent sexual inverts. As I explained above, Lombroso believed that the individual's ontogenic history retraced the phylogenic history of the human species. By linking sexual inversion to the concept of atavism, and accepting that there was a measure of hermaphroditism in human history and in embryological development, Lombroso was laying the foundations for a theory of latent homosexuality in each individual. But in 1881, his work was not as bold: he was merely attempting to substantiate his claim that same-sex desires were a form of atavism. Theories of latent homosexuality would be developed in years to come.

In 1883, one year before publishing the third revised edition of *L'uomo delinquente*, Lombroso released an article, 'Delitti di libidine' ['Sexual Crimes'], in which he developed some of the points cursorily addressed in the first edition of his book. Lombroso explained that civilisation affected the incidence of sexual crimes: as a result of evolution, there was an increase in psychological activity and general needs, which meant sexual desires and urges became more sophisticated.[34] In other words, Lombroso suggested that sexual habits were determined by the social context of the individual. Ancient and some nineteenth-century primitive populations had specific sexual customs that included, for example, institutionalised female prostitution and widespread sodomy.[35] In modern times, sexual habits presented a series of particularities, and so did sexual crimes. Lombroso, who rarely missed the opportunity to express his anticlerical sentiment, emphasised that in modern Catholic societies, priests would often commit sexual crimes related to pederasty, and so would famous writers.[36] Lombroso suggested that other social mores had further contributed to the spread of pederasty: because marriage was often a business arrangement rather than a choice made out of love, or out of the fear of having illegitimate children, people ended up hating sex within marriage and sought love 'against nature' instead.[37] Civilisation offered new opportunities to sexual offenders in places like schools and factories, where women and young children were vulnerable to sexual abuse.[38] Lombroso's proposed solution was to institutionalise female prostitution as a means to limit pederasty and rape.[39]

But historically contingent sexual moralities could not explain why certain individuals were inclined to any specific sexual crime: in other words, there was no socio-cultural factor that could explain why only some individuals were drawn to sodomy, rape, or bestiality. Lombroso believed that sexual offenders were constitutionally drawn to sexual abuse: as there were 'born criminals', so too there were 'born sexual criminals' and 'born rapists'.[40] Lombroso further associated sexual crime in general with physical and psychological gender inversion. Indeed, he remarked that most of the men convicted on charges of rape were insane, displayed a 'perverted sexual instinct', and felt 'sexual instincts as if they were female'.[41] These brief comments present an implicit contradiction they do not resolve: considering that this period's medical literature defined female sexuality as passive and responsive to the male's active stance, how could the rapist's sexual instinct be both female-like and aggressive? He then remarked that:

> Pederasty and tribadism are often a clear effect of mental sickness. Casper was the first to show this morbidity in Germany, and then Griesinger, Westphal, Gock, and Krafft-Ebing called it *Conträre Sexualempfindung*.[42]

Despite having analysed the concept of sexual inversion in '*L'amore nei pazzi*', Lombroso continued to talk about pederasty and only briefly mentioned the German term '*Conträre Sexualempfindung*. He stated that *Conträre Sexualempfindung* often caused the sexual arousal that led to pederasty and tribadism, and he accepted Casper's idea that these behaviours were often pathological. Yet Lombroso preferred to use the more classic term 'pederasty' to discuss both same-sex acts and desires, so that while he discriminated between pederasty and tribadism (meaning same-sex acts), and 'contrary sexual feeling', he also involuntary conflated them.

The same stance is evident in the third edition of *L'uomo delinquente*, published in 1884. In it, Lombroso further developed his concept of the 'born criminal', which he had introduced in his article '*Delitti di libidine*'; this concept was supposed to improve the earlier concept of atavism, which he acknowledged could not account for the presence of multiple anomalies in each and every criminal. This revision was in part a response to widespread criticism of his key concept of atavism. Lombroso explained that degeneration and atavism could combine to cause foetal diseases which in turn would result in physical and psychological malformation; this new formulation added a psychological and environmental element to his theory. The degeneration theory allowed the interplay of biology, environment, and psychology, since social factors like alcoholism, venereal disease, or malnutrition might debilitate the physical and mental health of a mother and therefore her foetus. But even if caused by external forces, degeneracy was thought to become a hereditary condition that progressively weakened future generations.[43] Lombroso also accepted the concept of 'masked epilepsy', a term used since the 1860s by British psychiatrists in general, and by Maudsley in particular, in their debates on 'moral insanity'.[44]

In the 1884 edition of *L'uomo delinquente* Lombroso argued that 'pederasts' felt the 'need to join each other in crime', and that they formed such a close 'confraternity [that] they recognise each other just through a glance'.[45] By this stage he had abandoned the view that same-sex desires were a symptom of arrested mental development, which he had put forward in *'L'amore nei pazzi'*. This reversal was so marked that he was now able to comment on the fact that pederasts have a 'high level of education and intelligence'.[46] Yet, this does not mean that Lombroso had abandoned his previous association between pederasty and atavism. Despite his admission that the notion of atavism was conceptually flawed, in the third edition of *L'uomo delinquente* he continued to talk about pederasts' atavism. In the context of his discussion about pederasty, Lombroso used the term 'atavism' to describe the 'vices' and proclivities of men, especially those from the upper classes, who engaged in same-sex acts, preferred women's clothes, and favoured ostensibly feminine activities.[47] Atavism also explained why the taste of pederasts was similar to the tastes of the ancient Greeks: pederasts belonged to an earlier stage of human development. Up until the 1884 edition of *L'uomo delinquente*, Lombroso continued to associate same-sex desires with criminal acts, rather than mental pathologies.[48]

As I have shown above, by 1881 Lombroso had already published a study on sexual inversion. Why, then, did he fail to expand his analysis and deploy the psychiatric category of 'sexual inversion' in his 1883 article, and especially in the third edition of *L'uomo delinquente*? Considering that in northern Italy, where Lombroso was working at the time, male same-sex acts were still a crime, it is conceivable that Lombroso preferred to analyse same-sex acts from a legal point of view so as to be in line with current laws.[49] His refusal to update his observations and employ the term 'sexual inversion' is nonetheless surprising, as are his remarks on the great intelligence of pederasts.

Since at least 1881, Lombroso had been aware of the new medical category, which he believed was similar to Casper's notion of pederasty as a form of 'mental hermaphroditism'. Although Westphal's and Krafft-Ebing's studies on sexual inversion had been known in Italy since at least 1878, it was not until the second half of the 1880s that Italian physicians started to embrace 'sexual inversion' *en masse*. It would be fair to assume that the concept of 'sexual inversion' was not widely acknowledged by the Italian medical community, and that Lombroso did not use it to avoid disconcerting his readers, especially northern lawyers. But it would have been unlikely for him to miss the opportunity to put forward medical novelties, especially if one takes into account that he considered himself a pioneer in the study of sexual perversion and was recognised as such abroad.[50]

There are at least two other plausible explanations for Lombroso's failure to mention 'sexual inversion' in the 1884 edition of *L'uomo delinquente*. The first one is that at this stage, Lombroso still considered 'sexual inversion' and 'pederasty' to be mutually related and possibly melded together. Lombroso associated effeminate behaviour and certain aesthetic tastes with pederasts, and he spoke of the ability of pederasts to recognise each other and create 'confraternities'. Therefore, 'pederasty' described not just sexual acts, but a broader set of behaviours. Just as sexual inverts had a number of psychological characteristics in common, Lombroso thought of pederasts as an identifiable group of people with shared psychological traits. In other words, the feeling of having a female sexual instinct made sexual inverts akin to pederasts, who were defined by their effeminate behaviour.

Another possible interpretation is that, in the process of updating *L'uomo delinquente* in 1884, Lombroso might not have considered it a priority to rework his observations on same-sex practices. The pattern of additions to the treatise indicates that Lombroso added new theories and observations to reinforce his principal arguments, rather than revising or refuting outdated explanations. Considering the scale of his output, it is reasonable to assume that when preparing the third edition of the book he did not elaborate on the concept of sexual inversion, whose novelty he also failed to grasp. Instead, he preferred to highlight other concepts such as the 'born criminal' and 'degeneration', which were the real conceptual novelties introduced by this edition.

While it is difficult to conclusively assert why Lombroso failed to elaborate on the idea of sexual inversion in 1884, an analysis of the fourth edition of *L'uomo delinquente* (1889) provides further insights into the original formulation and subsequent transformations of the concept. This version was greatly expanded and consisted of two volumes: the first dealt with 'the born criminal' and 'the morally insane', and the second with 'the epileptic criminal, furious criminal, the insane, and the *criminaloide*'. In the first volume, his treatment of pederasty and tribadism was no different from earlier editions, although there was a new section on psychiatric case-histories of 'sexual perversions', which dealt mainly with fetishism.[51] He highlighted the precocity of sexual perverts and noted that 'sexual inversion' was generally first experienced during childhood.[52] He was keen to underline what Krafft-Ebing had already pointed out: that in the morally insane, excessive sexuality was followed by impotency, as in the case of criminals.[53] Despite these new observations, he did not give a systematic account of sexual inversion in the first volume of the fourth edition of *L'uomo delinquente*.

In the second volume, Lombroso explored psychiatric categories that have 'parallels' in the criminal sphere, such as the pyromaniac and the arsonist, or the kleptomaniac and the thief.[54] Thus, the '[l]egal category of pederasty and rape has its double in *sexual inversion*'.[55] Finally acknowledging sexual inversion as an autonomous psychiatric category, Lombroso did not treat it as a symptom of other mental disorders, but rather as a mental pathology in itself. He further noted that:

> [In his *Treatise*, Kraepelin writes] that in individuals who normally show signs of psychological degeneration (especially emotional disorders); sometimes morbid impulses are displayed that once were

considered a special sickness (the so called Esquirol *monomania*); these [*monomanias*] are instead signs of mental pathology and incomplete mental organisation.[56]

Sexual inversion was a result of degeneration: a symptom of a faulty disposition of feelings and instincts.[57] Lombroso's acknowledgement that sexual inversion had its medical genealogy in Esquirol's monomania is highly significant in this context. Instead of resorting to the sexual behaviours of antiquity in his efforts to explain sexual inversion, Lombroso drew on contemporary medical traditions that considered same-sex desires as a derangement of feeling and instincts (Esquirol's monomania).[58] Referring to the psychiatric debate about so-called partial insanities also meant situating sexual inversion within the debate about criminal responsibility, an issue raised by a number of other medical writers such as Tamassia.

Adopting the standard psychiatric definitions, Lombroso then went on to explain the phenomenon in some depth: sexual inversion was not only a sexual perversion, but an 'odd anomaly' that determined the individual's psychological characteristics so that sexual inverts felt they belonged to the opposite sex. Sexual inversion could include platonic love and presented the same characteristics as heterosexual love but it was usually more emotionally violent.[59] Therefore Lombroso discriminated between same-sex acts and inclinations, as it was possible for sexual inverts not to perform same-sex practices, or any sexual practices at all, and yet remain sexual inverts. Sexual inversion was defined by the presence of same-sex desires and the feeling of not belonging to one's own biological sex.

A close reading of Lombroso reveals a conceptual continuity between pederasty and sexual inversion until at least 1889. Until then, in Lombroso's *L'uomo delinquente*, pederasty did not mean only same-sex practices: it also referred to inner predispositions. From the fourth edition of the book pederasty became more closely identified with a legal term to designate same-sex practices, while sexual inversion was identified with a mental disorder characterised by peculiar inner states and desires. In Italy in the second half of the nineteenth century there existed ideas of female same-sex desires that anticipated the concept of sexual inversion as the feeling of belonging to the opposite biological sex (see Chapter 3). The same continuity of ideas was evident in Lombroso's analysis of male same-sex desires. Pederasty was not just a sexual act because it defined a group of people with specific psychological characteristics. This brought it close to sexual inversion because both concepts were used to define a specific group of men with identifiable psychological characteristics. At the end of the 1880s, Lombroso began using two distinct terms depending on what field he was working in. The two terms do not describe exactly two different phenomena, but two different points of view: whenever Lombroso regarded male same-sex desires from a legal point of view he talked about 'pederasty' (at the time it was still considered a criminal offence in some parts of Italy), and whenever he regarded them from a psychiatric point of view, he talked about 'sexual inversion'. On the one hand, from a legal perspective, pederasty was a crime committed by a criminal who was characteristically effeminate, had specific aesthetic preferences, was keen to form confederacies with his fellow pederasts, and so on. On the other hand, one could look at same-sex desires from a psychiatric perspective: sexual inversion was a derangement of feeling that belonged to the same category as Esquirol's monomania, which made the invert feel like a woman trapped in a man's body.

I suggest that in talking about pederasty and sexual inversion, Lombroso was looking at both same-sex acts and desires from two disciplinary fields, and that he adapted his writings to suit his specific purposes. The emphasis was different because jurists and psychiatrists were concerned with different aspects of homosexual behaviour. Arguably, in *L'uomo delinquente*, Lombroso tailored his work on same-sex desires to fit conventional medical knowledge, which further explains why he did not deal with sexual inversion at any length until the 1889 edition of the book when this category was well known within the Italian medical community. His use of the same example – the sexual practices of the ancient Greeks – to explain two 'distinct' phenomena – sexual inversion and pederasty – also suggests that to him, they were one and the same phenomenon. In the nineteenth century there was a revival in classical studies, especially in relation to Greek culture, and some intellectuals invoked Greek sexual customs to defend their own same-sex desires.[60] This is not to say that Lombroso was championing same-sex desires. While intellectuals such as the British Symonds resorted to ancient Greece as a model for modern civilisation, in Lombroso, the rhetorical effect of referring to ancient times was rather different. For him, while ancient Greeks and Romans were highly developed for their own time, contemporary individuals' adoption of the psychological characteristics or sexual customs of the ancients bespoke an evolutionary failure, and should be considered as a sign of lack of progress.

Finally, the two case-histories of sexual inverts he published in 1881 can be interpreted as a bid to fashion a new subject in international medical literature. Because only a few such case-histories were known at the time, any new observation in the growing field of sexology was bound to get the attention of his peers overseas, which was precisely what happened.[61]

Popularising sexology

As I have mentioned, Lombroso was first drawn to the study of sexual deviancy in the course of his duties as a doctor and researcher in asylums and prisons. The case-histories he collected in the course of such research invariably featured mental pathologies and crimes. As a practising psychiatrist and lecturer, and later professor of mental pathology, Lombroso often acted as an expert witness in criminal cases in which the suspect was thought to be insane. This was because legal authorities required medical expertise to judge the mental capability of an individual in particularly violent or rare crimes. In this role as an expert witness, Lombroso came across a number of cases of sexual crimes and 'abnormal sexualities', which he used to substantiate his theories about sexual proclivities. To a certain extent, it can be argued that Lombroso's sexological research was prompted by extreme pathological cases such as the famous case of Vincenzo Verzeni to which I will return in Chapter 6.[62]

Yet Lombroso's numerous publications on sexual behaviour touched upon widespread modern anxieties: that civilisation gave rise to new diseases, that sexual crimes typical of the new urban environment may not be contained, that modern life stimulated moral corruption and vices. One of the reasons his analysis and practical solutions were so appealing was that he tackled issues that were at the core of late nineteenth-century society's debates and fears. But his success was also buttressed by his ability to mobilise people, both through his journal, the *Archivio di psichiatria*, and through international conferences.

The *Archivio di psichiatria* provided a forum for the exchange and discussion of ideas from disciplines as diverse as anthropology, psychiatry, medicine, jurisprudence, sociology, psychology, and pedagogy. While in the late nineteenth century the boundaries between different medical specialisations were not as strict and defined as today, Lombroso's journal was interdisciplinary in a very modern sense: it was open to contributions from different sources and institutions, including asylums, prisons, universities, and the law courts. In addition to topics such as criminality, which was appealing to all these audiences, the *Archivio di psichiatria* commented on urgent government problems. As has already been mentioned, Lombroso's followers, like many Italian physicians, were politically active.[63] The readership of the *Archivio di psichiatria* was composed not only of specialised psychiatrists, but also of the professional ruling elites preoccupied with essentially practical issues. Given its wide scope, the *Archivio di psichiatria* could be read by intellectuals and professional men of varied backgrounds, which explains how Lombroso's ideas became so popular beyond the discipline of psychiatry.

The *Archivio di psichiatria* had been concerned with the study of sexual pathology since it was first published in 1880; it regularly reviewed foreign works in the field and featured original contributions from Italy. From 1899 onwards, the *Archivio di psichiatria* devoted a specific section to the subject under the heading 'Psicopatie sessuali' ['Sexual Psychopathies']. As editor of the journal, Lombroso was especially concerned with popularising foreign sexologists such as Krafft-Ebing, reworking case-histories known to the international medical community, and publishing original case-studies and observations.[64]

Lombroso also helped to popularise sexology by supporting and contributing to the translation of major foreign works. In 1889, for instance, he sponsored the publication of the Italian translation of Krafft-Ebing's *Psychopathia Sexualis* through Bocca, a well-respected medical publishing house.[65] In his introduction to the volume, Lombroso drew a parallel between Krafft-Ebing's success and that of his own 'school' in Europe and South America. After asserting his pivotal role in the research of sexual psychopathologies, Lombroso listed some of the latest Italian works in the growing discipline, noting that all the researchers in the field were his own followers. Lombroso underlined the part he himself – and criminal anthropology in general – had played in establishing sexological research, as confirmed by his international fame and renown, and then went on to highlight Krafft-Ebing's own contribution to the study of sexual inversion.[66] Immodesty aside, criminal anthropology had indeed done much to promote sexological research in Italy.

Proof of the prestige sexological categories enjoyed within criminal anthropology was supplied by Lombroso's decision to give a paper on the topic of congenital homosexuality at the Sixth International Congress of Criminal Anthropology in Turin in 1906.[67] Lombroso's paper is interesting because it outlines the two perspectives – psychiatry and criminal anthropology – he had adopted in his study of same-sex desires, which he had conducted over the course of the last three decades. In this paper Lombroso argued that in a normal childhood there existed a 'kind of temporary homosexuality' akin to 'temporary criminality'. Recalling the well-known female homosexual phenomenon of *'fiamme'* to illustrate his point, Lombroso suggested that there was a homoerotic element in all childhood friendships. He then compared homosexuality with criminality: Lombroso reasoned that if there existed 'occasional criminals' and 'born criminals', there were also 'occasional homosexuals' and 'born homosexuals'. Occasional homosexuals could have 'normal' relationships, but when they were forced to live in single-sex environments such as prisons, colleges, or asylums, they would engage in obsessive relationships of a homosexual nature, whereas born homosexuals were sexually attracted to those of their own sex since childhood.

According to Lombroso, just as born criminals manifested a 'special physiognomy', born homosexuals displayed physical features typical of the opposite sex. Moreover, born criminals and born homosexuals shared some psychological traits: they were frivolous, vain, selfish, jealous, and mendacious, given to acting on impulse, and fond of gossip. All these characteristics were generally associated with women. Homosexuals, Lombroso explained, had sexual drives since early childhood, were prone to feigning madness, were immodest, and tended to be aesthetes, which is why artists, musicians, and actors were often sexual inverts. The aetiology of crime and illness was similar for born criminals and born homosexuals: they might be epileptics or neurotics, or have parents who were old and odd. They were consequently pathological, naturopathic, and incorrigible. Lombroso concluded that regardless of the similarities between the two groups, they had to be targeted by different social and legal solutions. And this because homosexuals were less dangerous than criminals: while criminals would continue to be socially dangerous for the rest of their lives, the sexual activity of homosexuals would eventually come to an end.[68] Lombroso did not indicate what kind of strategies would have to be used to treat homosexuality or limit its dangers, but he believed that society had to be defended and he perhaps contemplated locking up homosexuals in special institutions such as asylums.

Leaving aside the unflattering, if unsurprising, description of sexual inverts (after all, Lombroso considered homosexuality akin to criminality), the importance of this paper is that he conceded that transitory same-sex practices were normal in childhood. In doing so he was developing and making explicit what he had already suggested in his 1881 article '*L'amore nei pazzi*', that is: the presence of a kind of universal latent homosexuality in each normal individual. The implications of this paper go further and reveal the importance of homosexuality within medical research at the beginning of the twentieth century in Italy. By 1906, Lombroso was widely known and although many scientists had criticised the soundness of his research methods, his role as founder of criminal anthropology was unquestioned. His paper shows that homosexuality had become crucial within criminal anthropology. On the occasion of an important international conference, the very founder of the 'science of deviance' had discussed homosexuality. While same-sex desires were still associated with sexual crimes, Lombroso conferred a new status to the issue: homosexuality was no longer a footnote referring to a peculiar sexual perversion, but a central chapter within criminal anthropology.

Tribadism

Tribadism and asylums

In 1885, Lombroso made his first systematic contribution to the study of female same-sex desires by linking tribadism to specific environments like mental hospitals, prisons, *sifilicomi*, and harems.[69] Tribadism, he wrote, was 'a completely special vice', the 'pederasty of the fair sex', a 'dreadful practice', and was less common than male pederasty. Lombroso also thought that same-sex acts tended to flourish in female asylums. In the mental hospital at Pavia where he worked while preparing this study, out of a total of two hundred patients, 5 per cent were tribades.[70] Despite the fact that same-sex acts were generally less frequent in women than in men, Lombroso clearly stated that female sexual perversion was worse than male perversion. He provided some examples to substantiate this claim, which reveal how he evaluated the relative gravity of perversion. Amongst women he indicated two old female 'maniacs' who had been masturbating with a crucifix and eggs. Amongst men, he knew 'only' of a pyromaniac who had tried to rape his mother when he was seven years old, and a 'half idiot' who confessed to having raped a goat. His examples somehow confirmed Lombroso's theory beyond any doubt: women's sexual aberrations were much worse than men's.[71]

According to Lombroso, in a tribade couple there was usually a 'male part', and a tribade could change partners more than once in the same night.[72] Throughout his article, he refers to tribades rather than to 'sexual inverts', and uses the expression 'sad loves' – as opposed to 'normal' couple – to describe the union of two women. The essay included a case-history consisting mainly of a physical description of a patient without engaging in any psychological investigation. The virile fifty-year-old woman was a lame 'cretin' – in other words, she was mentally retarded and was affected by goitre. Her masculine appearance was 'nearly martial': she had a dolicho-cephalous skull, a low forehead, a marked mandibular prognathism, 'badly implanted' ears, dark skin, and atrophic breasts. Her genitals were abnormal: at four times the normal size and nearly as hard to the 'touch' as carti-lage, her left lip was hypertrophic, and her clitoris was larger at the base. Such deformations of the genitals, Lombroso explained, owed more to cre-tinism than to tribadism. Mental retardation in general was a condition that tended to produce abnormal erotic impulses, as well as degeneration in the genitals, or so it was supposed. Despite Lombroso's argument that, like any other sexual vice, tribadism was caused by organic anomalies and a certain degree of moral insanity, his article does not provide any specific and explicit indication about the exact physical origin of sexual perversion in women.[73]

His attempts to contain the spread of same-sex practices in the asylum of Pavia, however, suggest that Lombroso did believe female genital organs played a key role in the outbreak of sexual perversion. As Lombroso men-tioned, he used clitoridectomy to treat the aforesaid fifty-year-old 'cretin' tribade who 'infected' the other inmates with her 'vice'. It was all too apt to speak of an 'infection', since her tribadism – in the guise of a conta-gious disease – had eventually spread to nearly every other patient in the asylum. It had also been detected that the women in Lombroso's asylum rou-tinely masturbated, so it had been decided that at night their hands should be chained. But it was not until the 'cretin' tribade had been admitted to the madhouse that the first same-sex practices began. Lombroso saw the fifty-year-old woman couple with a 'maniac prone to furious excesses'; the former imitated the actions of a man by wrapping some clothes around her hand to function as a dildo and the latter was astride her. The two women were isolated for some months, but no sooner had they been returned to the common living quarters than tribadism spread to every nymphomaniac and a few other inmates. In order to suppress this 'wicked custom', Lombroso experimented with sedation – he used bromide of potassium, belladonna, and camphor – and imposed the strictest surveillance. But since the 'evil' seemed to be out of control, and assuming that female same-sex desires originated in diseased genitals, Lombroso resorted to 'cauterisation' of the clitoris for *'all those individuals'* who practised same-sex acts.[74] As Lombroso admitted, this measure was successful only for the shortest period of time: the practice of tribadism revived six days after the interventions had taken place. This is how Lombroso eventually came to believe that the most effective treatment would be to expel the 'cretin' tribade from the asylum before the 'contagion' could spread any further.[75]

It was not only in the treatment of female same-sex acts that Lombroso resorted to drastic surgical operations. In the 1881 article on sexual inversion analysed above, Lombroso made clear that surgical interventions were a common treatment for a variety of female abnormal sexual behaviours. On this occasion, Lombroso had noted that the mentally ill female patient exceeded her male counterpart in all 'sexual aberrations', which he supported with his observation that two thirds of the female patients presented anomalies in the genital organs. Sexual arousal and the 'sexual apparatus' affected more women than men. Hence Lombroso's comment that 'many centuries ago man used to call woman *racham*, or uterus in the language of the Semites'.[76] Lombroso indicated that removing the ovaries might prove an effective measure to curb progressive degeneration and abnormal sexual behaviours in insane women.[77] Lombroso was not alone in this way of thinking. As mentioned in Chapter 4, it was standard practice to remove the ovaries in order to treat virtually every gynaecological problem; far from being an exclusively Italian method, it was widely applied in Britain, France, and Germany.[78]

A range of other surgical operations – including clitoridectomy – could also be used to treat self-eroticism in women.[79] In surveying medical literature about sexual psychopathologies it is difficult to assess how widely Italian medical institutions resorted to this kind of surgery as a means to contain the spread of female masturbation, but one of Lombroso's case-histories might shed some light on the issue. In 1883 he published a study of two girls who had been onanists from a very early age. He resolved one case by applying 'some heat' on the clitoris, and cauterised the clitoris of his second patient.[80] The fact that in his 1883 case-study of female children Lombroso casually referred to the cauterisation of the clitoris as a means to cure abnormal sexuality in women suggests that such interventions were routinely performed, and their effectiveness taken for granted. On the other hand, the focus on removing or cauterising sexual organs has a wider implication: if these interventions sought to alleviate certain symptoms by removing the source of the disorder, it inevitably follows that female anatomy itself was being considered the cause of sexual perversion.

Tribadism and prostitution

Lombroso expanded his analysis of tribadism in *La donna delinquente, la prostituta e la donna normale* (1893), which he co-authored with his son-in-law

Guglielmo Ferrero. This book was the first of his treatises on criminality to be partially translated into English, only two years after the release of the Italian edition.[81] This is generally regarded as a key text produced in the context of Italian positivism which focuses on the topic of normal and deviant women. Published eight years after Lombroso's first extended contribution to the study of female same-sex desires, *La donna delinquente* articulated a general theory of female sexuality that also served to explain tribadism. The authors' portrait of the normal woman closely resembled the quintessentially good bourgeois mother: sexually passive and frigid, she had no autonomy, was naturally and organically monogamous, and depended on the father of her children.[82] Inspired by Darwin's theory of evolution through sexual selection, Lombroso described women as 'undeveloped men', whose inferiority in the evolutionary process explained why their bodies had less pronounced secondary sexual characteristics, and why their degenerative stigmata were less evident. Haeckel's law of recapitulation also helped Lombroso to explain the position of women in the evolutionary process: they were in an infantile and inferior state because their process of development in thousands of years of evolution had been halted at an earlier stage than men. In Lombroso's view, women were closer to nature and the primitive world than men. Unlike men, who spent most of their energy in a struggle for subsistence, women spent most of their energy in reproduction and child-rearing, which explained why their brains were underdeveloped.[83]

Some conclusions of *La donna delinquente* openly contradicted his previous studies on female sexual pathologies. His new position was that the rarity of specifically female sexual psychopathologies proved that women had a diminished sexual sensibility. The latter aspect was also proved by the fact that platonic love, although physiologically unnatural, was more accepted by women than men. The cultural value attached to virginity meant that women had their first experiences of sexual intercourse at a later age than men – although many people embraced chastity voluntarily, only women were required to refrain from having sex.[84] He then clarified his new position, which contradicted his earlier views: 'opposing views on the sensibility of women depend on the paradox that love is the most important thing in their lives'.[85] Love played such a crucial role in the lives of women not because they had a privileged relation to eroticism – on the contrary, normal women did not display any sexual feeling – but because they needed to be protected and had to satisfy their maternal instinct lest their existence be incomplete. A 'famous obstetrician' called Giordano had said to Lombroso that: '*Man loves woman for the vulva, while woman loves man as a husband and father.*'[86] In the female organism, the species' requirements prevail: the mandates of the biological world were more important than any individual desire or satisfaction the woman might experience. Not only did sexual desire occur only so that women may become mothers, but sexuality and maternity were incompatible; indeed, women who had an active sexual life were deviant and rejected maternity altogether. To sustain his theories, Lombroso exploited a study published by Harry Campbell called *Differences in the Nervous Organisation of Man and Women: Physiological and Pathological* (1891). Campbell had interviewed fifty-two poor patients at a working-class hospital on the subject of their wives' sexual instinct, and concluded that the sexual instinct of women was less intense than men's.[87]

Lombroso and Ferrero pointed out that female delinquency was less common than male criminality – a fact that was at odds with female inferiority. If criminality was a sign that a society had failed to progress, then women had to be considered more advanced than men as they did not commit crimes. But this was not Lombroso's thinking: rather, he reasoned that if one considered female prostitution as the typical female crime, male and female criminality were similarly widespread in society. Lombroso and Ferrero embraced this framework and explained that a woman became a prostitute not because she was poor, but because organic reasons compelled her to. In *La donna delinquente*, Lombroso departs from his earlier position on the subject of prostitution. In the first edition of *L'uomo delinquente* (1876), he had stated that female prostitution was caused by poverty and laziness; by 1893 he had decided it was caused by nature.[88] Moral insanity could also lead to prostitution: over the course of the centuries, moral evolution had made women more modest, so that the greatest degeneration that could affect them was a lack of modesty – a salient trait of the born prostitute. Lacking modesty, the prostitute was also a 'moral degenerate'.[89]

Physical descriptions remained a key element in the study of female deviancy, but in *La donna delinquente* Lombroso started to draw attention to the psychological features of his subjects. The psychology of the typical 'criminal woman' was characterised by 'excessive eroticism', which made her similar to men. This being the case, all women who were born criminals would also become prostitutes. Their masculine behaviour, coupled with their general dissipation, low cunning, and boldness, barred them from expressing maternal affection. To these virile qualities Lombroso added the 'worst qualities of women's psychology': a passion for revenge, deviousness, cruelty, the love of finery, and mendacity.[90] Women of this kind tended to dominate submissive persons, relish strenuous exercise, indulge in vices, and favour masculine-looking clothes. Thus the prostitute could be identified with, and compared to, the morally insane and the criminal woman. They shared a lack of modesty, a brazen attitude to vice, an irregular lifestyle, a love of idleness, a fondness for amusements, and a general inclination to indulge in orgies, alcohol, and vanity.[91] Above all, the prostitute brought to mind the 'primitive type' of woman. Lombroso returned to the concept of atavism to explain prostitution. Thus, the prostitute's atavism was evident in her virility, which essentially fitted the description of the criminal type because, as Lombroso wrote, 'what we look for above all in the female is femininity, and when we find its opposite, we usually conclude that there must be some anomaly'.[92]

Prostitutes and criminal women not only shared psychological characteristics: they were readily recognisable because they shared specific physical features. In their anthropometrical analyses, Lombroso and Ferrero had found prostitutes had limited cranial capacity, narrow or receding foreheads, prominent cheekbones, short arms (because they worked less), and prehensile feet (like monkeys); they were generally short, over-weight, and left-handed. Both prostitutes and criminal women shared a tendency to have particularly dark hair and eyes.[93] Lombroso conducted a number of sensitivity tests that found touch, taste, and smell were duller in prostitutes than in any other group of women. In prostitutes, such a 'dull' sensitivity even extended to the clitoris,[94] and he referred to the prevalence of sexual frigidity among prostitutes.[95] These observations about the sexual frigidity of prostitutes contradicted the theory whereby prostitutes were sexually unconstrained and sexually active and, as with so many of the logical problems embedded in Lombroso's work, he did not address this contradiction.

Lombroso's analysis of tribadism is presented in a chapter called *'Sensibilità sessuale'* ['Sexual Feeling'], which belongs to a section entitled *'Biologia e psicologia delle criminali e delle prostitute'* ['Biology and Psychology in Female Criminals and Prostitutes']. However, it is possible to find references to tribadism throughout Lombroso's *oeuvre*, as, for example, in his discussion of the insane, hysterics or *'donne di genio'* – talented women 'who look like transvestites'.[96] Lombroso observed that tattoos, which were a manifestation of atavism, were widespread amongst old tribades.[97] Lack of maternal feeling was a characteristic of the female sexual invert.[98] The mentally insane and hysterics were quite often sexually inverted.[99] Tribadism could also be observed in the animal world: in groups of cows, ducks, hens, and geese, females could replace males in sexual intercourse, and they would develop male secondary sexual characteristics as they grew older.[100] Throughout the book Lombroso and Ferrero used the terms 'tribadism', 'sexual inversion', 'homosexuality', and 'lesbianism' interchangeably to refer to female same-sex desires.

Lombroso believed that many tribades were both 'born criminals and epileptics', which explained 'the extraordinary violence of these loves'. This would also clarify why these homosexual loves were less stable, and why they were more 'physical' than normal love between men and women.[101] According to Lombroso, degeneration influenced the rise of sexual perversion in general, and same-sex desires in particular. Physical degeneration itself was caused by an 'atavistic tendency to return to a stage of hermaphroditism' that 'tends to bring closer and to confound the two sexes'. As a result, male criminals could exhibit the kind of 'feminine infantilism that leads to pederasty', which was mirrored by masculinity in female criminals. In women, this tendency to hermaphroditism often set in before puberty, with many dressing like men, deriving pleasure from looking at female organs, and avoiding female work.

In *La donna delinquente* Lombroso explicitly adopted the stereotype of the female sexual invert described in Krafft-Ebing's *Psychopathia Sexualis*:

> The female homosexual feels herself to be like a man; she delights in displaying courage and virile energy, traits that please women. She wears her hair and clothes in masculine style and delights in appearing in public dressed as a man. She has a liking only for male games, pastimes and pleasures; in her mind she feels longing for feminine personalities; in the circus and the theatre she desires only actresses; similarly in art exhibitions only portraits and statues of women awake her aesthetic sense and her sensuality. Her general appearance and clothes are male.[102]

The underlying assumption here is that female sexual inverts had a characteristic personality, a feature that Lombroso had not so explicitly articulated in any of his previous works. Krafft-Ebing was not the only medical influence in his description of sexual inversion. While studying sexual inversion, Lombroso also relied heavily on Parent-Duchâtelet's environmental observations on Parisian prostitution and considered the 'tribade-prostitute' a special kind of sexual invert.[103] It was not the first time Lombroso had noted this particular form of same-sex passion amongst prostitutes: he had already drawn attention to it in an 1874 anthropological article about the habit of tattooing amongst Italian criminals, which had been published in Mantegazza's journal – the *Archivio per l'antropologia e l'etnologia* [*Archive of Anthropology and Ethnography*]. On this occasion, Lombroso had observed that amongst 'the most dissolute'[104] female prostitutes, tattoos were very common: a young prostitute would often tattoo the name of her male lover, while older prostitutes frequently tattooed the initials of their 'tribade' lovers close to their pubis.[105] In 1874 Lombroso did not expand on the subject of female same-sex desires amongst prostitutes beyond these cursory comments.

Almost twenty years later, however, in *La donna delinquente*, he proposed a few social explanations to understand sexual inversion that were drawn from early nineteenth-century medical observations of the tribade-prostitute, such as those of Parent-Duchâtelet. In fact, Lombroso provided the reader with a guide to understanding female sexual inversion, for which he identified five causes:

1. Excessive lustfulness, which sought outlets in all directions, 'even the most unnatural'.[106]
2. Influence of the surroundings, in particular women-only environments like prisons – 'schools for lesbianism' – and asylums, where the appearance of a single lesbian was sufficient to 'infect' the rest of the inmates.[107]

3. Imitative conduct intensified the vices of each individual and increased collective vice. Women were naturally imitative creatures, even of abnormal behaviours, so that their congregation at boarding schools, or during carnival orgies and religious festivals, increased the diffusion of lesbianism. Prostitutes and lascivious women were more prone to be affected by female homosexuality.[108]
4. Maturity or old age tended to invert sexual characteristics, which further encouraged sexual inversion amongst women. Ageing itself was a form of degeneration, which explained why older females adopted 'masculine sexual habits'.[109]
5. Amongst prostitutes or immoral women, another cause of lesbianism was the 'apathy towards, and disgust for men occasioned by physical and sexual maltreatment'. As I have previously shown, this idea was already present in the work of the neo-Malthusian doctor, Drysdale.[110] Surprisingly, Lombroso sympathised with prostitutes who, after enduring years of abuse by clients, had become disgusted by men and consequently rejected heterosexual love.[111]

The first cause was an old medical stereotype that linked female same-sex desires with excessive sexuality, which had been described in 1862 by Tonini and others before him.[112] The second, third, and fifth causes were cultural and social explanations that Lombroso derived from his reading of medical writers such as Parent-Duchâtelet and Drysdale. The fourth cause listed was a physical interpretation linked to degeneration theory.

Despite his use of the new medical term 'sexual inversion', and his deep knowledge of the most recent work on the subject by sexologists such as Krafft-Ebing, Lombroso's explanation of female same-sex desires integrates theories from different sources. Lombroso's work on sexual inversion in *La donna delinquente* is a clear example of how earlier medical ideas of problematic desires were not easily superseded nor fully replaced by the introduction of the psychiatric category of sexual inversion. Interestingly enough, when summarising the phenomenon of sexual inversion and setting guidelines for its understanding, Lombroso did not do so in the terms of the 'new' category of sexual inversion; he did not suggest that female homosexuality was the expression of the inborn nature of certain women with precise psychological characteristics. Rather, in Lombroso's theories older medical notions of female same-sex desires coexist with female sexual inversion. Further, the concept of sexual inversion is not fully homogeneous in Lombroso's *La donna delinquente*: by the end of the nineteenth century, sexual inversion had become multilayered, containing anatomical, psychological, and sociological medical observations, themselves continuing and combining earlier theories.

Tribadism and murder

Lombroso linked female same-sex desires not only to prostitution, but to other forms of criminality: female homosexuality might on occasion serve to explain violent criminal offences, thus strengthening the old association between female same-sex desires and violent passion. In 1903, Lombroso published an article on 'the psychology of a tribade wife' who had murdered her husband in an atrocious manner. Celli was thirty years old when she poisoned and strangled this 'honest man', whom she later cut into pieces.[113] Celli's past was uncovered during the trial, and proved quite extraordinary: she had been educated in a convent, but expelled because of her 'obscene' relations with her fellow classmates. After leaving the convent, she had 'illicit affairs' with both women and men, including her own eighteen-year-old niece, who was an accomplice to the murder of the husband. The motive of the crime was Celli's 'pathological love' for a woman called Battalini whom she had met in the convent where both had been educated. Though still married to her husband, Celli celebrated a second marriage with Battalini on the church altar of her home town, in the presence of two witnesses. Since this second 'wedding' she had stopped sleeping with her husband, and slept with Battalini instead.[114]

Lombroso described Celli as 'degenerate' because her 'sexual habits' and body were masculine-like. The virile energy shown in killing her husband and the fact that she had worn men's clothes in a failed attempt to evade the police were signs of such degeneration. In turn, her homosexual love explained her virile disposition and her ability to murder her own husband. Since his first observations on sexual inversion, Lombroso had highlighted the 'violence' of homosexual love and its exceptional intensity, which bordered on the criminal. Judging by Celli's murderous act, tribades loved with such violence as was unknown to 'normal' women. Thus, Celli's second 'marriage' showed the extent to which homosexual love was excessive – according to Lombroso, such excesses were reminiscent of the oddities of the Roman emperor Nero. Even her convent life could help explain her homosexual behaviour, since it was there that Celli had experienced the excesses of religious mysticism, which fused with her sexuality.[115]

Physical virility had been the most salient feature of female sexual deviants throughout Lombroso's career, but by the 1890s he had also started discerning virile psychological features in lesbians, prostitutes, and criminals. Prostitution may have been the typical crime the 'fair sex' engaged in; lesbians and prostitutes lived in the very same places and were intertwined in the figure of the 'tribade-prostitute'. Female homosexuality could occasionally lead to criminal offences because it excited violent passions; hence female homosexuals were also extremely violent criminals, as in Celli's case. Once again, female sexuality provided an explanation for female criminality.

Stigmatising feminism?

Scholars have recognised that Lombroso's work on normal and abnormal women, in particular *La donna delinquente*, has been instrumental in cementing the link between female deviancy and prostitution that gives rise to the assumption that any female behaviour that does not conform to society's standard of normalcy entails some form of irregular sexuality.[116] Considering historian Mary Gibson's recent remark that Lombroso was the first to systematically theorise female crime in the Western world, it is not surprising that so many scholars have been drawn to study the birth of criminal anthropology and related scientific disciplines that engage with the body, gender, and women.[117] Gibson has also suggested that in writing *La donna delinquente*, Lombroso was trying to undermine Italian feminism.[118] She has noted that Italian feminists started to group together following Mozzoni's foundation of the *Lega promotrice degli interessi femminili* to promote female interests in Milan in 1881. In the 1890s, this trend had gained considerable momentum and there emerged organised groups of feminists who challenged their subordinate position in society by defying traditional categories of 'normal' and 'deviant'.[119] Gibson's interpretation is flawed in two substantial ways: on the one hand, it overstates the impact of feminism in Italy. On the other, it fails to locate Lombroso's work on female deviancy within a longer medical tradition of enquiry into sexual matters.

Like most of his colleagues, Lombroso thought that women were inferior to men, and he was certainly keen to distinguish the masculine from the feminine spheres. Women were better suited to maternity and family life, whereas men belonged in the public sphere and contributed to historical progress. Valeria Babini has carefully analysed the assumptions that underpin this idea of separate spheres to argue that in the nineteenth century, Italian medicine proposed a naturalistic stereotype of femininity that made it easier to justify social and political discrimination against women. She explains that differences between men and women were interpreted in biological terms so that 'natural' phenomena that led to both sexual differentiation and evolutionary differences were deemed normative. The physiological aspect of sexuality had a crucial function in these interpretations. It was thought that the role played by each of the sexes in reproduction explained and perpetuated the roles of the sexes in society: man sought pleasure in the sexual act, woman sought to become a mother.[120] In such a framework it was easy to conclude that the physiological differences between men and women fulfilled complementary biological purposes, which in turn gave rise to two dissimilar social fates: nature had designed men for an active role in the animal world and society, and women for a subordinate role of maternity and social sacrifice. As Babini notes, in the second half of the nineteenth century, few Italian people thought any differently. Even Italian feminists of the time associated femininity with motherhood and demanded civil rights on the grounds that maternity was related to the well-being of the family and the social sphere. Aside from marginal anarchist contributions, maternity was never construed as a choice in these political debates.[121]

In Italy as in Britain, the naturalistic interpretation of sexual differences served to legitimise social discrimination. Historical analysis must, however, consider local peculiarities. The central role of motherhood in Italian feminist discourses reveals that they did not aim to challenge the authority of science. In an 1877 letter to Josephine Butler on the subject of prostitution, the Italian feminist Mozzoni wrote that the problem of women selling themselves might be solved by eliminating male celibacy. She underlined the importance of the traditional family in curbing prostitution – an argument she supported by pointing out that in rural areas, where all men were married, prostitution did not exist.[122] Taking this position, Mozzoni, like Lombroso, considered the problem of prostitution inevitable because of men's natural sexual urges. Further, Italian feminists demanded equal access to education and the workforce, but unlike their British counterparts, they refrained from publicly discussing women's sexuality. Forums like the English Men and Women's Club, where traditional assumptions about sexuality were challenged, did not exist in Italy. Still, in 1908, when the well-known female physician and feminist, Maria Montessori, delivered a paper in support of 'sexual morality' education at the first Italian Women's National Congress in Rome, she was very cautious about approaching the topic of sexuality, which is a sign of the difficulty Italian feminists had discussing sexuality publicly.[123]

While positivist scientists certainly did not support feminism, the main goal of their writings was not a direct response to local feminists. If Italian positivist medicine was a response to any existing discourse, then it was against the sexual morality traditionally espoused by the Roman Catholic Church. While the study of women's physiology was certainly influenced by the political debates of the time and scientists never hid this (Chapter 4), Lombroso's contribution to the analysis of female sexual deviancy was much more influenced by the growing field of sexology. Within Italian sexology the general trend of the last decade of the nineteenth century displays a multiplication of works on female sexuality. Some of the treatises that belong to this wave focus on female prostitution or the biological nature of women, amongst them Giuseppe Sergi's *Le degenerazione umane* (1889), which devoted a chapter to the analysis of prostitution, and Mantegazza's *La fisiologia della donna* (1893), which focused on the defining features of female nature. Lombroso's interest in female sexuality is part of this medical expansion. Moreover, Lombroso's own view of female sexuality was not entirely new: Parent-Duchâtelet is an obvious precursor to Lombroso's analysis of female prostitution. In Britain, Acton's *Prostitution, Considered in its Moral, Social and Sanitary Aspects* (1857) could be seen as a British antecedent to Lombroso's work on prostitution. Both Acton and Lombroso were concerned with the causes of prostitution and the links between it and the environment in which the prostitutes lived. Indeed, both Acton's work on prostitution and Lombroso's *La donna delinquente* belong to the medical tradition that came from Parent-Duchâtelet, which drew attention to the socio-cultural analysis of prostitutes' lives. With Acton, Lombroso also shared some views on women's sexual desires. In *Functions and Disorders of the Reproductive Organs* (1857), Acton famously argued that for women, sex was only a means of satisfying their own maternal instincts and only a way to please their husbands, rather than their own sexual pleasure. In the course of his studies in Paris in the late 1830s, Acton had been much influenced by French physiology, which posed that male and female sexualities were fundamentally different. Men were defined by strong sexual urges, whereas normal women were characterised by a lack of sexual desire: the basic female instincts were motherhood, marriage, and domesticity. Lombroso shared these views. In Acton, the prostitute was deviant because she was defined by her sexual drive, which was in stark contrast with the normal woman's instincts.[124] As I mentioned above, in Lombroso, the characterisation of the prostitute as having an active sexuality and sexual frigidity at the same time was contradictory, but despite this, Acton could be seen as a forerunner to Lombroso's general views of women's sexuality.

Unlike some of his European colleagues, Lombroso did not stigmatise feminism by associating it with sexual deviancy. The historian Anna Rossi-Doria has highlighted that at the turn of the century feminists were classed as the 'third sex' – a category they shared with Jewish people.[125] Rossi-Doria's argument implies that in describing feminists as homosexuals, Italian science displayed an anti-feminist attitude, as such pairing led to the stigmatisation of feminists. Rossi-Doria is right to point out that there emerged a medical literature that popularised the stereotype of the feminist as a virile lesbian who threatened the stability of (male) social roles. Such medical stereotypes were indeed a reaction against feminism. However, as far as I have been able to see, in the Italian medical literature dealing with female same-sex desires, the first explicit association between feminism and homosexuality can be found in an article on bisexuality by the German physician Hans Kurella, published in the *Archivio di psichiatria* in 1896.[126] In Italian medical literature the stigmatising stereotype of the lesbian feminist would only gain momentum from the early 1900s onwards with the publication of *Geschlecht und Charakter* [*Sex and Character*] (1903) by Otto Weininger,[127] and *Das Sexualleben unserer Zeit in seinen Beziehungen zur modernen Kultur* [*The Sexual Life of Our Time in its Relations to modern Civilization*] (1907) by the German doctor Ivan Bloch.[128] This stereotype became more widespread towards the end of the first decade of the twentieth century, which coincided with sexology's increasingly systematised theory of sexual inversion, and in the context of a crisis in Italian positivism following Lombroso's death in 1909. In the writings of criminal anthropologists of the late 1890s there were no connections between feminism and homosexuality: a female homosexual would be more explicitly associated with some form of genius, and a woman's virility could also be considered a faint indicator of her superior intelligence. There is no comparison between the sexual invert and the feminist in any of Lombroso's writings published before 1900.[129]

Italian sexologists did in fact react against feminism and set out to stigmatise it, but not before the beginning of the twentieth century. By this time, their discipline was well established through systematic research into the subject of human sexuality. This is to say that, in *La donna delinquente*, Lombroso was not so much responding to feminism as contributing to the expanding the study of sexuality, which had been steadily growing in Italy and abroad over the previous three decades. It is easy enough to suggest that Lombroso felt threatened by the new roles women were playing in society and wrote *La donna delinquente* to forestall those changes. Scientists, and psychiatrists amongst them, create models to interpret reality and solve the problems – real or hypothetical – posed by society. Lombroso's solutions addressed a cluster of issues that arose in a specific historical context. For example, he drew attention to the fact that children were being sexually abused in factories and mines, and explained that because harsher punishment would do nothing to reduce the incidence of such crimes, female guardians should be appointed to supervise places where children worked, and that children should be stopped from working in mines altogether. Statistics had shown that children were abused not only in cities, but also in rural areas, a tendency he suggested could be modified by increasing female prostitution in those areas: men could not be restrained from looking for sex and would turn to children in the absence of adults.[130] While Catholic priests and later some feminists called for chastity, Lombroso's approach to social questions was eminently pragmatic: men's sexual instinct could not be curbed, so it was not possible to ask them to be celibate. In Lombroso's view, regulations dealing with sexual hygiene had to account for the fact that sexually anomalous behaviour would inevitably arise whenever these appetites were unsatisfied. His theory of course rests on the assumption that while women could be chaste without any risk of undesired consequences, men could not. On a related note, he acknowledged that there existed a moral double standard in regards to adultery, which at the time was a punishable crime for a woman: 'let me once again state quite plainly, setting aside any hypocritical prudery, that [adultery is also caused] by the prejudice that makes us condemn one sex for something that in the other sex is not even thought to be an infringement'.[131]

Lombroso also acknowledged that some professionals, including doctors, might be partially to blame for exacerbating the sexual abuse of women and children. Lombroso suggested that on occasions, there had been cases in which doctors either abused children or women, or were complicit when they were abused by a family member or carer. This problem could be solved by allowing women to pursue academic study; in this way Lombroso explicitly endorsed women's enrolment in university and in fact called for women to become doctors. In 1904, when Montessori obtained the chair in anthropology at the University of Rome, she was supported by Lombroso.[132] His own daughter Gina studied medicine and helped her father for many years, first as an editor and later popularising his work.[133] Driven by his fervent anticlericalism, he pronounced himself in favour of divorce and of *'ricerca di paternità'*.[134]

Historians have insisted that Lombroso's misogynistic theories were inspired by a strong anti-feminist sentiment, but in the late 1880s these statements were neither conservative nor misogynist. Gibson herself acknowledges that feminists Ellen Key and Anna Kuliscioff were regular guests at Lombroso's house in Turin, which suggests that his contemporaries did not perceive him as an advocate of staunch patriarchalism.[135] Kuliscioff in particular had a strong influence on the education and intellectual development of both of Lombroso's daughters.[136] Unlike Havelock Ellis, Lombroso did not associate feminist movements with any form of deviant sexual behaviour, including lesbianism. Lombroso was by no means a feminist, and unlike Ellis he did not try to advance equality between homosexuals and heterosexuals. Lombroso's writings are an example of how nineteenth-century medical work can combine 'progressive' and 'conservative' politics in widely controversial statements. Yet one should acknowledge that Lombroso was endeavouring to find practical solutions to some major problems faced by Italian society at the time. In writing about female sexual deviations, Lombroso was neither responding to nor trying to placate feminism; he was writing in the context of a medical and sexological tradition that had already formulated theories of femal sexuality.

Since very early in his career, Lombroso had been interested in degenerative illnesses; his enquiry into the physical and psychological deterioration of southern Italians cleared the way for later studies of other forms of degeneration. Lombroso's interest in same-sex desires and female abnormal sexuality stemmed from a more general concern with deviancy. Female criminals, prostitutes, and lesbians shared a series of characteristics: they were virile and they failed to manifest sexual dimorphism from men, which confirmed their atavism. Female offenders, prostitutes, and tribades were abnormal in displaying an excessive sexuality. At the same time, like many other scientists of the day, Lombroso brought different forms of female sexuality to the fore. He spoke about sex, and this conferred legitimacy on sexological research, which was perhaps his greatest contribution to the field of sexology in Italy. Having said that, the ways in which Lombroso, like a number of late nineteenth-century psychiatrists, repressed abnormal female sexual behaviour within asylums and prisons, remains a dark aspect of sexological research.

6
Pasquale Penta, 'First Class Sexologist'

If Lombroso's *Archivio di psichiatria* was the first Italian publication to foster the study of deviant sexuality, it was one of his disciples, Pasquale Penta, who should be credited with establishing sexology as an autonomous medical discipline. Penta had originally been trained as a psychiatrist and criminal anthropologist and, in the course of his professional duties, he spent a great deal of time observing the behaviour of prisoners and investigating the extremes of human deviancy. Like Lombroso, Penta had initially endorsed deterministic arguments based on organic theories of degeneration, according to which irregular sexual activities should not be considered immoral choices, but the expression of innate characteristics. Very early in his career, however, Penta began to challenge Lombroso and to favour more psychological explanations of crime and deviant sexuality, thus turning away from the biological explanations for deviancy typical of Lombrosian criminal anthropology.

While Lombroso gained international fame amongst his contemporaries and later historians, Penta remains vastly unknown. This has not always been so: at the turn of the century, Penta was well regarded in sexological circles.[1] In 1908, the German psychiatrist Paul Näcke wrote that the 'first class sexologist' Penta could 'claim the glory of having founded the very first journal for sexology' in an article entirely devoted to his Italian colleague published in the *Zeitschrift für Sexualwissenschaft* [*Journal of Sexology*].[2] He wrote that Penta's *Archivio delle psicopatie sessuali* could be considered the 'direct ancestor' of the German journal for which he was writing.[3] Despite the recognition that Penta garnered from his contemporaries for his pioneering contribution to the discipline of sexology, modern historians have credited Magnus Hirschfeld with founding the first sexological journal. Hirschfeld, who was a doctor and the leader of the first homosexual rights movement in Germany, founded the *Jahrbücher für sexuelle Zwischenstufen* [*Yearbook for Sexual Intermediate Stages*] in 1899 and *Zeitschrift für Sexualwissenschaft* in 1908.[4]

Penta died when he was only forty-five years old, which is perhaps why historians from Italy and elsewhere have largely forgotten him.[5] Despite his young age, his achievements were truly remarkable: he developed his career in Naples, a city with a strong neo-idealist cultural tradtion that rejected positivism; he established a network of psychiatrists and legal experts interested in the study of deviant sexuality, promoted the discipline of sexology, and challenged the prudery of the Italian scientific community as a whole.[6] His attitude towards sexual inverts was nevertheless somewhat ambivalent: on the one hand he argued that sexual inversion was a 'rudimentary' form of love or an acquired 'vice'; on the other, he was sympathetic towards sexual inverts, insisted they were neither worse nor better than normal people, and that they should be respected accordingly.

Historical interpretation often offers a dichotomised account that portrays nineteenth-century sexologists either as heroes of sexual reform, or agents of the stigmatisation of sexual minorities. Penta's sexological work, however, exceeds these categories. His is an example of a medical writer and practitioner who endorsed common contemporary medical views, but was nonetheless able to challenge the sexual morality of his time. As a result he was ostracised and forced to put an end to his sexological journal, the *Archivio delle psicopatie sessuali*. His opinions on sexual minorities were far from stable, and his stance cannot be accounted for by his personal or political interests. Instead, Penta's contribution to sexology reflected his determination to achieve professional eminence, which eventually resulted in him flying in the face of the sexual morality of the time.

Following Lombroso's path: the Verzeni case

The pioneer sexologist Penta took full advantage of the new research and career possibilities that Lombroso had helped establish. Born in Fontanarosa (Avellino) in 1859, he worked as an assistant to Professor De Renzi for a short period of time after completing his degree in medicine in Naples. Penta was later called up for compulsory military service and, like Lombroso at the start of his career, served as army doctor; his first posting was to a Naples hospital, after which he was assigned to the prison at San Stefano. It was there he first developed an interest in Lombroso's writings on criminal anthropology in general, and atavism in particular. Penta began combining Lombroso's ideas with Darwin's evolution theory to interpret the phenomena of criminality and deviancy. Once his stint as an army doctor had ended, Penta continued to study and practise criminal anthropology in the Criminal Asylum in Aversa and a number of other prisons in the south of Italy. In the 1890s Lombroso's *Archivio di psichiatria* published Penta's first works on criminality, which gave rise to a fruitful relationship, as the young doctor would go on to become one of the famous criminologist's closest collaborators.[7] In 1891, Leonardo Bianchi, who was one of the leading psychiatrists of the day, appointed Penta as his assistant psychiatrist in the Neapolitan mental clinic that he directed; this position allowed Penta to devote himself to the study of mental illness without fully abandoning his interest in criminology.[8] In the mid-nineties, Penta became Professor of Psychiatry at the University of Naples. Immediately after this designation had been made, Bianchi encouraged the university to separate the chair of Psychiatry from that of Criminal Anthropology because his protégé's profound knowledge of the area merited the development of specialised studies. As a result, the first Italian Chair of Criminal Anthropology was held by Penta at the University of Naples – and not by Lombroso in Turin, as it is often believed.[9]

Penta's starting point for his sexological studies was the infamous case of the 'Strangler of women', Vincenzo Verzeni, an Italian rural version of Jack the Ripper whose 'lust murders' made headlines in the 1870s. Italian newspapers displayed a morbid curiosity about the case, and Italian psychiatrists continued to study it in subsequent years. Verzeni's singular mental characteristics surfaced during the trial, and managed to attain notoriety beyond Italian borders. In *Psychopathia Sexualis*, Krafft-Ebing devoted a number of pages to Verzeni, who was described as a 'remarkable' sadist; a 'prototype' of lust murderer who omitted rape, and for whom 'the sadistic crime becomes the equivalent of coitus'.[10] The criminal inquest concluded that Verzeni had never raped his victims, and he even confessed that he did not know how women were conformed.

Born to a family of peasants the outskirts of Bergamo in 1849, by the time he was eighteen Verzeni had begun assaulting female members of his own family and trying to strangle them. Between 1868 and 1873, in his early twenties, Verzeni assaulted a number of women, which resulted in the death of two of them: a girl of fourteen and a woman of twenty-eight. The victims' bodies were found naked and so horribly mutilated that it was initially impossible to determine whether they had been raped before being killed. The genitals had been removed, the limbs smashed to pieces, the abdomen cut open lengthwise, the entrails pulled out and scattered on the road and mixed with the bloody clothing, or carefully hidden in basements and beneath nearby piles of straw. There was one more horrific detail: the killer had gathered up the victim's hairpins one by one and hammered them into symmetrical shapes in the victims' flesh and in the ground. The ensuing legal hearing revealed that Verzeni had also bitten and sucked on his victims' blood, which led local newspapers to dub him the 'Vampire of Bergamo'.

A total of eleven psychiatrists examined Verzeni in the course of the trial: for the prosecution Alborghetti, Fornasini, Galli, Manzini, Perolio, Previtali, and Tarchini-Bonfanti; for the defence Griffino, Quaglino, Terzi, and Lombroso. In an extraordinarily crowded courtroom, the psychiatrists for either party challenged each other by drawing their arguments from the same medical authorities, amongst them Morel, Pinel, and Esquirol. One of the critical points of the debate was whether Verzeni could be diagnosed with moral insanity: those psychiatrists intent upon proving that Verzeni retained his rational capabilities emphasised the murderer's depravity, while the defence psychiatrists tried to show that Verzeni was insane and unable to distinguish right from wrong.[11] In his deposition, Lombroso endeavoured to pinpoint the signs of Verzeni's insanity and supported his argument in the terms of his new theory of atavism. It was a bold decision on Lombroso's part to invoke such a theory during a trial, because Italian psychiatrists were not familiar with the concept yet. As a result, Lombroso was mocked by the other psychiatrists, and failed to persuade the Court of Justice in Bergamo. In 1873 Verzeni was sentenced to life imprisonment with hard labour.[12]

Lombroso's failure to convince the jury is an example of typical nineteenth-century clashes between legal and psychiatric interpretations of sexual perversion: in Italy, as in Britain, psychiatric theories were regarded with suspicion as soon as they set foot in the courtroom. The fact that Lombroso's study on Verzeni continued to circulate amongst the members of the international psychiatric community even though it had been thoroughly discredited in the legal proceeding is an index of how independent these two fields could be at the time. When Penta came across this case-history, he felt compelled to challenge his 'teacher' more than fifteen years after the actual trial.

Lombroso believed that Verzeni was suffering from necrophilia, 'otherwise called insanity due to monstrous or blood love'.[13] According to Lombroso, Verzeni carried out the assaults on women under the tyranny of a mounting delirium. As soon as he had grasped his victims by the neck, he experienced sexual gratification, which is why it never occurred to him to rape them. Verzeni had confessed to Lombroso that he felt spasms of sensual pleasure when squeezing a living throat and, above all, when mutilating bodies or biting and sucking the blood of his victims.[14] Later, in *L'uomo delinquente*, Lombroso described Verzeni as unbalanced and vague; he suffered from epilepsy and a morbid perversion of his feelings, which led him to develop moral insanity and confusion of the will. He exemplified the category of the inborn criminal who bore anatomical and physiological stigmata such as cranial asymmetry with overdevelopment of the left frontal lobe and the bony crest that is found in the great apes and primitive savages; ossification of the temporal artery, and a squint had also been observed.[15]

As mentioned above, Verzeni was the starting point of Penta's interest in human sexuality. In 1887, while working as a psychiatrist in the San Stefano prison in Naples, Penta came across Verzeni, who became his patient. The notes on his psychiatric examination were first published in 1890 in *La tribuna giudiziaria* [*Legal Tribune*]. Subsequently, Verzeni's case was the centrepiece of Penta's first sexology book, *I pervertimenti sessuali* [*Sexual Perversions*], which was published three years later. In line with the vast majority of sexological case-histories of his time, the study concerned itself with deviant sexuality, but it differed from other contemporary works in its extensive discussion of violent sexual excess. This interest in extreme sexualities would prove to be the distinctive mark of Penta's sexological research throughout his life.

Penta claimed that upon first interviewing Verzeni, he had found latent 'degenerative stigmata' that Lombroso had failed to notice: prior to the first interview, the patient had been asked to undress so that nothing could escape the doctor's observation. Penta applied this method with all the people he examined.[16] Despite differences in the actual physical observations, Penta's careful examination and measurement of Verzeni's skull and body illustrate the extent to which his method was in line with Lombroso's. What sets Penta's analyses apart from Lombroso's, however, is the amount of time he spent searching for the psychological traits of the 'Vampire of Bergamo'. Over the course of months in the controlled environment of a prison, Verzeni's life was recorded in great detail, incorporating both his own declarations and his fellow inmates'. Penta recounted how often and on what occasions Verzeni masturbated, what he thought when doing it, how he behaved with his fellow inmates, whether he had sexual relationships with them, and in what kind of sexual practices he engaged. In prison, where sexual perversion was widespread, Verzeni had lost his sexual drive, but was known to other prisoners as a '*cinedo*' [passive pederast].[17] Penta concluded that Verzeni's sexual perversion combined two opposing extremes: on the one hand, he was a violent man who killed for the sheer pleasure of exercising mastery over a human being; on the other, he was an 'urning', a 'female-male' who felt sexual pleasure where others would experience the 'most terrible outrage'.[18] Despite the fact that Verzeni's intellect was functional, Penta observed, he lacked any moral sense.

Penta's analysis was somewhat hesitant, so that while he was deeply influenced by Lombroso, he seemed to be at odds with his predecessor's overall diagnosis, particularly in regards to the explanatory power of psycho-social factors. Both case-histories concurred in the portrayal of Verzeni as a degenerate, but to Penta Verzeni's morbid heredity could not in itself account for his crimes. Penta specified that during puberty, poor or insufficient nutrition, an unhygienic environment, and corrupted social relationships could weaken human beings, who would become prone to acquire vicious and perverted habits. Verzeni's childhood was thus partially to blame for the sexual pervert to come. Penta noted that Verzeni had never attended school and that his parents taught him only the virtues of hard work and of going to church. He had started to masturbate as a child, and experienced great pleasure when, aged twelve, he strangled chickens for the first time. After this experience, he was sometimes compelled by a 'mysterious instinct' to slaughter all the chickens on the farm by wringing their necks.[19] Penta believed there was a psychological explanation for Verzeni's preferred method of murdering his victims – from a very early age, sexual craving and violence had been experienced simultaneously, so that sexual gratification must always be accompanied by a violent component.

Verzeni's passion for strangling, first chickens and then women, showed the extent to which sexual perverts associated sexual pleasure with an erotic image that had been a source of pleasure in early childhood. To Penta, sexual perversion was a 'rudimentary' and primitive form of love, because the normal sexual development of perverts had been thwarted by some accidental erotic experience in early life. Thus, Penta's explanation of sexual perversion is a synthesis of Lombroso's concept of atavism with Binet's theory of fetishism.[20] Penta's was neither the first nor the only psychological explanation of sexual perversion. In the last decade of the nineteenth century, there started to emerge a number of Italian medical writings that explained sexuality with reference to psychology and social context, and steadily less as a consequence of organic conditions. This turn-of-the-century development was accompanied by a growing uneasiness about degeneration theory, as has been discussed in Chapter 3.

Alongside psychological interpretations, Penta endeavoured to understand sexual crimes biologically by comparing human and animal sexual urges. In this comparative scale of sexual behaviour across the animal kingdom, Verzeni's conduct amounted to an 'animal throwback' analogous to fish that do not need penetration to get sexually excited, and toads that strangle females to death during sexual intercourse. Penta described at length the sexual habits of animal species that inflict wounds or kill their partners in the course of copulation, and pointed out that it was mainly the male of these species who derived pleasure from fighting and killing. Penta noted that, to a certain extent, it was common for animals to inflict pain during orgasm, which led him to wonder whether there was any relation between sexual intercourse and the beast killing its victim. This question encouraged Penta to explore the role of cruelty in humans. He believed that it was a well-established scientific fact that a lack of moral feeling and the presence of weak inhibitory ideas would explain why a psychopathic and ill-fitted individual would be sexually aroused by being cruel. But there was a further physiological condition that accounted for some male acts of monstrosity and sadism: according to Penta, even healthy men derive great pleasure from winning a woman who resists until the moment of surrender. A man encounters obstacles which he must overcome – this is why nature has given him an aggressive character. The aggressive components of a psychopath's personality are so accentuated that the subject develops a compulsion to destroy his object of desire. Thus, Penta implied that there was a close relationship between male sexuality and sadism, whereby the latter was nothing more than an exacerbation of otherwise normal aspects of the sexual life of males, which only became pathological because they were excessive.[21]

Penta's study of Verzeni was part of an overview of sexual perversion which distinguished mainly between rapists and sexual perverts, such as fetishists or sexual inverts. While rapists were morally insane and 'primitive', sexual perverts, like inverts, were 'degenerates', who on occasions had displayed a mental superiority that rapists never attained.[22] Overall, homosexual love was described as a 'simple and elementary [act], that is on the same low level as the initial experiments in love', often guided by an impulse almost as powerful as an emotional fixation, or the urge of animals in heat.[23] Thus, in the sexual invert the individual's development was incomplete, his arousal was tied to cortical stimulation of the brain at the sight of some specific objects, such as male nudes.[24]

While Penta's theoretical interpretations of sexual perversion diverged from Lombroso's, they both agreed on the issue of whether perverts were accountable for the sexual crimes they had committed: both of them denied free will in such actions, which went against the tenets of classic legal theories. Penta was all too aware of the existing conflict between legal and medical interpretations of sexual crime. Indeed Penta's earliest writings had attempted to broach the conflict between law and psychiatry when discussing phenomena such as sexual perversion, and continued to do so throughout his career. He reported that many morally insane people were locked in prisons instead of being placed in asylums because judges remained unconvinced by the results of psychiatric examinations.[25] Penta's approach to sexual perversion was mainly pragmatic: he thought that society had the right to defend itself from sexual crimes, and for this reason sexual criminals such as rapists had to be taken into custody in criminal asylums or penal colonies. Given their primitive nature, he wrote, the only way to discipline them was through hard labour and contact with animals. In the case of 'sexual perverts' like inverts or fetishists, who did no one any harm, Penta did not put forward any practical solution. He actually called them the 'oppressed of humanity', which implied not only a measure of sympathy, but also a certain disapproval of contemporary sexual morality.[26]

Penta's repeated castigation of jurists' great ignorance in psychological and psychiatric matters reflected not only the conflict between law and medicine over sexual issues as discussed above, but the extent to which Penta was conscious of the practical implications of sexological studies. The question of whether people suffering from disorders such as moral insanity, psychopathy, or sexual perversion were acting of their own free will or were simply mad remained a matter of debate within, and between, the legal and medical professions.[27] While lawyers were reluctant to challenge the view that individuals were autonomous moral agents, deterministic explanations of antisocial behaviours became crucial for psychiatrists, who appeared in court as expert witnesses and quoted biomedical theories. To psychiatrists sexual offenders suffered simultaneously from particularly strong sexual drives and thoughts, and markedly weak nervous systems which lacked the strength to control these virtually irresistible impulses. In a legal sense, this meant they could not be held personally responsible for what their drives compelled them to do, because their free will was impaired. As numerous historians have shown, many late nineteenth-century Continental psychiatrists believed that in many cases irregular sexual activities were not immoral choices, but symptoms of innate characteristics; they supported these arguments with deterministic theories of hereditary degeneration and neurophysiological automatism.[28] Medicine thereby challenged both religious and legal authorities, and advanced a new understanding of sexual deviance, transferring it from the realm of sin and crime, to that of health and sickness.

Archivio delle psicopatie sessuali

The decision to scrutinise Verzeni's case-history reflected Penta's readiness to address problems posed by sexual psychopathologies, and this case is also indicative of his particular interest in extreme sexualities. After publishing his studies on Verzeni, in January 1896, Penta launched the *Archivio delle psicopatie sessuali. Rivista quindicinale di Psicologia, Psicopatologia umana e comparata, di Medicina legale e di Psichiatria forense ad uso dei Medici, Magistrati ed Avvocati [Archives of Sexual Psychopathies. Fortnightly Journal of Psychology, Comparative Psychopathology, Forensic Medicine and Forensic Psychiatry for the Use of Physicians, Magistrates and Lawyers]*. This sexological journal initiated Penta's editorial career, which shows his willingness to challenge sexual morality, his readiness to compete with Lombroso's intellectual and editorial activity, a pioneering spirit, and a self-promoting interest. All these elements become clearer following his advancement of sexology in spite of the obstacles he encountered.

Penta was able to recruit thirty-three people – mainly psychiatrists and some lawyers from southern Italy – who were interested in collaborating on the project. Most of them were established professionals and the list of collaborators includes several professors in medicine or law, and directors of asylums. Some of them were: Leonardo Bianchi, who at the time was Director of both the Psychiatric Clinic and the Asylum of Naples; Guglielmo Cantarano, Vice Director of the Asylum of Naples and Professor of Clinical Medicine and Neuropathology at the University of Naples; Angelo Zuccarelli, Professor of Forensic Medicine and Psychiatry at the University of Naples;[29] Cesare Colucci, who at the time worked at the Psychiatric Clinic in Naples; Salvatore Ottolenghi, Professor of Forensic Medicine at Siena University; and Silvio Venturi, Director of the Girofalco Asylum. Some of these people had already published on the topic of sexual perversion and were thus well known in the field, amongst them Venturi, Cantarano, and Zuccarelli – the latter two had conducted and published pioneering research on female sexual inversion in the 1880s.[30] Penta also had a handful of foreign collaborators: Ellis and Raffalovich from London; Lombroso's famous rival Alexandre Lacassagne, Professor of Forensic Medicine in Lyon, and the psychiatrist Näcke, who at the time worked at the Hubertusburg

Asylum (Saxony). Ellis, Raffalovich, and Näcke were recognised for their tolerant attitude towards sexual inversion.[31]

Penta's network of contributors was remarkable. Although by the time he started the *Archivio* he had been Chair of Criminal Anthropology for five years, Penta was still very young when he launched his sexological journal, and yet he had enough connections to establish such a bold enterprise. His experience as a collaborator on Lombroso's journal, the *Archivio di psichiatria*, had certainly introduced him to a wide range of colleagues. He was also an active member of the medical community, participating in and organising conferences; at the 1898 Italian Congress of Forensic Medicine, for example, Lombroso was the president of the conference while Penta was vice-president.[32] This shows that despite his youth, Penta was already a reputable member of the medical community that was gathering around the new discipline of criminal anthropology.

The *Archivio delle psicopatie sessuali* was Europe's first sexological journal. Its goal was to advance the knowledge of sexual perversions because Penta was convinced that sexuality had a decisive influence on the individual's intellectual and moral life.[33] Penta's journal was published by Capaccini, a small Roman publishing house run by two brothers, who initially supported the *Archivio*'s engagement with sexology. In that same year, 1896, the Capaccini brothers also launched a series of scientific series called *Biblioteca dei pervertimenti sessuali* [*Library of Sexual Perversions*] which made available translations of texts by Tardieu, Moll, Krafft-Ebing, Ellis, and Jacques-Joseph Moreau de Tours.[34] During this same period, a larger and more renowned publishing house, Bocca, also published works on sexual perversion as part of a series called *Biblioteca antropologica-giuridica* [*Legal-anthropological Library*].[35] This illustrates that publishers had a growing interest in the field of sexology, that research in the area was proliferating in Italy, that there was a market for this kind of literature that had expanded beyond circles of physicians and lawyers to reach a larger audience, and finally, that at the end of the nineteenth century there was money to be made from books on sexual perversion. These phenomena certainly stimulated young physicians to engage in sexological research.

The Capaccini brothers themselves explained that they had been compelled to publish a journal on sexual perversion by 'many physicians, naturalists, lawyers and magistrates' who felt the need to have more information on what was published in Italy and abroad on such an 'important topic'. As the publishers noted, the study of 'sexual psychopathologies, and, in general, the sexual feeling' was developing apace, acquiring the 'most serious importance', beyond the specialised fields of forensic medicine and psychiatry, and into broader areas of interest like sociology and general literature. The Capaccini brothers specified that the *Archivio delle psicopatie sessuali* was not exclusively concerned with the scientific aspects of psychology, 'human and comparative psychopathology', and their practical application in forensic medicine; by making 'new knowledge' readily available, it would contribute to a better understanding of human history, art, and literature.[36] They claimed that sexual matters greatly influenced the experience of men and women, their actions, and their intellectual production.

The staff of the journal also stressed the 'highly scientific and humanitarian' goal of the *Archivio delle psicopatie sessuali*, namely: to render less painful the 'fate of many sad people born anomalous and degenerated'. Although they would sometimes be obliged to recount obscene stories, their intention was far from promoting 'particular forms of human weakness'. In reporting shocking cases, they merely wished to 'shine a light' on the 'darkness of the human heart'.[37] This stress on the scientific character of the publication is most likely a hint that its staff anticipated some negative (moralistic) responses. The success of Krafft-Ebing's *Psychopathia Sexualis* had elicited two critiques from the international scientific community: first, that works on sexual perversion had somehow justified sexual deviancy, and thus recognised the right of sexual inverts to exist, so much so, that some inverts had written to Krafft-Ebing and other psychiatrists justifying and legitimising their feelings; and second, that sexual fantasies narrated in scientific works were contributing to the spread of sexual perversion amongst laymen who were able to access such works.[38]

The *Archivio delle psicopatie sessuali* appeared every fortnight, and was around fifteen pages long. It contained original articles, clinical cases, and detailed reviews of recent works. The reviews section was usually grouped around a main topic such as marriage, neurasthenia, sexual impotence, fetishism, or masochism. Articles and case-histories were contributed by different authors and focused on abnormal sexuality. While some medical writers – notably Ellis – occasionally argued that sexual inverts were normal individuals, the majority of the pieces dealt with cases of sexual inversion as observed in mental hospitals, which indirectly emphasised the pathological nature of sexual inversion.[39] Interestingly enough, none of Penta's case-histories discussed male inversion on its own; he did, however, publish a number of articles that dealt exclusively with female same-sex desires.[40] Male homosexuality was just another abnormal sexual behaviour in a constellation of other, more unusual, sexual practices. Other researchers also tended to analyse extreme forms of sexual perversion such as a case of auto-pederasty in which a man inflicted injuries upon himself, or a masochist pederast who engaged in necrophilia.[41] The journal devoted a considerable amount of space to literary representations of sexual perversion: Penta was specifically interested in the ways in which writers described their own sexual life and in the manner in which autobiographies tended to anticipate abnormal sexualities. Such materials, Penta said, would illustrate the 'universal character, origin and meaning' of sexual perversion.[42]

Penta was the soul of the journal. He regularly wrote original articles and produced bibliographic surveys covering the latest works on sexual perversion by Italian and foreign authors. Näcke heaped praise on Penta's reviews and bibliographical essays, and with good reason:[43] Penta had a deep knowledge of the most recent and pioneering German, French, American, British, and occasionally Russian, sexological works. Raffalovich's and Ellis's participation in the project, together with Penta's keen interest in Ellis's latest sexological work, explained why the *Archivio delle psicopatie sessuali* followed the British sexological scene so closely. In July 1896, Raffalovich published a provocative article on the current state of British research on sexual psychopathologies, accusing British society of hypocrisy. He highlighted the difficulties English scientists such as Ellis had when publishing on sexual inversion.[44]

While in the late 1890s the British scientific community did not show much enthusiasm for Ellis's work on sexuality, the Italian medical community responded favourably to Ellis's contribution to the field. Penta would continue to write detailed and prompt reviews of Ellis's work for the rest of the *Archivio delle psicopatie sessuali*'s short life. For example, in one of the July issues, there was a review of a work Ellis had co-authored and published with E. S. Talbot in the *Journal of Mental Science* in April of the same year, just a few months earlier.[45] The original work was based on an American case-history of insanity associated with sexual inversion and melancholia following the removal of the testicles. In one of the September issues, Penta reviewed another article Ellis had published in the American *Alienist and Neurologist* only three months before.[46] Ellis himself contributed an essay on the artistic skills of male sexual inverts to the *Archivio delle psicopatie sessuali*. Building on Moll's observations that many sexual inverts had to act all the time to disguise their sexual preferences, Ellis argued that sexual inverts had mental characteristics that brought them close to artistic genius.[47] He further argued that sexual inverts' artistic skills came from an inborn nervous inclination,[48] an argument he would revisit in *Sexual Inversion*, where he likened congenital sexual inversion to a form of talent (Chapter 7). Apart from displaying a considerable degree of internationalism, this small journal illustrates that Italian and British sexologists were interacting in unexpected ways, by reviewing and disseminating each other's work, and actively engaging in joint projects that sought to popularise the new discipline amongst the members of the international medical community.

Penta's role as director of the *Archivio delle psicopatie sessuali* was not restricted to his ability to mobilise a number of physicians, lawyers, and even amateur sexologists like Raffalovich; his tireless personal contribution to the journal often took the form of original articles. His most innovative studies in the *Archivio delle psicopatie sessuali* betrayed an ever-growing interest in the psychological causes of sexual perversion and a willingness to move beyond current medical assumptions. For all his tentative explanations, Penta was still unable to offer a final account of the causes of perversion in 1896.

According to Penta, scientists would often describe the clinical phenomena of sexual perversion without trying to uncover its origin, or when they did approach the question of aetiology, they would only do so superficially, by invoking a 'functional sign of psychic degeneration'. As he explained, degeneration had proved to be a medically insufficient explanation and, by the turn of the century, more and more European scientists denounced its shortcomings and complained that it was being used too vaguely to explain all sorts of not-yet-understood phenomena.[49] Although he was certainly aware of the inadequacy of degeneration theory, Penta continued to use it as a working concept in the essays he published in the *Archivio delle psicopatie sessuali*. In 1896, he still thought that every sexual perversion should be considered a 'manifestation of human degeneration', as attested by the fact that all sexual inverts had a 'naturopathic or alienated' temperament and were quite often completely insane and needed to be treated in an asylum.

Another bodily source of sexual perversion was further identified in the role of the reproductive system. Penta cited clinical histories which proved that the removal of the sexual glands before puberty prevented the individual fully developing a sexual feeling.[50] Yet, while he identified physical origins for sexual perversion and placed it within the all-encompassing phenomenon of degeneration, he was attempting to move beyond bodily explanations. An example of this is Penta's effort to explain the exact process through which sexual perversions arose. He believed that in certain individuals some accidental pleasant external sensations, like the sensation felt at the sight of a person of the same sex, were felt along with the first stirrings of sexual excitement. These sensations could then become stable and be permanently linked to sexual excitement. If this was the case, the individual could only be aroused when recalling – in other words, regressing – to these first pleasant sensations. This was a 'regressive tendency' that occurred only in degenerates.[51] Unable to craft an original comprehensive explanation of sexual perversion at this stage of his research, Penta relied on the well-worn degeneration framework, which he combined with Binet's theory of psychological causation and an organic component.[52] In his early works published in the *Archivio delle psicopatie sessuali*, he struggled to unravel the interactions of his twofold account to no avail.

'Natura non facit saltus': pursuing the study of sexuality

After only one year, Penta's journal folded without giving any warning or explanation. In a letter to general practitioners dated 28 April but inserted in one of the November issues, the publishers announced that the final number of the *Archivio delle psicopatie sessuali* would appear at the end of the year; they also noted that previous issues had sold out, and that in order to satisfy the demands of physicians who kept asking for copies, the first three numbers of the journal would be reprinted. It is reasonable to assume from this that the journal sold well, as did other publications about sexual perversion.[53] Indeed, in another supplement, the Capaccini brothers highlighted that German, French, and Russian doctors were taking an active interest in the study of psychopathologies and that Italian physicians were increasingly following suit. The *Biblioteca dei pervertimenti sessuali* had proved widely successful in medical circles, and for this reason and because they received some complaints about the price of the original *Biblioteca*, the entire collection would be made available in a second, cheaper edition; the treatises would consist of sixteen, thirty-two, or forty-eight pages, and would appear fortnightly. The Capaccini brothers were keen to point out that this budget publication was for doctors only.[54]

If it was neither for financial reasons nor due to lack of popularity of medical studies on sexual perversion, why did the publishers close down the *Archivio delle psicopatie sessuali*? There does not seem to be enough historical evidence to answer this question conclusively. However, Bianchi observed that to some, the *Archivio* had seemed 'shameful'.[55] Näcke suggested that the journal – which is probably to say Penta himself – had clashed with the publishers.[56] Later in his career, Penta briefly remarked that the difficulties he had encountered with the *Archivio delle psicopatie sessuali* had been the result of the prudery of Italian society – which is probably a suggestion of why he had been forced to put an end to his sexological journal.[57]

Far from being discouraged in the pursuit of his editorial activity, Penta set out to establish a broader framework that would allow him to reach a larger audience. From 1898 until his death in 1904, he published the *Rivista mensile di psichiatria forense, antropologia criminale e scienze affini* [*Monthly Journal of Forensic Psychiatry, Criminal Anthropology and Related Sciences*]. This time, Penta's enterprise was supported by a small Neapolitan publishing house called Tocco. The *Rivista di psichiatria forense* offered original essays and articles on sexual matters, just as its predecessor had done, but by providing extensive coverage in the field of forensic psychiatry, criminal anthropology, and related subjects, it managed to attract the interest of a wider audience. Special attention was paid to criminality in southern Italy, mainly amongst the mafia. Unlike the *Archivio*, however, the *Rivista* did not have any foreign collaborators, even if Penta continued to review the latest sexological developments and update his readers on the activities of his ex-collaborators. For instance, Penta continued to review Ellis's work promptly and closely followed the Bedborough trial, which elicited his comment on the oddity of the British situation. According to Penta, Britain was the only country to reject sexological research, which was well established elsewhere, especially in France and Germany.[58]

The new journal's first issue included a platform in which the editorial board explained its 'humble purpose' was to present the best works on forensic psychopathy from around Italy (especially the south) and overseas. They would approach their subject from a social, juridical, and biological point of view, but more specifically, the editors hoped the journal would advance research in three distinct directions.[59] First, they wanted to examine the 'natural history of the criminal and insane' in its social and legal aspects, while conducting an analysis of artistic works dealing with such topics.[60] Second, they wished to promote the study of sexual pathologies and crimes. They highlighted that, to a large extent, the subject had been neglected by other journals due to 'excessive and false prudery'. The high incidence of sexual pathologies had been established beyond any reasonable doubt, and such a widespread phenomenon merited scientific scrutiny in its own right. This new editorial board was keen to emulate the *Archivio delle psicopatie sessuali*, which had been successful both in Italy and abroad, and had the support of a wide reading public.[61] Third, they hoped to advance the study of sociology and jurisprudence following the methods of the natural sciences.[62] The journal's target-audience comprised doctors, judges, and lawyers, as well as the educated public who took an active interest in the scientific and legal aspects of 'highly social' topics like criminality, insanity, and neuropathologies.[63] Aside from Penta, none of the contributors came from the *Archivio delle psicopatie sessuali* and, in fact, most of them were lawyers. One of the psychiatrists involved in the new project was Augusto Luzenberger, Professor of Psychiatry at the University of Naples.[64] The new journal resembled Lombroso's *Archivio di Psichiatria* in its focus on criminality, but Penta's second journal took a special interest in southern Italian criminality and usually devoted one section to sexual pathologies. The latter section was by no means less prudish than the *Archivio*, but in the new format sexual pathologies were just one topic amongst many. The *Rivista* might thus appear to be less preoccupied with sexual perversions than Penta's earlier journal had been.

In the *Rivista mensile di psichiatria forense*, Penta relied mainly on environmental and psychological theories to explain same-sex desires, before finally abandoning the concept of congenital inversion altogether. In the first year (1898) he published 'A Case of [male] Sexual Inversion' which, despite the title, dealt mainly with transvestism and masochism.[65] The patient was effeminate and loved to dress as a woman, to the extent that he had been photographed in women's attire. Penta held that in this case the abuse inflicted by the father and the resulting pain might have combined with the first signs of sexual feeling, which had initially turned the son into a masochist.[66] To substantiate its claims, the article cited some of Moll's recent studies, which had shown masochism to be the beginning of male homosexuality – hence Penta's remark that inverts were often masochists, and in turn, masochists often became homosexuals.[67] Penta concluded the analysis of this case-history by stating his new conviction that homosexuality was more an effect of environment and education than the expression of an inborn tendency.[68]

When presenting other case-histories, Penta seemed to further endorse a view of homosexuality as an acquired condition. In 1903, he published an article about a 'paedophile fellator'; even if the article's main concern was the paedophilia of the subject, it was interpreted as a case of homosexuality because he had sex with only male children. Additionally, the man suffered from alcoholic dementia, which had led to a pseudo-paralysis. Penta believed that this case showed that paedophilia was not inborn, but acquired. He further stated that every sexual perversion had deep roots, so that once it had been established it became such a powerful impulse that it never failed to be elicited by specific stimuli and was always experienced in the same fashion.[69] In general terms, even when a sexual pervert was aware of what he was doing, it was impossible to talk of free will.[70]

Penta's popularisation of the study of (abnormal) sexuality was not restricted to the work published in journals. Indeed, his lectures on psychiatry at the University of Naples covered the field of sexuality to a significant extent; they are highly relevant to historians because they illustrate the influences that shaped Penta's ideas about sexual perversion in the later years of his career. In his courses for the 1899–1900 academic year, Penta started by teaching his students how to conduct a clinical psychiatric examination: he first explained how to examine the vegetative (bodily) functions, mobility, sight, and sensitivity, and then he drew attention to the psychological functions. Amongst these, erotic and moral feelings, as well as behaviour patterns, had to be closely observed – especially those aspects which might be linked to instincts.[71] Insanity mainly affected people between thirty and forty years old, and the insane began evincing sexual perversions from an early age.[72] The unmarried and widows or widowers were more prone to insanity than married people. Spinsters and bachelors were considered to have escaped family responsibilities, which explained their accordingly weak moral sense and underdeveloped personalities. Penta explained that sexual feeling gave rise to the psychological personality: the individual's personality was developed in sexual phases that ranged from masturbation, to sexual attraction towards certain kinds of women, and thence to a specific woman. Thus, according to Penta, single people were prone to insanity and were often more unbalanced than the married because they would never attain the final stage of development in sexual life; namely, the exclusive love of one specific person.[73] Psychopathies and neurosis could be consequences of excesses, especially dietary excesses, so widespread amongst the 'fat, greedy, vain, tedious and foolish bourgeoisie'.[74] Even if Penta never declared it openly, his statements on poverty, wealth, and the exploitation of the poor by the bourgeoisie bordered on a kind of socialism.

Penta spent a fair amount of time explaining Freud's essays on neurosis, which had been published in the French *Revue Neurologique* (1896) and the German *Neurologisches Centralblatt* (1899). While Italian positivist psychiatrists explained madness as a bodily disease and, broadly speaking, tended to reject Freud's psychology, Penta introduced his students to nascent psychoanalytic theory, which emphasised a causal relationship between an individual's sexual life and the development of neurosis.[75] In these same years, Lombroso's *Archivio di psichiatria* was reluctant to popularise Freud's theories and the journal continued to be generally hostile to psychoanalysis in later years. According to Morselli, Italian positivist psychiatrists rejected Freud's theories because of their excessive focus on sexuality, the absence of organic explanations, and their overall pessimism.[76] The positivist scientific community set great store by the rationality of the world, and in its quest to find it had made no allowances for the irrationalist aspect of Freud's position.[77]

According to Penta, Freud's ideas about sexuality were not in themselves original, because previous researchers had pointed to the defining importance of sexuality in the outcome of mental diseases; it was only the degree to which sexuality was highlighted that was new.[78] Both Penta and Freud conceived of sexuality as a central aspect of an individual's personality, which is perhaps why Penta was willing to popularise Freud's ideas. Penta underlined that what Freud was doing for the study of neurosis, other psychiatrists such as Albert Eulenburg, Albert von Schrenck-Notzing, Luzenberger, and Morton Prince had already done for sexual inversion. Schrenck-Notzing, like Freud, had explained that early in life, a pleasant erotic sensation was indelibly associated with a specific image and idea. Sexual inversion took place when the link united the image of a person of the same sex with a pleasurable erotic sensation; Schrenck-Notzing thus conceived of homosexuality as acquired rather than inborn. Penta went even further: he deemed Krafft-Ebing's theory of sexual inversion as an inborn disease outdated, and stated that neither sexual perversion nor criminality were inborn. In espousing these views, Penta was indirectly criticising Lombroso's criminal anthropology.[79]

Penta focused on sexual inversion in the second part of his lectures, which dealt mainly with mental pathologies and clinical psychiatry. When introducing what psychiatrists traditionally called 'psychological degeneracy', he touched on sexual inversion. Penta explained that the nervous system of some individuals with a morbid heredity was weak, so that they were easily influenced by suggestion. Consequently, following Binet, sexual perversion was explained in terms of the association between pleasurable feelings and specific ideas. Despite the fact that homosexuality thus conceived was acquired, Penta stressed that it was not possible to condemn sexual inverts: their nervous system was weak and the outcome invariably morbid. Further, because such a frail nervous organisation could not counter the force of unmanageable sexual impulses, sexual inverts were not fully in control of their actions. Penta regretted that not even doctors were prepared to accept that a sexual 'vice' was also as 'morbid' as a 'primitive psychological degeneration', or at least was beyond the control of the individual. In Penta's terms, 'vices' were not the effect of a moral choice, but an effect of the environment, like poor food or hygiene, or harsh economic conditions. For instance, a poor diet would weaken an individual's body and nervous system, thus enabling the appearance of sexual vices. Society and the struggle of modern life caused nervous breakdowns, which led to the development of vicious and irresistible tendencies. Living in society required constant confrontation, which was often nerve-shattering, and further, the wealth and pleasures of contemporary times triggered a collapse of the nervous system, which in turn increased propensity for vice.[80] It thus emerged from Penta's explanation that sexual perverts could not be held responsible for their condition, because the latter was a product of society.[81]

Penta's attitude towards deviant sexualities was decidedly ambivalent, particularly in regard to homosexuality. He thought homosexuals were not responsible for their acts, and insisted they were characterised by a weak and sick constitution. The underlying premise was that sexual behaviours were natural phenomena that had to be understood rather than judged or morally condemned. Despite his attempt to dissociate personal responsibility from sexual inclination and thus suspend moral condemnation of such tendencies, Penta initially seemed unable to break away from the assumption that sexual inversion was pathological. Even his attempt to explain sexual inversion was far from fully articulated: if the nervous system of the homosexual was weak, was the body, and the brain in particular, still the cause of sexual inversion? How did environment and the nervous system interact?

Despite not being able to answer these questions conclusively, Penta kept a remarkably open approach to homosexuality: he continued to read widely on the topic and was not afraid of changing his opinions. Thus, in 1903, he wrote a long review of Hirschfeld's 1899 *Jahrbuch für sexuelle Zwischenstufen*, noting that it was interesting and 'philanthropic'. The German sexological journal had the purpose of deepening the knowledge of all the physical and psychological characteristics which, despite belonging to the one sex, can be found in the other, that is, of the intermediate grades between the two sexes, such as sexual inversion.[82] 'These people [homosexuals] are not monsters', Penta wrote, 'and must not be looked at with contempt, because they are neither worse nor better than other people.'[83]

Penta wished Hirschfeld's journal a better chance of success than the *Archivio delle psicopatie sessuali*. As he wrote, in 1896 he too had sought to 'challenge the prejudices and the false or excessive prudery of society'.[84] He explained that when the publishers had forced him to close down the *Archivio*, there emerged a gap, 'which [was] now bridged by the excellent German yearbook'.[85] Penta went on to say that:

> Even if [Hirschfeld's journal] does not feature genuinely observational and clinical studies, it has the great merit of popularising certain ideas, dispelling the contempt and disgust that surrounds these natural phenomena, and of causing us to love and respect those individuals who suffer – *comprendre c'est pardonner* – because throughout these pages filled with facts, quotations, memoirs and pictures the whole, so to speak,

history of homosexuality [...] is presented in a clear and simple style, with its causes, characteristics, and biological meaning, and it is demonstrated that ultimately complete virility and outright femininity are types, almost artistic abstractions, or exceptions. Whereas Linnaeus's dictum, that *Natura non facit saltus* [nature does not show interruption] is also true of sexual characteristics, homosexuality is not an atavistic phenomenon, a degeneration, a monstrosity, but a natural and common phenomenon which must be respected legally, like heterosexuality.[86]

In 1903, about ten years after his first study on sexual perversion, Penta regarded homosexuality and heterosexuality as equally natural phenomena, and further acknowledged there was no clear boundary between normal and abnormal sexuality. No longer classed as an example of pathological sexuality, same-sex desires had become normal. Remarkably enough, Penta supported Hirschfeld's political programme, comparing it with his own sexological activity, which he trusted would result in the obliteration of society's prudery and eventually win acceptance for uncommon sexual 'types'.

Female sexual inversion

In the initial stages of its development on the Continent, sexology was strongly linked to the study of sexual inversion, but the discipline's object of study diversified as sexologists expanded the taxonomy of sexual perversions to include sadism, masochism, zoophilia, paedophilia erotica, and others in the 1890s.[87] Both the *Archivio delle psicopatie sessuali* and the *Rivista mensile* reflected this expanding interest in a wide range of sexual perversions. In his early 1890s work on Verzeni, Penta had noted that there were at least ninety case-histories of (mainly male) sexual inversion on record in the international scientific community.[88] The topic had become reasonably well established and he perhaps thought it was time to move on to other sexual categories. After all, professional prestige is often associated with discovery, or in other words, being a pioneer in a new field is what appeals most to a scientist.[89] Penta was interested in the study of male sexual inversion as long as it afforded him the opportunity to engage with lesser known forms of abnormal behaviour. Penta did not publish any case-histories of male sexual inversion in the *Archivio delle psicopatie sessuali*, and only two in the *Rivista*. As it has been mentioned above, the first case-history was published in 1898, and did not centre so much on sexual inversion itself, as on the subject's accompanying perversions – transvestism and masochism.[90] The second case-history, as has also been discussed above, was published in 1903, and focused on the paedophilia of a man who had sex with young boys.[91]

Penta observed male same-sex practices not so much for their intrinsic interest, but as part of a wider range of sexual behaviours in which prisoners, sadists, and murderers could engage. Penta did not think that the mere fact that a man had sex with another man was representative of the medical category of sexual inversion, and he pointed out that many poor young men in Naples sold themselves to foreign men, that sodomy in prison could be used as a form of violence or retribution, and that in this context same-sex practices often implied power relationships with active men dominating passive prisoners.[92] On the other hand, from the first very number of the *Archivio delle psicopatie sessuali*, Penta had drawn attention to female sexual inversion both as an independent phenomenon that shared some characteristics with male inversion, and as a phenomenon with its own specific traits.

Penta thought that women were less developed morally and physically than men, not only as a result of natural selection, but also because of men's domination over them. Along with many contemporary scientists, Penta believed that natural selection impinged more on men than women because of the more active role men played in the natural and social environments. Interestingly enough, Penta pointed out that in the course of history men had always chosen as partners those women who submitted to them and were faithful; men had exacerbated the psychological and physical weakness of women through these choices, and so had played a role in reinforcing those sexual differences that had acted in their favour. Women's inferiority was thus 'an artificial product' created and sustained by men.[93] According to Penta, women were weak because their nervous system was less developed than men's, which also explained why they normally experienced 'less sexual excitement'.[94] Their 'inferiority' did, however, offer women a greater measure of protection against suffering from madness and engaging in criminal acts: if women were less affected by these phenomena, it was because they were less involved than men in the struggle for existence.[95] At this time, Lombroso's influence on the field of Italian criminal anthropology was at its peak, and his work on female criminality encouraged his followers to conceive of female prostitution as the equivalent of male criminality. Penta, however, never approached female deviancy through the analysis of female prostitution, and he did not consider the latter to be the typical female crime or sexual perversion. Instead, he focused on female inversion as an extreme form of female sexuality and noted its frequent incidence in schools.

In the first issue of the *Archivio delle psicopatie sessuali*, Penta commented on a scandalous case from a small town in southern Italy. The focus of this case was F.R., a woman who had acquired sexual perversion as a result of a pathological nervous heredity. F.R. was forty-eight years old and was originally from the north of Italy. Her grandfather had committed suicide and her parents no longer lived together. She had spent six years in an asylum because of excesses of lypemania with signs of delusion.[96] In 1892, when she was working as a teacher in an elementary school in southern Italy, F.R. was discovered having sex with three female pupils aged eleven and under. From the ensuing investigation, which presumably took into account the pupils' depositions, it transpired that in the past she and her pupils had engaged in oral sexual intercourse and genital contact without penetration. It was also discovered that a pupil once brought her brother to school when he was still a baby and F.R. had abused him.[97] F.R. was imprisoned and Penta was summoned to conduct a psychiatric examination.

Penta's examination found F.R.'s genitals to be normal, as was the rest of her body, which he described as 'masculine-looking'. While in prison, F.R. was inspired by religious mysticism: she spent most of her time alone, prayed a great deal, and wanted to wash other prisoners' feet. It was observed that she suffered severe genital pains as a result of habitual insertion of objects into her vagina.[98] When Penta questioned her about her past sexual experiences, she revealed her first sexual experience had occurred when she was twelve years old, and was aroused when her genitals came in contact with a shaking train. Soon after that she fell in love with a young boy, who reciprocated her feelings, but they were forced to break up because she could not bear '*immissio penis in vaginam*'; as a result, she developed a fear of ejaculation and started avoiding men altogether. Subsequently, she was affected by erotomania upon reaching menopause.[99] Penta described the woman's crimes as 'rudimentary loves': the kind of sex she had with her female pupils resembled the first sexual experiences that everybody has as a baby – a pleasant feeling resulting from touching and sucking.[100] She also displayed a form of 'paradoxical love' when abandoning all caution in her search for sexual pleasure from her pupils. According to Penta's analysis, the source of her acquired sexual perversion was the sexual fantasy of being with the young man she had loved earlier in her life.[101] Aside from a brief remark on F.R.'s nervous heredity, Penta moved away from strictly physical explanations with emphasis on signs of degeneration, providing a psychological interpretation of F.R.'s sexual crime instead. He explored the woman's childhood and first sexual experiences, demonstrating the extent to which F.R.'s sexual life was undeveloped. Her fear of men, for example, explained her preference for children as sexual partners. Above all, this case was evidence that women, as well as men, could abuse children. Penta thus highlighted the fact that women could display the same extreme sexual perversions as men.

Fear of men and the idea of an undeveloped form of sexual feeling also explained other cases of same-sex desires in women. In 1896, Penta and Alfredo d'Urso, a physician at the Naples asylum, co-authored a study on a young female sexual invert who developed sexual feelings while still at school.[102] Like other physicians at the end of the nineteenth century, Penta was particularly interested in pupils' sexual awakening in all-girls schools. A.V. was twenty-two years old and came from a wealthy Neapolitan family. Since some relatives were insane, she had a psychopathic heredity, but overall she was good-looking and did not display any stigmata of degeneration. However, ever since childhood she had suffered from epileptic crises, had derived pleasure from the company of ladies, and avoided boys. When she was nine years old she fell in love with a female schoolmate, becoming ill-tempered and sad; since then her only source of happiness was the time she spent with her beloved friend, whom she liked to kiss and hug. When her friend died, she no longer went out, ate, or spoke a word, and from the age of thirteen her intellectual abilities had visibly declined. She subsequently fell in love with a few other young ladies, who in turn avoided her.

To support his claim that she was fully aware of loving women in a sensual fashion, Penta quoted passages from the letters A.V. had sent to two of those she loved. A.V. also showed a strong resistance to the idea of getting married, which was what her parents wished for her. Penta observed that A.V. was a modest young woman and that her parents preferred him not to ask her questions of too intimate a nature. Accordingly, he did not dare explain to them the nature of their daughter's sexual feelings and perceived they did not quite know what to make of his scientific interest in A.V.'s sexual life. However, her parents' readiness to hand over their daughter's letters reflects, at least to a certain extent, a willingness to cooperate with Penta in exploring her inner world.

According to Penta's analysis, A.V. was an example of 'rudimentary love' with a real 'phobia' of the male sex, and her sexual desires were more psychological than physical. Her sexual inversion had very peculiar characteristics: she seemed to feel a 'sweet feeling', as well as a 'cult' and 'adoration' for other women, which Penta described as 'instinctive organic characteristics, not much developed, undifferentiated and inferior'. This had all the characteristics of 'amore fiammesco' [fiamma love].[103] Penta stressed how this undifferentiated love was common during puberty, especially in all-girls colleges.[104] It was so frequent indeed, that later that year, when reviewing the work of his American colleague William Lee Howard, Penta noted a very similar case on the other side of the Atlantic involving a woman who developed her sexual inversion at school.[105] Reflecting on A.V.'s case, Penta pointed out that at that time all psychiatrists agreed on two features characterising sexual perversions: that 'sexual feeling' was critical in the development of the psychology of the individual, and that it was generally accepted that any 'psychic degeneration' had an effect on sexual life. Yet, more work was necessary to clarify how different individuals manifested different forms of perversion.[106]

As the A.V. case-history shows, it was customary for late nineteenth-century psychiatrists to use patients' letters when diagnosing mental diseases. Both Krafft-Ebing and Ellis made extensive use of male sexual inverts' writings in their studies, quoting passages from the letters sexual inverts had addressed to their doctors. Penta adopted this methodology and started to use letters sent by female sexual inverts to their lovers, although without the cooperation of the authors.[107] In 1900, Penta published a remarkable article focusing on female sexual inversion in which he used letters written by a woman. He claimed that Mantegazza,

Lombroso, and Krafft-Ebing had never been able to collect letters from such women, which in turn made him the first to make female inverts' desires speak for themselves.[108] Penta was thus claiming to be a pioneer.

Nicola Tramontano, a lawyer, had given Penta some letters Maria M. had written some time earlier. While he was not allowed to disclose how Tramontano had come across these letters, he thought they merited publication, regardless of whether some people thought they were obscene and immoral, as it was equally necessary to know human beings' virtues and their vices. No matter how revolting or obscene those 'vices' might be, if the physician could understand their source, he might find the means to heal and prevent them.[109] Penta stressed he pursued a moral purpose in publishing Maria M.'s letters: he wanted to warn fathers of the risks their daughters ran by being on too intimate terms with their girlfriends.[110] Excessively long and tender kisses, warm embraces, and overly sensual sentences in daughters' letters were signs that they were liable to be 'contaminated' by 'abnormal loves'.[111] It was necessary to act promptly, not only because those individuals who were prone to mental diseases could 'suffer from sexual perversions', but also because normal individuals could learn by example, imitation, and habit.[112] Penta believed that silence and ignorance about sexuality were more dangerous than being informed about the mechanics of normal and abnormal sexual passions. Knowledge was thus necessary to prevent the spreading of sexual deviancy, and awareness of its pathological nature might deter people from such behaviour.

In order to do this, Penta summarised for the reader some important points put forward by recent studies that were relevant to the case-history he was introducing. Clinical cases of the *fiamma* phenomenon had proved that sexual inversion was acquired by 'moral contagion', and that it was possible to treat same-sex desires through a change of milieu. Schrenck-Notzing, Freud, and Luzenberger had recently shown that in most cases of sexual inversion, sexual perversion was acquired.[113] According to this argument, some sexual stimuli acted on the 'unconscious' during childhood, linking sexual pleasure to a specific idea, and this unconscious association re-emerged during puberty. This phenomenon was very dangerous, as indicated by the fact that anybody working in juvenile delinquent prisons, colleges, or seminaries was familiar with sexual inversion: at least a third of all prisoners practised pederasty;[114] most of the young boys in gaol had their first sexual experiences at an early age, had seen pederasty in and outside the prisons, had seen how their parents made love with each other because they grew up in crowded rooms, and had shared a bed with their sisters. Neapolitans were only too well aware that visitors came to their city in search of sexual intercourse with boys, who later blackmailed their foreign lovers.[115] Physicians knew that most educated sexual inverts had their first sexual encounter in colleges or seminaries. In colleges, boys and girls started by writing love letters to each other and ended up engaging in same-sex acts.[116]

After this digression on the means of diffusion of homosexuality and the threats posed by boarding schools, Penta warned that the letters he was publishing contained vivid depictions of 'Sapphic love' and other forms of female sexual perversion. From the context it is possible to discern that Maria M., the author of the published letters, replied to a newspaper advertisement written by another young lady, Giovannina. Maria was a theatre singer who lived on her own, although a man kept her in exchange for sex from time to time. Maria was very keen to say to her new friend that this was the only man she had 'known', and that she had loved women since childhood. Because she could not make a living from her singing, she agreed to become the lover of a rich man who was willing to help her. The two girls corresponded for a while, using *posta restante* addresses and collecting letters from their respective post offices. After exchanging photographs and reinforcing their mutual sympathy, they tried to organise a rendezvous. After agreeing that it was critical to avoid gossip, they decided to meet at Maria's house because she was freer than Giovannina. Reading the letters that Penta published, it is clear that the two women were both looking for the same thing: a sexual encounter with a person of the same sex. They had been very explicit about sexual matters since the first letter, and they described what they liked and wanted. Maria, for example, openly wrote that she masturbated while gazing at Giovannina's photo and decided to send some of her 'wet' pubic hairs to Giovannina. Maria also said she wanted to be Giovannina's 'slave'.[117]

Penta was right to claim that he was the first to have published such a document. The content of these letters is unique in scientific literature because Maria M. described her sexual fantasies openly. As I have mentioned above, Penta believed in principle that women were more protected than men from madness and criminality because of their marginal participation in evolution. Yet on the basis of his research, he admitted that women were just as exposed to sexual deviancy as men were. Penta insisted and showed that, just like men, women could be sexual perverts, could commit sexual crimes, and had strong sexual drives.

Penta's introductory remarks on the scientific purposes of his research on Maria M. and his insistence that the journal was not an obscene publication should not come as a surprise. Today's reader might even wonder whether Penta's success owed mainly to the scientific merits of his writings, or to his work's titillating qualities – in particular the long descriptions of perverts' sexual habits. The question remains whether it was only doctors and lawyers who read Penta's journal. The answer is probably not. In the 1903 volume of the *Rivista di psichiatria forense*, the physician Filippo Saporito published a study on criminality amongst soldiers. The case-history focused on a young soldier who ended up in a military prison for disobeying orders. While in prison, he started to write his autobiography, which included a brief account of his sexual encounters with other men; he called himself a 'pederast', was familiar with medical terms, and was evidently well acquainted with Mantegazza's work on same-sex desires.[118]

Even if the *Archivio delle psicopatie sessuali* was only meant to circulate within medical circles, it probably had a much wider audience; Penta's *Rivista di psychiatria forense*, on the other hand, explicitly aimed to reach laymen. But did women have access to such works at the turn of the century? Did women read medical works? Lombroso was certainly very popular in Italy and he was read not only by scientists. The social conventions of the time, however, made it hard for women to gain access to works such as the *Archivio delle psicopatie sessuali*, which targeted men, either for scientific or for titillating purposes.

Penta's repeated warnings about the scientific nature of his work and his emphasis on the importance of extending the knowledge of all natural phenomena, including sexual behaviours, also betrayed his anxieties about possible misunderstandings of his sexological studies. He might have been aware of the highly erotic quality of his writings and it is possible he exploited this feature for maximum effect. Yet, so far as I have been able to tell, there is no evidence that Penta used his sexological studies as a vehicle for financial gain. What is certain is that his editorial activity in the field of sexology was ostracised.

Penta's positions throughout his career were far from static or coherent; when assessing the overall development of his ideas, it is possible to discern a move from physical to increasingly social and especially psychological explanations. It is not fortuitous that Penta promptly assimilated and popularised Freud's ideas on sexuality at a time when most Italian scientists rejected them. Indeed, Penta's own ideas were akin to Freud's, and in a general sense mirrored the transition from Krafft-Ebing to Freud that was taking place in European psychiatric debates. His change was also motivated by practical reasons. Biological explanations offered little hope of improving a patient's situation, and Penta was an optimistic practitioner, as suggested by his advocacy of a better education for criminals on the grounds that they could be reformed. He also had socialist leanings. Penta escaped the deterministic limitations of degeneration theory by assembling a psychological interpretation of sexuality that opened up the possibility of therapeutics, although he never wrote about treating sexual inverts.

Ultimately, his shifting views on homosexuality show that when dealing with sexologists, the villain/hero dichotomy is limiting. Like Ellis in Britain, Penta saw himself as a scientist whose purpose was to liberate others from sexual prudery, but whose work was restricted by hypocrisy and social conventions. Both experienced difficulties in publishing their sexological research in their own country. Nonetheless, Penta challenged contemporary views on homosexuality and was prepared to express sympathy for Hirschfeld's political programme. Italian society met

Penta's sexological activities with controversy, which shows the extent to which Penta was challenging its moral values and customs.

There is no proof that his 'tolerance' derived from his own personal or political beliefs. Penta's bold sexological activity acquires a clear meaning when understood in the context of the medical community of the time. Penta is an example of a professional scientist who, in a country dominated by Lombroso's criminology, made sexual perversion his own specialisation. In the course of the nineteenth century, medicine became increasingly professionalised; in this framework, Penta is an example of how a young physician cultivated sexological research as his own field of expertise. To him, debates on sexual inversion were a means to secure professional prestige, rather than a political instrument to achieve social acceptance for, or stigmatisation of, sexual minorities.[119] He was an ambitious scientist who wanted to be the first to deal with new developments in his discipline. It is not a coincidence that he did not focus on male inverts, despite having a deep knowledge of the specialised literature that discussed them. Instead, a large proportion of Penta's case-histories focused on female inverts, and he approached the study of their perversion in an original way: he sought to let these women speak for themselves. While physicians had published letters written by male inverts, Penta thought that collecting letters written by female inverts was an important step for sexology because little had been done in that area. Not only did Penta manage to find and publish letters from such women, but he did it so effectively that his readers might have been easily shocked by their explicit descriptions of sexual desires and practices. Rather than viewing him as an oppressive sexologist, historians might gain more by considering Penta as a practitioner who was unveiling female sexuality, because despite the fact that they were written for a male professional or general audience, his studies on female sexuality reveal vivid aspects of late nineteenth-century women's sexual desires.

7
Havelock Ellis and Sex Psychology

In 1897 Henry Havelock Ellis published *Sexual Inversion*, the first English monograph on homosexuality. It took him five years to collect all the data and case-studies; it also took two years of collaboration with a man of letters, John Addington Symonds, and help from various American and Continental medical writers, as well as his personal friends.[1] Ellis's aim in publishing his study of same-sex behaviour was to demonstrate that same-sex desires were just a 'natural' expression of the sexual instinct: he proposed that homosexuality was a common biological manifestation in human beings and animals alike. He also used examples from both anthropological and historical studies to show that homosexuality was present across a wide range of different cultures. *Sexual Inversion's* radical proposition rested on the broader implications of the book: if sexual inversion was neither a sin nor a sickness, it followed that the difference between heterosexuality and homosexuality was simply in the choice of object of desire. Its argument that homosexuality should be treated as a natural phenomenon, subject to no religious or legal constraints, meant that *Sexual Inversion* was pitted against the morality of its time. It fostered sexual tolerance, proposing that individuals had a right to follow their sexual inclinations and desires.

Earlier works on the subject of 'sexual perversion', such as Krafft-Ebing's *Psychopathia Sexualis*, dealt with sexual inverts from the perspective of pathology. The importance of Ellis's treatise is that, probably for the first time, a scientific authority opted not to portray sexual inverts as degenerates. Further, he had a specific political agenda in mind when he started to write *Sexual Inversion*: it was to be a plea for the decriminalisation of 'gross indecency between men'. This he achieved by presenting a range of case-histories illustrating that sexual inverts were normal people and not insane or criminals.

In 1897, talking about sex with tolerance and permissiveness, and advocating for its study on rational grounds, was as radically subversive and progressive a project in Britain as anywhere else. The wider implications of the project, however, were even more complex. Ellis believed that in the course of history, society had grossly exaggerated the differences between the two sexes by imposing different gender behaviours. Nevertheless, many of his arguments rested on the assumption that there existed profound differences between the sexes, which led him to argue that due to their very difference, women should be given a greater voice in the ordering of society. His case-studies were based on the conviction that 'masculinity' and 'femininity' were qualities rooted in deep biological differences: male sexuality was seen as essentially active while female sexuality was theorised as passive and responsive to male sexuality. As a result, feminist scholars have often pointed out that Ellis left the Victorian stereotypes of women's sexuality unchallenged, and that in fact he reinforced negative stereotypes associated with female homosexuality.[2] Indeed, Ellis's *Sexual Inversion* has been seen as responsible for the consolidation of the 'stigmatising' notion that lesbians are masculine, criminal, insane, and almost monstrous creatures.[3] Ellis's essentialist approach to sexual differentiation, however, did not impede the portrayal of lesbians in a more positive light than is often acknowledged. The chapter that follows analyses Ellis's work with close attention to the internal and reciprocal relationship between its constituent elements and the medical context within which it emerged. I will argue that Ellis's work on female sexual inversion was much less a matter of stigmatising lesbians than of negotiating what homosexuality was within specific medical debates about sex and sexual inverts, and women's role in society.

First, I will depict *Sexual Inversion* as the outcome of an unusual collaboration between Ellis and Symonds, and examine their underlying political agenda. Second, I will show how difficult it was to pursue the study of sexuality in Britain through an analysis of British physicians' response to the Bedborough trial (1898), which led to *Sexual Inversion* being banned on charges of indecency. Finally, Ellis's conception of lesbianism will be examined. Through a close reading of *Sexual Inversion*, it is possible to gauge the extent to which Ellis considered lesbians independent, intelligent, and even remarkable women.

Writing *Sexual Inversion*

Despite his fame as a sexologist, Ellis was an outsider to the established British medical community. Born in 1859 in Croydon, Ellis left England in 1875 and joined his father, the sea captain Edward Peppen Ellis, for a world trip. He stopped in Australia due to bad health, and in 1878 was employed as an elementary school teacher at Sparke's Creek, a remote town in New South Wales. He moved back to England in 1879 and in 1881 started his medical training at St Thomas's Hospital, London. Ellis obtained his medical degree after seven years of training, although his qualification – the Licentiate of the Society of Apothecaries – was the lowest level required for embarking on a medical career, and he only practised medicine as a *locum tenens* physician. During the course of his studies, Ellis began to frequent the Fellowship of the New Life and other progressive London circles, and shortly after became a regular contributor to the magazines these groups put out. In these years he also met Edward Carpenter and feminists such as Olive Schreiner and Edith Lees, whom he would go on to marry in 1891. He started his career as a writer by tackling both literary and scientific subjects before focusing exclusively on the psychology of sex in the 1890s. By 1928, the last tome of his encyclopaedic seven-volume *Studies in the Psychology of Sex* was completed. It was only in 1938, a year before his death, that he was admitted to the Fellowship of the Royal College of Physicians in an explicit recognition of his contribution to research on sexual knowledge.[4]

The genesis of *Sexual Inversion*, which was originally the first and subsequently second volume of his *Studies in the Psychology of Sex*, is contained in Ellis's and Symonds's correspondence, which was initiated in 1890. They never met personally, but were nonetheless able to collaborate by letter in the study of sexual inversion.[5] In July 1891, Symonds wrote to Ellis alerting him to the importance of same-sex desires in his research on the Italian Renaissance, which Symonds had studied extensively; in the same letter he reported that he had privately printed two essays on same-sex desires.[6] Indeed, in 1883, Symonds had published a closely argued defence of ancient Greek boy-love, *A Problem in Greek Ethics*, and then, in 1891, a plea for the sympathetic reconsideration of contemporary sexual inverts, *A Problem in Modern Ethics Being an Inquiry into the Phenomenon of Sexual Inversion*. In the latter work, Symonds engaged with medical research on same-sex desires, displaying a fair knowledge of specialised Continental literature.

In July 1892 Symonds approached Ellis – through their mutual friend, the poet Arthur Symons – with the proposal that they should co-author a book about sexual inversion to be published in the Contemporary Science Series, which Ellis edited.[7] In his reply, Ellis explained that he was 'at work on a study of the present state of knowledge of the secondary sexual characters in man and woman', which appeared in 1894 as *Man and Woman*. He added that his interest in the subject was of much longer standing: 'for fifteen years I have been more or less occupied in preparing myself to deal with the question of sex'. Ellis was not trying to impress Symonds: his interest in sexual knowledge can effectively be traced back to his Australian trip of the late 1870s.[8] However, it was not until 1892 that he began enquiring into the nature of sexual inversion, and this for two reasons: partly because he had observed its existence to a greater or lesser extent in many people he knew, or knew of, and for whom he had 'much love and respect';[9] partly because he attained a clearer grasp of the relation between sexual inversion and the law after having read a summary of a pamphlet on this subject in an Italian journal, the *Archivio di psichiatria*.[10]

Some of his most esteemed friends indeed had same-sex relationships, amongst them his wife Edith, and Carpenter. Moreover, Ellis was interested in recent developments in Italian science: he was perhaps the main English supporter of criminal anthropology and had helped to popularise this kind of Italian research in Britain. The correspondence between Ellis and Symonds reveals that both were familiar with Italian sexological research about same-sex desires. *Sexual Inversion*, the most famous British turn-of-the-century text on the topic, not only grew out of a dialogue and negotiation between Symonds and Ellis about the nature of homosexuality, it is also the product of their engagement with Italian sources.

Sexual Inversion as a negotiation of scientific knowledge

The correspondence between Ellis and Symonds reveals that they had a shared political agenda, and that they were both equally determined to change the legal situation of sexual inverts.[11] Ellis thought that the law against male same-sex acts in England was 'outrageously severe', and lamented how 'a perfectly beautiful form of inversion' could be shamefully tainted by current legislation.[12] Symonds fully endorsed this view and both of them agreed it was time to call for a change.[13] The real obstacle was settling the question about the nature of same-sex desires. This was not a secondary problem, and they both realised that if the book was to trigger a change in British attitudes towards sexual morality it was important to reach a mutual agreement on what kind of image of sexual inversion it was going to promote. From the outset, it was clear to them that the success of their project was reliant on their negotiation of concepts of disease and normality as applied to homosexuality and, consequently, the strategy they used to present their results had to be carefully weighed. This was not an easy task as initially their use of words such as 'abnormal', when applied to same-sex desires, contained a certain degree of ambiguity that disappeared in subsequent letters.

In an 1892 letter to his friend Carpenter, Ellis explained that the goal of the book he was writing with Symonds's help was to obtain 'sympathetic recognition for sexual inversion as a psychic abnormality which may be regarded as the highest ideal and to clear away many vulgar errors preparing the way, if possible, for a change in the law'.[14] Ellis's adoption of the term 'psychic abnormality' did not imply he supported a form of pathologisation of, and consequent discrimination against, sexual inversion. On the contrary, Ellis openly indicated that sexual inverts could well be considered 'the highest ideal', meaning that it is possible to hold inverts in high regard. Writing to Carpenter, however, Ellis did not clarify what he meant. For his part, Symonds, who was homosexual and the son of an eminent physician, referred to homosexuals as persons 'abnormally constituted', but whose 'congenital abnormality is not vice or crime, but imperfection, aberration from the standard'.[15]

Despite the fact that Symonds was suspicious about Ellis's possible approach to homosexuality, it was Ellis who removed any uncertainty from their position. Symonds's initial scepticism was due to Ellis's medical background. Symonds was well aware that general medical knowledge regarded the phenomenon of sexual inversion as a form of psychopathic or 'neuropathic' derangement, inherited from morbid ancestors and developed in the patient by early habits of self-abuse.[16] He recognised, however, that medico-legal scientists had taken a considerable step forward when they acknowledged that homosexuality was inborn. Indeed, this concession made it possible to speak of homosexuality not as a vice or a crime, but as something that was innate in an individual, and hence not punishable by law.[17] According to Symonds, 'the theory of morbidity [applied to homosexuals] is more humane, but is not less false, than that of sin or vice'; the diagnosis of the homosexual instinct as pathological was perforce an error on the part of psychiatrists: the passions of many of the best and noblest Greek men had been guided by such instinct, and it was simply not possible to assert that ancient Greece, the cradle of Western culture, had been in some sense diseased.[18] Symonds sought to promote a positive image of sexual inverts by establishing a close association with ancient Greek culture.

Although for his part Ellis also wished to show that sexual inverts could be normal individuals, he did not want to pick a quarrel with medical psychologists, choosing instead to concentrate on the production of a medical treatise with the potential to transcend current medical assumptions. At the time it was easier for physicians to write about homosexuality in medical texts than it was for authors addressing a general readership in, for instance, literary novels. Non-specialised writers were limited by the Obscene Publications Act (1857), which restricted the subject-matter of books published in England.[19] Thus for Ellis, working within the medical tradition entailed a strategic compromise: 'I do not wish to put myself in opposition to the medical psychiatrists, the people who have most carefully studied the question; to do so in any case would be a bad policy; I simply wish to carry their investigations a step further.'[20]

Replying to Symonds's criticism about the sanctioned medical position in regard to sexual inversion, Ellis reminded his correspondent that in *A Problem in Modern Ethics*, he had described sexual inverts as having an 'abnormal nature', a term other psychiatrists were also fond of using.[21] Ellis's main objection to the British psychiatrists was less their use of a pejorative phrase than their unwillingness to even discuss the question, wrapping 'a wet blanket round it, with averted eyes'.[22] Forced to clarify his position, Symonds then responded that when referring to sexual inverts as 'abnormal' and 'morbid', he had simply meant that they were 'in a minority', and 'exceptions to the large rule of sex'. Sexual inversion was never morbid in itself, but on occasion might coexist with morbidity.[23] Symonds eventually found a middle ground and admitted that sexual inverts were frequently 'neurotic'; he nevertheless doubted that neurosis should be regarded as the cause of inversion. He also recognised that the word 'sport', which he had also used to define homosexuality in *A Problem in Modern Ethics*, was misleading, although it was the preferred alternative to 'neuropathic', a widespread term in the field of psychiatry.[24] 'I have been growing to regard these anomalies as sports, that is to say, as an occasional mal-arrangement between the reproductive function and the imaginative basis of desire.'[25] While struggling to define his position, Symonds had used the current medical terminology, arguing on the same grounds and adopting the same assumptions as doctors. After discussing the issue with Ellis, he concluded that sexual inversion would eventually be regarded as:

> a comparatively rare but quite natural and not morbid deflection from the common rule, due to mental imaginative aesthetical emotional peculiarities of the individual in whom it occurs. It would then be neither a morbidity nor a monstrosity – except in the same sense as colour blindness be termed either.[26]

Therefore, eventually Symonds grew inclined to abolish the neuropathic implication. Considering that all boys are exposed to 'the same order of suggestions (sight of a man's naked organs, sleeping with a man, being handled by a man) and that only a few of them become sexually perverted', he argued 'it is reasonable to conclude that these few were previously constituted to receive the suggestion. In fact, suggestion seems to play exactly the same part in the normal and abnormal awakening of sex.'[27]

After reading Symonds's letters, Ellis replied stating 'that the difference in point of view is very slight – not greater than is desirable in order to obtain an all-round presentation of the matter'.[28] There could be no doubt that scientific opinion agreed with Symonds's initial characterisation of sexual inversion as a neuropathic condition, but after pondering the various available stances, Ellis wished to avoid any assumption of the necessity of psychopathic conditions. He wrote:

> [t]hat sexual inverts are often neurotic persons there can be little doubt: that suggestion is sometimes a factor in the causation I am quite prepared to believe; the fact that both neurotic condition and suggestion are so common would certainly not prove that they have no connection with sexual inversion. The causation is probably complex.[29]

Ellis went even further, and pointed out that 'to call the sexual invert a "sport" [...] seems another way of calling him – in the uncoloured scientific sense of the word – a "monster"'.[30] Thus, the use of the term 'sport' would bring the homosexual close to the field of teratology.[31] Later in the same letter, Ellis seemed quite pleased with the comparison with colour-blindness put forward by his correspondent, but proving to be more astute and bolder in his approach than Symonds, he judged the analogy with 'secondary sensations' – for example coloured hearing – to be still better:

> such sensations may be either congenital or acquired, and it is not easy to distinguish one class from another; the subjects of such sensations sometimes, though by no means invariably, show minor neurotic characters or a neurotic heredity, while at the same time it is possible to argue (as some of the subjects of such sensations do argue) that coloured hearing indicates a further step of human development. I think we may regard sexual inversion as a psychic abnormality in just the same way as coloured-hearing is.[32]

In the final draft of *Sexual Inversion*, Ellis wrote that the sexual invert may be 'roughly compared to the congenital idiot, to the instinctive criminal, to the man of genius', who are not all strictly 'concordant with the usual biological variation', but who become 'somewhat more intelligible to us if we bear in mind their affinity to variations'. In *Sexual Inversion* Ellis cited Symonds's comparison of inversion to colour-blindness, and while he conceded that 'such comparison is reasonable', he insisted it was better to refer to homosexuality as an analogue of colour-hearing.[33] This condition could be considered an 'abnormality,' but not a 'diseased' condition, a claim he supported by reiterating the points of the letter quoted above.[34] Inversion as such was a mutation, an 'anomaly' which after all could be regarded as a sort of talent, so that it was possible to assume that the invert was not only an acceptable member of society, but a gifted one.[35]

With this in mind, it becomes clear why, at the very beginning of *Sexual Inversion*, Ellis stressed that none of the individuals whose case-histories he published had ever been charged with misdemeanours or were degenerate, and that the subjects of his research belonged to the most cultivated circles of the community. Indeed, Ellis took great care to include no subjects from asylums or prisons; he was concerned with 'individuals who live in freedom, some of them suffering intensely from their abnormal organisation, but otherwise ordinary members of society'.[36] His originality resided in his decision to publish case-studies of 'normal' individuals. Like Krafft-Ebing, Ellis was convinced of the hereditary basis of inversion, but differed in his attitude towards inverts so that while for Krafft-Ebing they were virtually pariahs, 'step-children of nature', Ellis accorded them some dignity as complete human beings.[37]

Both Ellis and Symonds appealed to the authority of science as the only viable path that might lead to changing the law against 'sodomy' in late Victorian times. Ellis suggested they 'adopt a rather austere style in this book, avoiding so far as possible a literary or artistic attitude towards the question – appealing to the reason rather than to the emotions'.[38] To be effective, their work had to be rational and scientific in tone. Just as for Ellis's generation science represented liberation from primitive forms of social organisation and repressive forms of government, so in Ellis's own view a scientific attitude was a form of emancipation from the mental restrictions that had been applied to sexuality in the past.

As Chris Nottingham has suggested, Ellis 'used science in the sense grafted onto British culture by Jeremy Bentham and his followers: as a body of applicable knowledge which must by reason and of necessity become the basis for understanding society and developing the agenda of modern government'.[39] Science appeared to many of Ellis's contemporaries as the practical remedy to false beliefs and ignorance. Science thus understood offered a guarantee for progress to the governing elite. Ellis's conviction that science could play a critical role in society and politics was equivalent to the faith in science that animated Italian positivist doctors from Lombroso to Penta.[40] Although Lombroso supported the view that sexual inversion was a disease, he believed that science could improve cultural attitudes towards sexuality, a point shared by Ellis. Italian sexologists like Lombroso used science to counter the cultural influence of the Roman Catholic Church in sexual matters, but since the approval of the Zanardelli penal code in 1889, which decriminalised private male same-sex acts between consenting adults, there was no reason to appeal to science to change the law as there was in Britain. Here the law remained something that sexologists had to confront – as in the case of Ellis – or accept – as in the case of the majority of physicians.

Symonds also recognised the symbolic prestige of scientific authority. He knew that if his ideas were to be considered sympathetically at all, he must collaborate with a respectable figure with a medical background. Like Ellis, he believed that an emotional-literary approach to the question of sexual inversion would have appeared 'eccentric' and ineffective.[41] Symonds had initially assented to the medicalisation of sexuality, and he argued within the bounds of entrenched assumptions about physical, mental, and moral health. He proposed that homosexuality was natural, that there were many noble souls amongst inverts, and he recalled the great achievements of ancient Greek culture to counter the idea that homosexuality was a symptom of moral decadence.

Sexual Inversion and the Italian sources

Ellis and Symonds read and extensively discussed current writers who engaged with the subject of same-sex desires, ranging from the strictly scientific Lombroso and Moll, to avant-garde poets such as Paul Verlaine and Walt Whitman. The correspondence between Ellis and Symonds shows that both of them had links with Italy: Symonds had published a work on the Italian Renaissance, studied Italian literature, visited Venice regularly, and had travelled all over the country, where he had the chance to observe different social conditions that seemed to stimulate homosexuality.[42] There is no doubt that Ellis kept abreast of new developments in Italian science: he had introduced criminal anthropology in general, and Lombroso's ideas in particular, to the British scientific community in the 1890s.[43] In 1890, Ellis launched a new section within the *Journal of Mental Science* entitled 'Criminal Anthropology', in which the 'Italian School' featured on a regular basis. He translated Lombroso's *The Man of Genius* (1891), which was published in the Contemporary Science Series, and he further contributed to the field of criminal anthropology with his own *The Criminal* (1890). Despite his admiration for Lombroso, Ellis criticised him on the grounds of his 'indiscriminate procedure in collecting data':

> Lombroso's work is by no means free from faults. His style is abrupt; he is too impetuous, arriving too rapidly at conclusions, lacking in critical faculty and in balance. Thus he was led at the beginning to over-estimate the atavistic element in the criminal, and at a later date he has pressed too strongly the epileptic affinities of crime.[44]

Nevertheless, Lombroso is still quoted more often than any other authority in *The Criminal*, so it is probably fair to say that he is the book's chief influence. Ellis wrote to Symonds:

> Nothing too severe of Lombroso's lack of critical judgement and historical insight and accuracy; one forgives it all because he has opened up so many new lines of investigation and set up so many good men to work.[45]

When attending the International Medical Congress in Rome in 1894, Ellis had the chance to meet Lombroso personally; they first met in Ellis's hotel room, and on the occasion Lombroso appointed his British colleague to the virtually honorary post of secretary of the Congress's Psychiatric Section.[46] Then, in 1901, when Lombroso's methodological shortcomings had become all too evident to the scientific community, Ellis continued to praise him as a 'Columbus' of science.[47]

Aware that Symonds was working on Michelangelo's life, Ellis suggested that he read Lombroso's notes on Michelangelo's 'psychic anomaly' and sent him the article in question.[48] Symonds was highly critical of Lombroso's work and judged the article 'worthless', saying that Lombroso was 'entirely untrustworthy from the historical-critical point of view'. In a letter to Ellis he wrote that Lombroso 'stuffs into his book at second hand whatever suits his purpose',[49] not realising that in their efforts to describe sexual inversion, Ellis and himself were doing much the same thing. For instance, in passages where he wanted to reinforce his opinion, Ellis used the views of different scientists without discussing methodological issues. Collecting data was an integral part of the scientific model of the time: British scientists were particularly keen on their natural history, as Italians were proud of their positivism. However different, natural history and positivism shared a common methodology: the base of their epistemological edifices was the collection of 'facts'. Thus, in Ellis's and Symonds's co-authored work, anthropological, sociological, medical observations, everything was grist to the mill.

In his private correspondence with Ellis, Symonds had been particularly harsh towards Lombroso, but he made a point of moderating his public criticism. In *A Problem in Modern Ethics*, Symonds had listed Lombroso alongside Moreau de Tours, Benjamin Tarnowski, and Krafft-Ebing as the author of one of the four 'most recent, authoritative contributions' to medical research on same-sex desires and ranked his work 'upon the whole most sensible studies'.[50] On the one hand, Symonds acknowledged that Lombroso had precipitated a 'revolution of ideas, which gives new meaning to the words sin and conscience, which removes moral responsibility, and which substitutes the anthropologist and the physician for the judge and jury'.[51] On the other hand, he criticised Lombroso for making no distinction between 'innate crime' (same-sex desires) and 'moral insanity', and indistinctly linking them to epilepsy:

> This introduces a certain confusion and incoherence into his speculative system; for he frankly admits that he has only gradually and tardily been led to recognise the identity of what is called crime and what is called moral insanity. [...] Criminal atavism might be defined as the sporadic reversion to savagery in certain individuals. It has nothing logically to connect it with distortion or disease – unless we assume that all our savage ancestors were malformed or diseased, and that the Greeks, in whom one form of Lombroso's criminal atavism becomes established, were as a nation morally insane.[52]

According to Symonds, Lombroso's arguments led to the identification of same-sex desires with disease. His criticism was based on the assumption that ancient Greeks could not be a population of sick men because the art of ancient Greece had realised the ideal of health and sanity.[53] As he did in his letters to Ellis, again Symonds adopted historical 'facts' to disprove medical 'facts'.

Ellis wrote to Symonds about another Italian scientist, Silvio Venturi, who had published a treatise on the subject of sexual inversion, *Degenerazioni psico-sessuali* (1892).[54] Symonds found Venturi's book 'tedious' and 'stupid', aside from his view that the absorption of male semen through the mucous membranes of women might explain the flourishing physical state of young brides.[55] At that time Symonds was speculating with Carpenter on the possibility of transferring one man's virility into another through the absorption of semen. He had no doubt that the absorption of the semen generates 'a real modification in the physique' of the person who absorbs it, a fact that Symonds believed Venturi had proved.[56]

While Symonds was sceptical about Lombroso and Venturi, he showed a certain sympathy towards another Italian, Mantegazza, the 'distinguished anthropologist' who had paid particular attention to the physiology and psychology of what he called '*pervertimenti dell'amore*' [perversions of love].[57] This was due to Mantegazza's observation that 'pederasty', or 'infamous abomination', was not confined to the vilest classes of society, but might also be found in the 'highest spheres of wealth and intelligence'.[58] Mantegazza's 'striking passages' indeed supported Symonds's view that sexual inversion was present even in 'quite intelligent, talented, and highly-placed persons, of excellent and even noble character'.[59] According to Symonds, Mantegazza's anatomical explanation of same-sex desires was also worthy of attention and ought to be tested by empirical research.[60] As noted in Chapter 3, Mantegazza had explained that 'anomalous passions' occurred when the nerves responsible for pleasurable sensations were displaced to the rectum from the genital organs, where they are to be found in normal cases.[61]

Not only did Ellis and Symonds draw attention to a number of Italian scientists; they also observed that some social conditions in Italy seemed to prove that sexual inversion could be a matter of custom. What Symonds had experienced in Italy led him to attach 'very great influence to customs and example' regarding the phenomenon of homosexuality.[62] In his social observations Symonds also distinguished between northern and southern regions, and reported to Ellis that they regarded themselves as essentially different from each other:

> A male prostitute whom I once saw at Naples told me that he was Venetian, but he had come to Naples because at Venice he only found custom with Englishmen, Swedes and Russians whereas at Naples he could live in excellent Italian society and be abundantly supported.[63]

Ellis was interested in Symonds's cultural observations:

> It would be worth while to give some attention to the question of homosexuality in southern Italy. Would not the same be true, though in a less degree of the Italians generally? And if so, one wonders if there is any connection between this fact and the marked inferiority of Italian women to Italian men which experimental psychology seems to show.[64]

Ellis elaborated on these observations in the final draft of *Sexual Inversion*. He stated that there was a special 'proclivity' towards homosexuality amongst certain races and in certain regions. On the whole, this tendency seemed more common in the hotter regions of the globe. In Europe it was probably best illustrated by the case of southern Italy – which was completely unlike northern Italy in this respect. Moreover, according to Ellis, when it came to discussing their sexual practices, Italians were generally franker than men of the northern 'race'. Ellis also suggested that it was possible to explain the prevalence of homosexuality in southern Italy in connection to 'Greek influence and Greek blood'.[65] In northern countries like England, Ellis thought, homosexual phenomena could not be present in the same way as in southern Italy or ancient Greece because all traditions and all moral ideals, as well as the law, were firmly opposed to all public expressions of homosexual passion. It required strength to counter a compact social force that coerced the individual into the path of heterosexual love on all sides. In ancient Greece, as in southern Italy, it was possible to identify a number of cases in which there existed an 'organic and racial disposition to homosexuality', but the state of social feeling also induced a large proportion of the ordinary population to adopt homosexuality as a 'fashion'.[66] Thus Ellis, at least apparently, came to think of homosexuality as part of the Italian cultural tradition.

The idea that hotter regions were more inclined to pederasty than colder regions came from Richard Burton, whose ten-volume translation of *The Arabian Nights* was followed by a 'Terminal Essay' addressing a number of interpretative issues, amongst them 'Pederasty', examined in Section D.[67] His essay represents one of the earliest modern efforts to collect and make known both cross-cultural and historical information about same-sex practices.[68] Burton regarded the 'vice' of pederasty as 'geographical and climatic, not racial'. According to Burton there existed a 'Sotadic Zone' which included southern France, the Iberian Peninsula, Italy and Greece, plus the coastal regions of Africa, from Morocco to Egypt. Within this area pederasty was considered:

> popular and endemic, held at the worst to be a mere peccadillo, whilst the races to the North and South of the limits here defined practice it only sporadically, amid the opprobrium of their fellows, who, as a rule, are physically incapable of performing the operations, and look upon it with the liveliest disgust. [...] Within the Sotadic Zone there is a blending of the masculine and feminine temperament, a crasis which elsewhere occurs only sporadically.[69]

In *A Problem in Modern Ethics*, Symonds briefly expounded Burton's theory, noting that some medical writers held this hypothesis to be empirically sound, but he argued that the phenomenon of sexual inversion could not be regarded as geographical and climatic because it was registered across the globe. The 'problem', according to Symonds, was social. He also recognised that Burton did not consider the 'vice' of pederasty as being against nature, but rather a natural phenomenon.[70] What Symonds had not grasped was that Burton collected data from literature just as Ellis and Symonds themselves had, and as Lombroso was doing in Italy at the same time. The same data were used over and over in different medical accounts.

One must be cautious when suggesting a strong Italian influence on Ellis and Symonds, for while they most certainly drew from existing Italian sources, they also expanded on them substantially. Their collection-based methodology makes it difficult to speak of influence in general. The fact that Ellis was engaged in an encyclopaedic compilation is often overlooked: much of his study consists of more or less direct quotations from his sources – which he often does not acknowledge as such, and so pass as his own opinions. To take only one example, the entire section of *Sexual Inversion* concerning the link between female prostitution and lesbianism was based on Lombroso, but Ellis mentions the Italian anthropologist only briefly at the end.[71] Yet the final version of *Sexual Inversion* was in part a result of Ellis's and Symonds's engagement with Italian sexology which is plainly disclosed by their correspondence. Ultimately, the fact that both Ellis and Symonds, who to a certain extent adopted different approaches to sexual inversion, knew and engaged with Italian research shows the popularity of Italian research in Britain.

Ellis's use of Italian sources did not end with the publication of the first English edition of *Sexual Inversion*. In the second edition, published in 1901, Ellis added an appendix titled 'The School-Friendships of Girls'. The section was explicitly based on the Italian study of the phenomenon of *fiamme* by Obici and Marchesini, which Ellis expanded with new material based on some American research.[72] Obici and Marchesini published their book in 1898, and Ellis referred to it and elaborated on their results in 1901. Ellis had only had three years to learn about, get hold of, and digest a lengthy study in a foreign language. This goes to show that he had been reading Italian medical and anthropological research since the 1890s, and continued to be up-to-date with new publications through the first decades of the new century. He could easily have relied on British medical literature dealing with female same-sex desires as observed in schools, which, as has been shown in Chapter 4, was available on his side of the Channel. Remarkably, however, Ellis expanded on Italian sources, which suggests that he actively engaged with, and trusted, Italian sexological research.

The Bedborough case

Symonds died of tuberculosis in Rome in April 1893, but Ellis persevered with the writing of *Sexual Inversion*. By the time the book was in its final draft, the Wilde trials and associated scandals had made its publication almost impossible. Ellis decided to have his manuscript translated into German by his friend Kurella, and at the end of 1896 the book was published in Leipzig as a part of the Library of Social Science that Kurella himself edited.[73]

The *Journal of Mental Science* reviewed the German edition but not too enthusiastically, despite the fact that Ellis had worked regularly for the journal in the preceding years. According to the reviewer, the study contained 'some interesting observations' but did not 'throw special light on inversion as a congenital sexual abnormality'. The *Journal of Mental Science* judged the case-histories to be no 'more conclusive than those already published by others'.[74]

In the meantime, individual chapters of *Sexual Inversion* were published abroad without raising any particular outcry: in America in *The Medico-Legal Journal* and in *Alienist and Neurologist*, and in Italy in *Archivio delle psicopatie sessuali*.[75] After the book's favourable reception in Germany, Ellis decided to pursue the possibility of a British publication, and he submitted the book to the respected medical publishers Williams and Norgate, which in turn forwarded it to Hack Tuke for review. Tuke, who was editor of the *Journal of Mental Science*, advised against publication not because he had any intrinsic objection to the material, but because he felt that the distribution of the book could never be confined to specialists.[76] Williams and Norgate subsequently rejected the manuscript.

Ellis's friend, F. H. Perry Coste, directed him to Dr Roland de Villiers, the German director of an obscure and only marginally reputable publishing house called Watford University Press, which was to be responsible for the first English edition of *Sexual Inversion*, in April 1897.[77] However, Symonds's literary executor, Horatio Brown – who originally gave the publication his blessing – asked Ellis to remove Symonds's name from the title page to protect the good name of the family. He then bought the entire edition from de Villiers and destroyed it. The book was published again at the end of October 1897 without Symonds's name. In the subsequent six months *Sexual Inversion* created scarcely a ripple amongst British medical readers.

Amongst de Villiers's publishing projects was *The Adult*, a monthly magazine issued by the London-based Legitimation League, one of a number of organisations that advocated the 'New Sexuality' in both Britain and the United States during the 1890s.[78] In May 1898, a Scotland Yard agent called John Sweeney attended some Legitimation League meetings posing as an anarchist and asked George Bedborough, who was secretary of the League, to sell him a copy of *Sexual Inversion*. Bedborough indeed used the front room of his home in John Street to sell *The Adult* and several other publications, including a few copies of Ellis's book. On 27 May 1898 Bedborough was arrested and brought before London's Central Criminal Court for selling a 'wicked, lewd, impure, scandalous, and obscene libel, *Studies in the Psychology of Sex: Sexual Inversion*, written by Mr. Havelock Ellis'.[79]

A Free Press Defence Committee was formed by such luminaries as George Bernard Shaw, Frank Harris, George Moore, and the socialist H. M. Hyndman, amongst others. Bedborough's committee-appointed counsel was the prominent Horace Avory, who had represented the Crown at the second and third trials of Oscar Wilde; he went to great lengths to have the Bedborough case transferred from the Criminal Court to the High Court of Justice, arguing that Ellis's work was not 'obscene' but scientific and hence unobjectionable if published for persons with a scientific interest in the subject. Bedborough's trial ensued, and lasted from May to 31 October 1898, when he eventually struck a deal with the police, claiming to have sold *Sexual Inversion* in innocence and laying all the blame on de Villiers.[80]

During the trial *Queen v. Bedborough* the Recorder, Sir Charles Hall, informed Bedborough that although he might have been 'gulled' into believing that Ellis's work was scientific, 'it is impossible for anybody with a head on his shoulder to open the book without seeing that it is a pretence and a sham, and that it is merely entered into for the purpose of selling this filthy publication'.[81] This deposition sums the spirit of the trial, which never addressed the issue of whether *Sexual Inversion* was obscene, but was concerned only with who was responsible for its publication: its licentious nature was assumed from the onset and there was no chance to present an argument to the contrary. Consequently, *Sexual Inversion* was banned as an obscene work. The Bedborough trial and all debates that it generated at the time are interesting and overlooked historical evidence that illustrate the conservative and prurient attitude of the official British scientific community in dealing with sexual matters, and the extent to which Ellis's work was radical and groundbreaking for its time.[82]

The charges had been pressed against Bedborough and not Ellis, who was thus prevented from defending his work in the trial. In order to vindicate *Sexual Inversion* from the charges of obscenity, Ellis published *A Note on the Bedborough Trial* (1898) and wrote in *The Lancet*. For Ellis, the whole course of the case was 'profoundly unsatisfactory', because despite receiving spontaneous support and encouragement from colleagues in Belgium, Britain, Germany, France, Italy, and the United States,[83] not a single prominent British scientific or medical figure came forward publicly for the defence in the course of the legal proceedings. Ellis regretted that his own country was 'almost alone in refusing [him] the condition of reasonable intellectual freedom':

> In this country it is a sufficiently hard task for any student to deal with the problem of sex, even under the most favourable circumstances. He has entered a field which is largely given over to faddists and fanatics, to ill-regulated minds of every sort. He must, at the same time, be prepared to find that the would-be sagacity of imbeciles counts him the victim of any perversion he may investigate. Even from well-balanced and rational persons he must at first meet with a certain amount of distrust and opposition.[84]

Foreign physicians seemed surprised by the outcome of the Bedborough trial, which in essence represented a challenge to Ellis's scientific credibility.

The American physician James Kiernan wrote to Ellis that regarding sexual matters 'the cant element is stronger in Great Britain than in the United States'.[85] Kurella observed that for Continental scientists 'such a proceeding is altogether incomprehensible. What would become of science and of its practical applications if the pathology of the sexual life were put on the Index?'[86] When Penta remarked that in countries like Germany and Italy works such as Ellis's found their most numerous readers amongst magistrates themselves, he was criticising British law for stopping sexological works from reaching the same level of development as they had on the Continent. He was also making a point about the peculiar situation of Britain, which set it aside from other European countries:[87]

> Having with Darwin overthrown hieratic superstition and scientific prejudices, can England be now frightened of your book, which only opens another page in the unprejudiced, serene, and objective study of Nature? Such moral hypocrisy as this does wrong to the land of scientific positivism, a land which is certainly one of the most civilised countries in the world.[88]

English 'hypocrisy' was criticised not only by foreign scientists, but also by a few English commentators. For example, Conolly Norman wrote to Ellis that the 'vagaries of the law are of course unaccountable, and British hypocrisy which makes us the laughing-stock of the Continent is a great force in things insular, but I must say it seems to me preposterous to interfere with the writing or publication of your work'.[89]

In Britain, the medical establishment's response to *Sexual Inversion*, however respectful, did not accept its conclusion, namely, that sexual inverts could be normal individuals. In his *A Note on the Bedborough Trial*, Ellis praised his professional audience for accepting his work 'in the serious spirit in which it was put forward' and for acknowledging the 'scientific tone and temper' in which he had executed it.[90] However, these cautious remarks masked the important way in which the book had failed: the 'stigmata of degeneration' which Ellis had sought to remove persisted tenaciously.

This is exemplified by the two foremost British medical journals' reaction to the trial. Five days after Bedborough pleaded guilty, the prestigious *British Medical Journal* informed its readers that it had examined Ellis's work and disagreed with the Recorder's view on the unscientific character of *Sexual Inversion*. According to the *British Medical Journal*, the book was scientific and contained nothing 'to pander to the prurient mind'. They stressed that the book treated 'a subject which is to most people extremely disagreeable [...] but is one of those unpleasant matters with which members of the medical profession should have some acquaintance'.[91] Then they went on to quote a letter written by Norman that Ellis himself had submitted to the *British Medical Journal*: 'In its relation to insanity, to degeneration, and to the neurotic

state, the subject of sexual inversion has much medical interest. It is, therefore, a matter which must be discussed and written [about].' It thus emerges
that Ellis was defended on account of his being a member of the medical
profession, which sanctioned his study of sexual perversion. Norman, however, agreed with the editors of the *British Medical Journal*, who took it for
granted that sexual inversion was linked to degeneration and sickness, and
consequently rejected Ellis's conclusions about sexual inverts.

Two weeks later, *The Lancet* put forward an even bolder judgement. In a
defensive editorial attempting to explain why it had not reviewed Ellis's work
before, it admitted that the book was concerned with sexual inversion, an
aspect of the psychology of sex that did indeed exist, and that Ellis 'seem[ed]
to have been badly treated in the [Bedborough] matter'.[92] As a matter of
fact the trial had 'clos[ed] his mouth and prevent[ed] him from making any
defence'. However, *The Lancet*'s editors emphasised that the topic of sexual
inversion 'touches the very lowest depths to which humanity has fallen', and
although it certainly was an adequate subject for medical discussion, they
admonished the debate should be limited to qualified medical researchers.
'Scientific' though the book was, Ellis's choice to publish for the general
public was sharply criticised.[93] *The Lancet*'s editors made it especially clear
that they did not agree with Ellis's conclusion:

> [Ellis] has failed to convince us on these points [referring to the mode
> of publication]; and his historical references and the 'human documents'
> with which he has been furnished will, we think, fail equally to convince
> medical men that homo-sexuality is anything else than an acquired and
> depraved manifestation of the sexual passion; but, be that as it may, it
> is especially important that such matters should not be discussed by the
> man in the street, not to mention the boy and girl in the street.[94]

If the righteous indignation over the plight of the book as a scientific contribution was not accompanied by comparable indignation over the plight
of the invert, it was because the medical class felt under attack by the law.[95]
Medicine and jurisprudence were two opposing fields competing for the control of sexuality during the nineteenth century. Physicians defended Ellis as
a member of their class, and in doing so they were defending their privilege
to deal with deviant sexuality and with its moral implications. As for Ellis's
views on sexual inversion, no member of the medical community was prepared to endorse them, not even a member of the *Journal of Mental Science* –
for which Ellis worked for many years. The reviewer of *Sexual Inversion*
expressed his qualms:

> The whole crux of the question seems to us to be whether this is or is not
> to be fairly called congenital sexual inversion. Is it 'a sexual desire directed
> by a constitutional abnormality towards a person of the same sex,' or

is it so directed by suggestion, by vicious education, or other external agencies?[96]

In the same issue, the *Journal of Mental Science* gave a somewhat more sympathetic review of another work on sexual inversion. In his discussion of Raffalovich's *Uranisme et Unisexualité*, the reviewer stated that it was not yet proven that sexual inversion was an inborn condition, and further:

> That a certain or uncertain proportion should adopt the evil is perhaps inevitable, but the fact that this proportion varies according to climate, race, education, habits, moral atmosphere, and a hundred other influences tends to show that in no individual case is the worst choice inevitable.[97]

The editorial of the January 1899 issue of the *Journal of Mental Science* did not dispute the status of *Sexual Inversion* as a 'scientific study', but they took the trial as an appropriate occasion for reiterating their scepticism regarding viewing inversion as a congenital condition, and they insisted that only the scientific community was fit to deal with such an 'immoral' subject, warning that care should be taken concerning the 'mode of production' and the sale of such works:

> Observers in the field of sexual depravity will bear in mind the lessons of the Bedborough case. The perpetual repetition of the theory of Ulrichs, that some people are naturally possessed of a perverse sexual feeling, is tiresome. We are never favoured with an atom of proof, and writers seem to imagine that they advance their arguments by heaping up unsavoury details. [...] We are sorry for Mr. Ellis [...] but we are of the opinion that he should have exercised more care with regard to the mode of production and sale of his volume, in its English form.[98]

The *Journal of Mental Science* warned physicians not to provide details about sexual inversion because if certain accounts fell in the hands of the 'vulgar', they could be treated as 'very dirty stories', and as such, were 'liable to become part of the stock-in-trade of the pornographic bookseller'.[99] According to the *Journal of Medical Science*, it was only Ellis's 'well known reputation as a criminal anthropologist' that was a 'sufficient guarantee' of his motives in writing about inversion.[100]

Almost the same attitude is shown by those physicians who wrote to Ellis, all of whom sympathised with him and agreed that the trial enacted an 'abuse of the form of law' against a doctor. Notwithstanding loyalties to their colleague, some of these physicians failed to make their view on sexual inversion clear; amongst them Norman, Urquhart, Mercier, and Clouston.[101] Rayner[102] spoke openly of his disagreement with Ellis's viewpoint, but was against the legal interference and the fact that his book was banned. Of all the correspondents who expressed their support for Ellis, only Goodall and Savage showed some sympathy with the ideas put forward in *Sexual Inversion*.[103]

The medical community was outwardly united in defending Ellis's reputation, but there was conflict amidst its ranks on what the view on sexual inversion should be. The official position was that sexual inversion was a sexual vice. This is proven, for example, by the fact that the *British Medical Journal* was considerably less charitable when reviewing Krafft-Ebing's *Psychopathia Sexualis*, saying it belongs to 'the most repulsive group of books of which it is the type'.[104] In response to Krafft-Ebing's hope that his book would bring comfort to those who read it, the editors wrote: 'We should prefer that the book should convey solace by being put to the most ignominious use to which paper can be applied.'[105] In the same issue there was an equally scathing review of volumes one and two of Ellis's *Studies in the Psychology of Sex*.[106] There exists a telling contrast between those who had written about *Sexual Inversion* in 1898 and more recent reviewers, who strongly disagreed with Ellis. Now that the social pressure around the Bedborough trial had cleared, the medical community was free to openly dismiss Ellis's work as disgusting and not quite scientific.

According to the historian Paul E. Stepansky, on the basis of the Bedborough case it is reasonable to assume that for physicians, as for others, the defence of heterosexual behaviour at the turn of the century had political overtones relating to Britain's *moral fibre*, and by implication, the role Britain could and should assume in world affairs. The growth of German sea power, the unfavourable reaction to the Boer Wars, and the possibility of a future European conflict, all prompted a moral reassessment of Britain's strength and served to foster a close identification between moral and martial strength. Moreover, the reaction to *Sexual Inversion* was tied to the social status of the medical profession itself. As leaders of social opinion, physicians accepted the Victorian belief that privileged social status must be justified by superior morality, and in all likelihood, they saw Ellis's depiction of the inverts as incompatible with the standard of sexual normality on which their own social standing was predicated.[107]

These are certainly parts of the story, but the pivotal interest of the Bedborough trial is that it epitomised the struggle between law and medicine in dealing with homosexuality and sexual matters in general. Faced with legal constraints, physicians defended their own right to speak about sexuality, and as a matter of fact they had closed ranks to support Ellis as a man of science in 1898, even if they did not agree with his position on inversion, or they deemed his choice of publisher highly objectionable. Physicians claimed that they had the prerogative of dealing with issues surrounding sexual perversion on the grounds that they were qualified to handle them from a scientific viewpoint, with expert and disinterested detachment.

Subsequently, medical men dismissed Ellis's arguments that sexual inversion was an inborn, natural phenomenon, an abnormality but not a diseased condition. They believed that Ellis's arguments were not supported by medical evidence and, further, that his findings were incompatible with the existing clinical literature, which asserted the relationship between inversion, insanity, and degeneration.[108] On the other hand, the Bedborough trial shows that sexology, as synthesised by Ellis's experience, could prove to be a radical enterprise, because given its purposes and area of enquiry, the British medical profession was inclined to find it altogether disagreeable; conversely, some inverts like Symonds and Carpenter saw in the pursuit of sexological research an opportunity to improve their social status. It is true that Ellis worked within the margins of science, but he did so in the hope of changing existing scientific assumptions.

The Bedborough case also amounts to clear evidence that, in Britain, disseminating knowledge about sex was a difficult – even perilous – undertaking. With the publication of *Sexual Inversion*, Ellis was saying that sexuality should be discussed while opening up new ways of doing so. This was the crucial meaning of Ellis's work, which was along the same lines as his colleagues from the Continent. Moreover, if the comments made by foreign observers about this case are to be relied on, British prudery was stronger than that of the Germans, the French, or the Italians. Compared with some of the sexological works published in Italy, *Sexual Inversion* was very prudish indeed. Yet it was not just a question of British prudishness and Italian openness to sexuality. I have shown that Penta's activities were met with considerable opposition, how Italian sexological research still had titillatory qualities even after the *Archivio delle psicopatie sessuali* had folded, and how there was nonetheless no major legal opposition.[109] But it was not a coincidence that both Ellis and Penta faced problems in their careers due to their stance on deviant sexuality. Even in Italy it was difficult to pursue sexological studies if one displayed an attitude that challenged contemporary conventions.

Psychology of sex: female sexual inversion

Unlike Symonds, who had previously ignored the phenomenon of female inversion, Ellis pointed out to him in an 1893 letter that he had 'been impressed by the frequency of homosexuality – both congenital and acquired – in women'.[110] Despite the fact that psychiatrists were curious about female sexual inversion, they usually complained about the difficulty of gathering case-histories. Krafft-Ebing's seventh edition of *Psychopathia Sexualis* (1892) contained forty-one male case-studies, but only eleven female case-studies.[111] For the 1897 edition of *Sexual Inversion* Ellis was able to collect twenty-seven case-histories of men, but just four of women; he eventually added two other female cases to the later edition.[112] Ellis's wife had

homosexual relationships and was actively involved in feminist and socialist groups; it was in fact through Edith Lees's network of friends that Ellis
obtained his female case-histories.[113]

Independent women

Ellis's particular engagement with the subjects of female sexuality and sexual
inversion has led to debates amongst historians. At the root of the matter is
the feeling that, by adopting a biological approach to the question of sexual
differentiation, Ellis embraced a staunchly conservative ideological stance.
For these critics, Ellis's suspect handling of sexual differentiation was only
compounded by his eulogies to motherhood, a tendency which is alleged
to have become more marked as eugenics came to play a larger part in his
analyses and prescriptions.

Thus, far from being hailed as a prophet of sexual emancipation and as
an ally of women, Ellis is presented as a reactionary figure. As a result, some
researchers have unduly emphasised Ellis's failure to challenge stereotypes
of masculinity and femininity in relation to both sexuality and the social
roles of men and women. Further, this line of criticism has portrayed Ellis's
sexological model of sexuality as 'patriarchal' and hence instrumental to the
domination of men over women, and responsible for the consolidation of
the 'stigma' of lesbians as masculine creatures.[114]

Ellis was, without a trace of a doubt, a biological essentialist who took the
existence of scientific 'facts' for granted, but he also argued that society had
built cultural differences and unnaturally exaggerated superstructures upon
these 'facts'. The underlying premise of *Man and Woman* (1894) is the fundamental 'equivalence' of the sexes. The book went through several editions,
but the general concept of sexual equivalence remained unchanged: men
are neither better nor worse than women, but the two sexes are not equal in
the sense of identity. Ellis insisted that there was no equality in the world of
living things. It made no sense, he observed, to speak about inferiority and
superiority amongst men and women, because the sexes were basically, as
Ellis puts it, 'equivalent'.[115]

In order to support his theory of equivalence, Ellis presented an historical
and anthropological analysis: he wrote that the division of labour amongst
primitive people was sharply drawn, independently from race and climate.
Tasks requiring power and intermittent energy with corresponding periods
of rest generally fell to man; the care of the home and children – which
required less energy but constant exertion – fell to women. It was during the
early industrial age that men began to specialise in crafts that had formerly
been the realm of women, leaving them only with household chores. The
eighteenth century, Ellis argued, had entailed a call to reason and the casting
off of prejudice, which contributed to a redefinition of the social status of
women. At the same time, there was an economic revolution that brought
women into labour 'independently' of men. Work came to be organised in

urban centres. Machinery allowed men and women to work side by side, and they began studying in the same schools. These changes meant that the cultural – artificial – differences between men and women tended to fade, as did signs of male superiority.[116]

Ellis noted that the transition was still in progress and that only after the cultural differences were removed could biological differences be properly considered. These, Ellis thought, were natural differences:[117] just as women's primary sexual characters and reproductive functions were unlike men's, their psychological characteristics could never be expected to be man-like.[118] However, in the conclusion to his work, Ellis was quite sceptical about the chances of clearly identifying what those natural differences between men and women were:

> It is abundantly evident that we have not reached the end proposed at the outset. We have not succeeded in determining the radical and essential characters of men and women uninfluenced by external modifying conditions. [...] We have to recognise that our present knowledge of men and women cannot tell us what they might or what they ought to be, but what they actually are, under the conditions of civilisation. By showing us that under varying conditions men and women are, within certain limits, indefinitely modifiable, a precise knowledge of the actual facts of life [...] forbids us to dogmatise rigidly concerning the respective spheres of men and women.[119]

Bio-medical studies tended to presuppose that sex characteristics were mutable. This led Ellis to conclude in *Man and Woman* that it had been impossible to determine 'the radical and essential characters of men and women uninfluenced by external modifying conditions'.[120] There was, in short, constant interaction between nature and culture.

A further complication of Ellis's biological essentialism is his use of a late nineteenth-century embryological hypothesis. Ellis thought that the foetus was originally bisexual and that traces of this phenomenon were manifested in adult life, for example in male nipples and in the clitoris – that 'rudimentary penis'.[121] It has already been shown that at the end of the nineteenth century it was suggested that in the earliest period of intra-uterine life human embryos were sexually undifferentiated.[122] This hypothesis allowed scientists to think that the original anatomical bisexuality of the embryo could leave a physiological trace in the human being, and that these traces could be the starting point of sexual inversion. It was possible to conclude, as Ellis did, that homosexuality was a result of the original bisexuality present in all human beings.[123] Sexologists like Ellis and, as has been shown, Lombroso, used embryology to argue in favour of a normal universal bisexual disposition in the sexual sphere. Within this construction, homosexuality was seen as a natural outcome for some individuals.

In addition to the ramifications of the notion of mutable sexual characteristics illustrated above, Ellis was fully cognisant of gradations between
the ideal masculine and feminine types, although he strove to maintain
a clear distinction between them and considered sexual behaviours to be
gendered.[124] In an effort to distinguish between the two ideal types, he identified active sexual attributes as masculine and passive ones as feminine.
Thus men were expected to be dominant while women were constitutionally
modest. Lesbianism was a sign that a woman's masculine attributes over-
powered the feminine aspects of her constitution and inclinations.[125] Some
historians have noted that Ellis stressed the absence of effeminate characteristics in male sexual inverts, while emphasising the masculine traits in
female inverts, which is why historians have frequently charged him with
having stigmatised lesbianism.[126] This is, however, an oversimplification for
two main reasons: first, some male sexual inverts were in fact described as
effeminate, both physically and psychologically.[127] Second, in Ellis's work
'masculine' attributes exceed purely (negative) physical characteristics.

I suggest that in *Sexual Inversion* Ellis's use of the term 'masculine' refers
to sexual autonomy. Ellis adopted a biological model according to which
the male type was active and the female passive. Female sexual inverts in
particular did not fit this model because they did not depend on masculine
proactivity to initiate sexual acts, which is to say that they were altogether
independent from men. The only way to explain female inversion within
Ellis's theory of ideal gender types that are expressed in behaviour and personality traits was to underline the incidence of the 'masculine' element,
that is, the 'active' element. Ellis was clear on this point:

> The actively inverted woman differs from the woman of the class just
> mentioned [...] in one fairly essential character: a more or less distinct
> trace of masculinity. She may not be, what would be called a 'mannish'
> woman, for the latter may imitate men on ground of taste and habit
> unconnected with sexual perversion, while in the inverted woman the
> masculine traits are part of an organic instinct which she by no means
> always wishes to accentuate. The inverted woman's masculine element
> may in the last degree consist only in the fact that she makes advances
> to the woman to whom she is attracted and treats all men in a cool,
> direct manner, which may not exclude comradeship, whether of passion
> or merely a coquetry.[128]

Although lesbians' sexual autonomy was biologically rooted, it did not
make them 'mannish' or monstrous. Ultimately, the main characteristic
these women displayed was independence from men, as his case-histories
illustrate.

Case-studies were central to Ellis's work. Indeed, they were what set
his writing apart from amateur English sexologists such as Carpenter and

Symonds, and made him more acceptable in psychiatric and medical circles. Case-studies condensed all the results of his research, the data upon which his theories were based. As already mentioned much of *Sexual Inversion* consists of more or less direct quotations from other sources, but the theoretical speculations upon case-studies are the book's most original contribution.[129] Because Ellis did not practise medicine full-time himself, his case-histories did not feature his own patients; rather, the histories were provided to him by other doctors or by inverts themselves, who wrote to him to submit information about their own experiences. Some of the case-histories in *Sexual Inversion* are also taken from existing medical literature, which Ellis reworked specifically. As Ivan Crozier has pointed out, Ellis asserted the biological base for the psychology of individuals in his presentation of case-histories.[130]

Ellis's cases of female sexual inversion described women physically and he admitted that the conformation of their bodies could be feminine. Ellis adopted the same strategy towards women as he did towards men: in his case-histories he tried to highlight the fact that these individuals were normal, if not high-minded, and were fit for normal society. In the 1897 edition of *Sexual Inversion*, Case XXVIII, Miss S. is a 'business woman of fine intelligence, prominent in professional and literary circles. Her general health is good.'[131] In Case XXIX, Miss M. is a musician 'of good intelligence, and always stood well in her classes'. The development of the intellectual faculties is 'somewhat uneven' because while she is 'weak in mathematics, she shows remarkable talent for various branches of physical science'.[132] In Case XXX, Miss B. is 'perfectly healthy', and 'her person and manners, though careless, are not conspicuously man-like'.[133] Case XXXI portrays Miss H. in a less positive light without turning her into a kind of monster or madwoman: she is 'energetic, and with a somewhat neurotic temperament', and 'has suffered much from neurasthenia at various periods, but under the appropriate treatment it has slowly diminished'.[134]

In *Sexual Inversion*, the term 'masculine' designates the autonomous personality, the active attitude of these women who of course had to be independent from men; and to Ellis's late Victorian eyes this kind of woman was comprehensively not common or 'normal' in society. When writing to Symonds on the topic, Ellis had observed that 'congenitally inverted women are nearly always to some extent masculine in character'.[135] Such is the key assumption underlying the book's description of the female invert, who is characterised by her 'brusque, energetic movements, the attitude of the arms, the direct speech, the inflections of the voice, the masculine straightforwardness and sense of honour, and especially the attitude towards men, free from any suggestion either of shyness or audacity, will often suggest the underlying psychic abnormality to the keen observer'.[136] Ellis was probably also observing a change in women's customs, which were more evident in sexual inverts than in other women. Ellis's friend, Carpenter, wrote in the late nineteenth century that the 'modern woman is a little more masculine'. 'The growing sense of equality in habits and customs – university studies, art, music, politics, the bicycle, etc. – all these things have brought about a *rapprochement* between the sexes.'[137]

It might seem that the expanded edition of *Sexual Inversion* stresses female sexual inversion as a morbid condition characterised by masculine aspects due to its discussion of more recent medical studies that had reported, for example, the story of Alice Mitchell, an American woman who had killed her female lover by cutting her throat.[138] However, the apparent emphasis on diseased aspects of female homosexuality was only superficially so. As I remarked above, Ellis's cumulative method has to be taken into account, for, like a practitioner of natural history, Ellis collected many different kinds of information about those subjects in whom he was interested. Ethnographic, anthropological, literary, and historical data were lumped together with medical observations. In the revised and expanded third edition of *Sexual Inversion*, Ellis reported cases of female inverts observed by other psychiatrists. Although a violent case such as Alice Mitchell's could not be sidestepped, Ellis started his chapter on female sexual inversion with a lengthy treatment of Sappho rather than with a criminal female sexual invert. Sappho was intended to exemplify Ellis's view that 'inversion is as likely to be accompanied by high intellectual ability in a woman as in a man'.[139] Ellis moved on to recount how female homosexuality was portrayed by French writers such as Zola, and as a current example of a female invert he cited Renée Vivien, a French poet whose verses were, he said, amongst the 'finest in the French language', and who was described as 'very beautiful, very simple and sweet-natured, and highly accomplished in many directions'.[140] Only subsequently did Ellis deal with the famous case of Alice Mitchell and other psychiatric cases.[141]

Feminism and female homosexuality

Ellis's analysis of the relationship between female homosexuality and feminism is more problematic than his description of female inverts in terms of 'masculine' characteristics, which can be partially clarified by situating Ellis's comments alongside medical ideas of the period. Historians have insisted that Ellis was convinced that the women's movement encouraged a 'spurious imitation' of lesbianism – not lesbianism itself – and the acquisition of masculine characteristics.[142] Yet Ellis's use of the phrase 'spurious imitation' was consistent with the general sense of his analysis inasmuch as he explained sexual inversion as a congenital state, all 'acquired' lesbianism was a faint reproduction of the inborn inverted conformation, and hence a 'spurious imitation'. Laura Doan goes still further and argues that Ellis saw female homosexuality and its relation to feminism as a threat to the survival of both civilisation and the race.[143] This is an extreme conclusion to draw: Ellis believed that women fought to attain the freedom, education, work and responsibility they were rightfully entitled to, but that these battles were being waged in an atmosphere that continued to discourage female sexual rights.[144] He wrote: 'marriage is decaying, and while men are allowed freedom, the sexual field of women is becoming restricted to trivial flirtation with the opposite sex and to intimacy with their own sex'.[145] Further, he stated that women were no longer forced to sit by and wait for a husband, and sometimes found love where they worked, amongst other independent women.[146] It did not follow for Ellis that civilisation was doomed, or that lesbianism was directly caused by the women's movement; what Ellis said, rather more plainly, is that at the turn of the century there were fewer opportunities for a woman to express her sexuality in the traditional manner, and more opportunities to find intimacy amongst those of her own sex.

This being said, it is clear that Ellis's discussion of the link between female sexual inversion and 'modern movements' was shrouded in ambiguity. For instance, Ellis wrote that the women's movement carried with it certain 'disadvantages': feminine criminality and insanity were steadily rising, he thought, to match the masculine standard. According to Ellis, the influence of modern movements might not directly cause sexual inversion, though insofar as they promoted 'hereditary neurosis', the development of the 'germ', and a 'spurious imitation' of sexual inversion, they were indirect causal agents.[147] On the whole, involvement in the women's movement was viewed as the result, rather than the cause, of homosexuality. Lesbians possessed characteristics from birth, such as independence and an active mind, which equipped them to be involved in the women's movement.

Ellis did not explain how modern movements promoted the development of hereditary neurosis: he simply took current medical assumptions for granted. In the Victorian period nervous disorders were perceived as a common occurrence. Although hereditary causes were thought to be predominant, the medical profession tended to agree that nervous derangements could also be acquired within a lifetime, with or without hereditary predisposition. Socio-cultural factors could make an inherent predisposition become manifest. The nervous system was understood in terms of nervous economy, and its malfunctioning was attributed to abnormalities in the production, regulation, or exertion of 'nerve force, power, or energy'. Both sexes were vulnerable to the risks of over-straining the nerves, when they wore themselves out by studying too much, when they indulged in 'luxurious living', or when businessmen worried excessively about their work and spent too many hours without pausing to rest or exercise.[148] Nineteenth-century medical discourses had embraced the notion that women's current social roles could be detrimental to their health. As it was, women had to tend to the demands of their domestic responsibilities, disappointment in love, and abusive husbands, and it was feared that their undertaking new jobs and study opportunities led to a higher incidence of nervous symptoms. The new social conditions of women were thus providing more opportunities for nervous neuroses to develop.

Environmental conditions altered the biological make-up of women. In the nineteenth century, the environment, second only to constitutional factors, was seen to affect individuals' health and determine which diseases they would suffer from. This applied not only to geographical and climatic conditions, but also to circumstances created by human culture. Because education, occupation, and lifestyle were equally important in shaping the physiology of individuals, questions of women's social position were inextricably tied up with preoccupations about female health.[149] In consonance with the medical principles of his time, Ellis believed that sexual inversion was inborn, but that environmental factors could intensify the phenomenon, and therefore assumed that new social conditions might well favour female homosexuality, which could therefore develop more easily than in the past.

In his study on female sexuality, Ellis reproduced some of the typical contradictions of late Victorian medical science, but insofar as he was able to go further in contesting contemporary medical writers to argue that sexual inverts were normal individuals, he challenged some of the major prejudices of his own culture. While giving a paper at the British Society for the Study of Psychology of Sex in 1918, he passionately advocated for the 'erotic rights of woman'.[150] Ellis argued that established customs had a 'repressive and unnatural' influence on women's erotic life; he highlighted the importance of 'the play-function of sex', conceived of female masturbation as a normal and even healthy sexual manifestation, argued that individuals were entitled to be sexually satisfied, and even had the right to be actively involved in debates on sexuality.[151] All this must have struck contemporary readers far more forcibly than the biological or environmental medical assumptions, which he took for granted.[152]

Moreover, as Nottingham has rightly pointed out, in the 1930s and especially in the USA, Ellis was afforded an almost totemic status in the development of feminism. Prominent feminists embraced him as an ally and, in some cases, as a mentor as well. His delineation of sexual differentiation became a progressive orthodoxy and was still being cited as authoritative in sociology textbooks widely in use in the 1960s.[153] Some first-wave feminists supported Ellis's emphasis on essential differences between men and women, and his belief in the importance and centrality of motherhood in women's lives. For Ellis, as for many contemporary feminist writers, one of society's major evils was its failure to allow women to fulfil their specific mission. According to this view women were, amongst other things, prevented from exercising their 'natural function' as the civilising influence on society and from maintaining their role as the guardians of order against the disrupting depredations of the male instinct.[154] The Swedish feminist Ellen Key advanced a very similar agenda. The British campaigner for women's rights, Marie Stopes, had a chequered relationship with Ellis, but her most influential book, *Married Love* (1918), included a great deal of respectful quotations from his work. It is well known that the American birth control activist, Margaret Sanger, was influenced by Ellis. Key, Stopes, and Sanger all derived ideas from Ellis's writings. In the contemporary feminist debate, his appeared to be the most reasonable position on how the cause of women might best be advanced.[155]

Schools and homosexuality

Ellis's social observations were not restricted to his analysis of feminism and its relationship to homosexuality; they were also mobilised in his investigation into the nature of relationships in all-girls schools. This study presents some affinities with late nineteenth-century Italian research on the *fiamma* phenomenon. According to Ellis, in girls as in boys, homosexuality made itself apparent during school days, at the age of puberty, and its origin could be either peripheral or central. In the first case, when close to each other in bed, two children may mutually be aroused by touching and kissing more or less unintentionally. This was a 'spurious kind of homosexuality', as it was merely the precocious play caused by the normal instinct, and had no necessary relation to true sexual inversion. In the girl who was congenitally predisposed to homosexuality such feelings would continue to develop, whereas in the majority of cases these explorations would be soon forgotten. In cases in which the source was 'central', on the other hand, the schoolgirl or young woman formed an ardent attachment to another girl, probably somewhat older than herself, often a schoolfellow, and sometimes her schoolmistress, upon whom she lavished an astonishing amount of affection and devotion. The girl who bestowed this wealth of devotion was overwhelmed by emotion, but was often unconscious or ignorant of sexual impulses, and thus would not seek any form of sexual satisfaction. Kissing and the privilege of sleeping with the friend were, however, actively sought, and when the occasion of physical closeness arose, it was not uncommon for the comparatively unresponsive friend to feel a more or less definite sexual emotion. Physical manifestations of the sexual emotion consisted of a 'pudendal turgescence with secretion of mucus and involuntary twitching of the neighbouring muscles'.[156]

Ellis argued that this phenomenon usually went unnoticed, and given the common ignorance of girls about sex matters, the girls themselves might fail to understand what they were doing. In some cases there is an attempt, either instinctive or intentional, to develop the sexual feeling by close embraces and kissing. This rudimentary kind of homosexual relationship is, Ellis believed, more frequent amongst girls than amongst boys, and this for several reasons: more often than not a boy would have some acquaintance with sexual phenomena, and would thus regard such a relationship as 'unmanly'. The girl has a stronger need for affection from another person than a boy has, and while under the existing social conditions she has less opportunities to find an outlet for her sexual emotions than a boy does, social conventions allow a considerable degree of physical intimacy between girls. Thus, social conventions encouraged and cloaked some common manifestations of homosexuality amongst women. According to Ellis, such relationships, especially when formed after school days, are fairly permanent. The actual specific sexual phenomena experienced in such cases vary greatly, and although in certain relationships the emotion may be 'latent or unconscious', such cases are on the borderline of true sexual inversion, but according to Ellis cannot be included within its boundaries. Sex in these relationships is seldom an essential or fundamental element.[157]

In his work on the school environment, as with the rest of his observations on sexual inversion, Ellis was reluctant to force his conclusions and kept an open-minded approach to his materials. He clearly wished to liberate sexual activity from certain contemporary restrictions; his ideas were unmistakably intended to serve a radical purpose and were accepted as such by his contemporaries. His reformist and unconventional attitude towards female same-sex desires acquires full meaning when Ellis's *Sexual Inversion* is considered alongside the broader context into which it emerged. His collaboration with Symonds makes plain that his political agenda was to contribute to changing British law, which considered male same-sex practices to be a criminal offence. In order to achieve this, both Ellis and Symonds believed that they had to challenge the law through medical arguments, even if Ellis wanted to go beyond accepted psychiatric assumptions. The difficulty of such enterprise is well illustrated by the Bedborough case, which ultimately shows just how difficult it was to pursue the study of sexology in Britain, and how the case-histories presenting 'normal' sexual inverts in *Sexual Inversion* were bound to be rejected by the British medical establishment.

Some of Ellis's feminist contemporaries, such as the radical sexual reformer Stella Browne, took him as a source of inspiration in their attempts to promote discussion about lesbians and sexual liberation. Feminist writers today, however, believe that Ellis was responsible for undermining feminism and stigmatising lesbians. Ellis is ultimately blamed because his essentialist viewpoint on sexual differentiation is considered incompatible with current feminist political critique. These criticisms repeatedly fail to account for the context in which Ellis worked and, most dangerously, they overlook how his sexological work went against the grain, such as his stress on the importance of women's sexual pleasure, alongside his belief that female masturbation was a normal and even healthy sexual manifestation. All this must surely have caught the attention of readers far more than the biologistic assumptions he took for granted, which so alarm current readers.

Ellis was not an enemy of women and lesbians, nor did he plot against their sexual freedom. Quite to the contrary, Ellis's case-histories presented lesbians who were refined, educated, intelligent, and feminine in appearance. Sexually autonomous and independent lesbians appeared as bold as men, and it is for this reason that Ellis described lesbians as women who exercised 'masculine' prerogatives. This was in line with his political agenda: with the aid of science, he sought to release sexual activity from certain contemporary restrictions, pleading in turn for a reorganisation of sexual relationships. Just as he argued that sexual relationships should be recast on the basis of mutual respect, so too did he advocate for the reconstruction of gender relationships on the basis of equality.

8
William Blair-Bell and Gynaecology

Almost twenty years after the publication of Ellis's *Sexual Inversion*, the prominent English gynaecologist, William Blair-Bell, revealed in *The Sex Complex* (1916) that femininity was determined by internal secretions. The body of the higher type of woman was naturally suited to – and geared towards – perpetuating the race. Deviations from the procreative norm occurred whenever there was an imbalance in the internal secretions; the sexual invert, a freak of nature, was affected by such an irregularity, which explained her typically masculine physiognomy. Bell's work thus belongs to the British tradition of explaining same-sex desires by reference to the body, only instead of doing so in the 'old' languages of anatomy, he privileged the new vocabulary of endocrinology. In contrast to Ellis's progressive stance, Bell exemplifies British medicine's more conventional view on female sexual inversion as a diseased condition. Further, the analysis of his work on sexual dysfunctions makes it possible to chart the emergence of same-sex desires as a major gynaecological concern that now deserved full chapters in the discipline's treatises.

In examining Bell's conception of the 'normal woman' as opposed to sexual abnormalities like the intellectual woman, the hermaphrodite, or the sexual invert, this chapter first shows the extent to which Bell adopted the new language of endocrinology to reinforce older medical tenets that linked a woman's normality with maternity. The chapter goes on to analyse Bell's studies on hermaphroditism to illustrate how, although he played down the role of genital anatomy in defining an individual's true biological sex, he continued to believe that there existed only two opposite sexes: male and female. Finally, while he did exploit some psychiatric categories, Bell maintained a considerable distance from Continental psychiatric research on sexual inversion. Unlike other European theories that defined sexual inversion in psychological terms, his explanation of this pathology relied heavily on bodily explanations. As I will show, on the whole, Bell's endocrinological explanations did hold some affinities with earlier anatomical understandings of same-sex desires, but interestingly they superseded the anatomical models because Bell based sexual inversion on a proto-chemical perception of the body.

While Bell's interpretation of sexual inversion is not explicitly allied to Continental medical models, his physical description of the invert's body implicitly echoes the criminal anthropological stereotype. In Bell's account internal secretions replaced the stigmata of degeneration, and women's masculine physiognomy became a mark of sexual deviance. The refusal of Bell's gynaecology to conceive of same-sex desires in terms of psychology further illustrates how bodily oriented understandings of sexual inversion persisted in British medical thought. Despite having adopted the psychiatric term 'sexual inversion' and being familiar with Ellis's work, Bell's inverts did not have a psychological dimension and were characterised in strictly physical terms. According to Bell, a sexual invert was a woman who was compelled to engage in sexual acts due to a dysfunctional endocrinological system.

The normal woman and her internal secretions

Bell was born in 1871 in New Brighton, Cheshire. In 1890 he entered King's College, London, where he was trained as a doctor. After obtaining his qualifications as Bachelor of Medicine and Bachelor of Surgery, he practised in the Wallasey Hospital from 1898, and became assistant consulting gynaecologist in the Liverpool Royal Infirmary from 1905. In 1921 he was appointed to the Professor's Chair and was a leading figure in the struggle to establish the Royal (formerly British) College of Obstetricians and Gynaecologists, which he presided over for many years. Bell campaigned tirelessly for the recognition of gynaecology as a true science; he appealed to the new experimental studies of sex glands to support his claims that gynaecology and obstetrics should be considered a single discipline, independent from medicine and surgery. For Bell, the scientific authority and innovative aspects of endocrinology would transform the old practice of gynaecology and obstetrics into one genuine, unified science.[1]

Unlike Ellis, Bell followed an institutional path in establishing his professional prestige, but like him, he encountered some obstacles in this process. Indeed, the medical elite of the time did not approve of the specialisation in gynaecology because the gentlemanly ideal of generalism was valued over specialism. The British medical elite held the view that the ideal doctor was one who could treat the whole body; as a result, gynaecology continued to be regarded as a manifestation of the pernicious rising trend of medical specialisation well into the twentieth century.[2] Notwithstanding this debate, the gynaecological market expanded rapidly from the second half of the nineteenth century onwards due to new technological developments in surgery; these made it possible to perform abdominal operations such as ovariotomy and hysterectomy, as well as removing the gravid uterus, repairing vescicovaginal fistulae, and relieving uterine and vaginal prolapses.[3] Gynaecologists

supplemented these technological developments with new theoretical interpretations of women's physiology, all of which contributed to establish the
scientific validity of their discipline in the eyes of the medical elite.

A major discovery prompted nineteenth-century scientists to reformulate
theories of the functioning of the female reproductive apparatus. With the
discovery of the mammalian ovum in the first half of the century, scientists
came to understand the role of the ovaries in reproduction and to articulate
the ovular theory, according to which the process of cell growth, which took
place inside the ovaries, constantly stimulated women's nerves. This stimulation periodically reached a threshold, which led to an arterial congestion
of the genitals by means of nervous reflex. It was this periodical congestion
which was seen as the cause of ovulation and menstruation, which were
thought to take place at the same time. The implicit assumption was that
the female nervous system was particularly weak and irritable.[4] By the mid-
nineteenth century, gynaecologists promoted a view of the ovaries as the
very essence of a woman: to the extent that they were the seat of sexual
instinct and automatic behaviour, they also regulated the female nervous
system. The importance of the ovaries for women's nature was synthesised
in the French physician Achille Chéreau's famous medical motto: '*Propter
solum ovarium mulier est id quod est*' [A woman is only what she is because of
her ovaries].[5]

With the discovery of the production of internal secretions there emerged
a framework for the explanation of women's nature that further reinforced
the medical belief that women's moral traits were a reflection of the characteristics of their reproductive system.[6] The term 'internal secretion' was
coined in 1855 by the French physiologist Claude Bernard, who believed
that endocrine or ductless glands such as the thyroid, testicles, and adrenals,
released their secretions directly into the bloodstream.[7] In 1889, the
French physiologist Charles Édouard Brown-Séquard suggested that the male
gonads produced a secretion that controlled the development of the male
organism, and the ovaries secreted substances regulating the growth of the
female organism. As early as 1896 and 1900, two Viennese gynaecologists,
Emil Knauer and Josef Halban, identified chemical substances secreted by
the ovaries, suggesting that these – and not nervous impulses – determined
female characteristics.[8] In the early twentieth century, the British physiologist, Ernest H. Starling, reformulated this theory of 'internal secretions' to
introduce the concept of hormones, which were understood as chemical
messengers that travelled through the body in the blood. Starling believed
that these chemical messengers, designated 'sex hormones', originated in
the gonads: the male sex hormone was present in the secretion of the testis,
and female sex hormone in the ovarian secretions. The first results in iso-
lating the steroid structure of the female hormones were obtained in the
late 1920s, and by the mid-1930s both the female and male hormones were
completely isolated.[9] However sketchy these studies were at the time, Bell
embraced them and based his gynaecological theories on the mysterious

and elusive substances known as internal secretions. Contradicting earlier tenets of gynaecology that situated the ovaries at the very basis of feminine nature, Bell proposed a new gynaecological dictum: '*Propter secretiones internas totas mulier est quod est*' [A woman is a woman as a result of all internal secretions].[10]

Bell's best-known work, *The Sex Complex* (1916), was based on the conviction that the 'internal secretions' that existed in the endocrine glands made a woman's psychology feminine.[11] Bell recognised that instincts and mental disorders could be influenced by external factors such as class, education, social environment, and even intellectual pursuits, but his central theme was that the psychological make-up of women was mainly dependent on their metabolism – itself regulated by ovarian secretions.[12] Before the onset of puberty, 'the metabolism of girls does not differ appreciably from that of boys, the maiden is often self-reliant and somewhat a "tom-boy"'.[13] In Bell's opinion, the reason boys and girls eventually adopted socially sanctioned gendered behaviours was because of the influence of 'feminine' or 'masculine' bodily secretions. There were, he wrote, two stages in the development of a woman's secondary sexual characteristics: from birth to puberty 'the mind [of the girl] is often not pronouncedly feminine', although sometimes girls 'show a marked liking for dolls and sewing', whereas boys like 'balls and fighting'. Puberty, on the other hand, was marked by a remarkable physical and psychical change: at this stage a girl would normally become shy, reserved, and 'essentially feminine in her pleasures and in her relations with men'.[14]

Despite his adoption of such a new model based on internal secretions, Bell's endocrinological framework did not fundamentally contradict what many nineteenth-century physicians and psychiatrists held to be true: that a woman's mind was at all times and irrevocably influenced by her special functions, which is to say the reproductive system.[15] For instance, he held that the 'characteristic functions [...] of the female are those associated with genital activities – menstruation, gestation, parturition and lactation'.[16] Typical female characteristics included dependence on males, associated with an ardent desire to be loved.[17] 'Sexual and maternal capacity' depended upon the 'proportion of femininity' in the individual woman's hormonal make-up. Thus, 'the normal woman' found a certain pleasure in sexual intercourse, but was centrally defined by her desire to have children. In other words, female sexuality was subordinated to maternity. In the 'truly feminine' woman, feminine psychological characteristics such as the maternal instinct were mirrored by physical signs that suggested that all natural functions were 'perfectly coordinated and correlated'; a woman who fitted this description would, for example, menstruate regularly and her breasts would be 'well formed'.[18]

According to Bell's analysis, the human female had developed highly complex mental processes in order to assist with sexual selection, the care and rearing of offspring, and in general to keep pace with the general mental evolution of the species. Human males, on the other hand, had evolved intelligence 'to sustain their mates and offspring in an environment compatible with advanced civilisation' and to deal with the 'intricate processes' of modern life.[19] It was senseless, Bell wrote, to consider the male intellect as necessarily superior to the female intellect, which was 'a source of personal pleasure and pride'.[20] He noted, however, that 'it must surely be recognized by all that the male mind and masculine form are suited for the business life' while 'the central motive of a normal woman's existence is the propagation of the species'.[21] Given that reproduction made fewer demands on masculine metabolism, the nerves of men were 'steady, the mind stable and the physical strength great'.[22] Men's more diverse intellectual skills, however, were in no way superior to women's 'patient and absorbing work': in modern conditions this was a 'far greater self-sacrifice'.[23] As the male and female sexual apparatuses complemented each other, so too woman complemented man.

It was not only his medical conviction that moved Bell to declare that feminine psychology was informed by women's reproductive function. Throughout his life, Bell strove to elevate the status of gynaecology: the fact that women were wholly and exclusively governed by their reproductive apparatus was the best and highest justification for the existence and the advancement of the gynaecological profession. He also attempted to raise the prestige of gynaecology by broadening its immediate field of expertise. Indeed, Bell's gynaecology was not restricted to treating the diseases of the female reproductive system: a considerable part of his work crossed the disciplinary boundaries between gynaecology and psychiatry, and dealt with the wider implications of female disorders by touching on definitions of normal and abnormal mental characteristics. It was in the context of such a discussion that Bell denounced that civilisation had produced women who wanted children but not sex, and vice versa:

> The woman who delights in sexual pleasures, but has no maternal instincts, is not strictly speaking normal, although she may be as much a product of civilisation as is the female who abhors the idea of both intercourse and maternity. [...] The woman who desires children but has no sexual inclinations is a product of civilisation. [...] Her physical and psychological functions are not perfectly adjusted.[24]

By the same token, women whose 'reasoning faculty' was more developed had 'less need for the lure of sexual gratification to ensure the perpetuation of race'.[25] Women who did not have sexual reproductive inclinations had an ovarian insufficiency that could be treated by means of 'organotherapy'.[26] Those who had an excessive ovarian secretion, however, were the hardest cases, because although it could affect women who were 'extremely feminine in appearance and character', these disorders could cause excessive sexuality, culminating in 'sexual insanity'.[27]

The abnormal woman type

Bell portrayed the abnormal woman as a type, a product of modern times that did not conform to the medical idea of woman whose life was organised around the event of procreation. He admitted that 'it is difficult to establish what may be described as a *normal standard* of the mental attitude towards coitus. Civilisation, with its social exigencies supported by religion, has undoubtedly raised an artificial standard.'[28] Despite these acknowledged difficulties, Bell did describe a specific kind of woman characterised by her masculine appearance in the section entitled 'psychoses and neuroses' of his classification of gynaecological disorders.

Bell thought that masculinity and femininity were attributes of the body as well as of the psyche, and noted that in certain women masculinity was present in all bodily tissue. Psychological and biochemical elements were correlated and depended to a considerable extent on the distribution of the 'sex characteristics':

> The larger the proportion of femininity in a woman, and the less the proportion of masculinity, the greater will be her inherent sexual and maternal capacity. If a woman have [sic] a comparatively large proportion of masculinity in her composition, it will be reflected in every cell of her body. The characterisation of the organs of internal secretions will be modified, and the activity of her ovaries reduced.[29]

In abnormal women, the internal secretions were responsible for the development of masculinity and, further, with the onset of menopause some women became even more masculine and displayed 'acromegaly'.[30] Bell explained that in extreme cases of hormonal disorder menstruation could cease, the skin might become coarse, the voice deep, the breasts shrunken, and last, and most interesting of all, the clitoris could hypertrophy until it resembled a small penis.[31] The category of abnormal women included those who suffered from ovarian diseases and what today are called hormonal diseases, as well as women with 'excessive' sexuality or without sexual instinct at all, 'self indulged' women, and sexual inverts. All of them had masculinity in their external bodies caused by their internal secretions.

While Italian criminal anthropologists such as Lombroso had scrutinised the stigmata of degeneration in abnormal women, Bell looked for some signs of the disease under the skin. Yet internal secretion dysfunctions became visible on the female body. In both Bell's gynaecology and in Lombroso's criminal anthropology, abnormal women were recognisable because of their masculine looks. In both cases the prolific mother was considered normal and contrasted with the virile woman, who was deemed pathological. Unlike Lombroso, however, Bell never discussed criminal women; in his gynaecological writings there emerged three main types of deviation: intellectual women, hermaphrodites, and inverts.

Intellectual women

According to Bell, a woman who took pleasure in the sexual act but had no maternal instinct was 'the result of the conditions in which we live' because her intellectual exertion had led to her reproductive functions being hindered. The type of woman who shunned both sexual relations and maternity was 'on the fringe of femininity', to the extent that her femininity was 'almost neutralised by her masculinity'.[32] These intellectual women were usually flat-chested and plain. Although they did menstruate, for the most part their metabolism was masculine in character and they were 'ill adapted' to the female role in reproduction.[33]

The physical description of intellectual women is similar to those women whose ovaries did not work properly. Indeed, Bell proposed a correlation between ovarian insufficiency, a woman's masculinity or reduced femininity, and the female mind. When a woman's masculinity-producing internal secretions became abnormally active, her entire metabolism and body became more inclined towards masculinity, and the mind became 'less feminine in its outlook'. Although cases of so-called 'true hermaphroditism' were rare:

> there is little doubt that women with a larger share of masculinity than is normal are extremely common. These unnatural individuals are easily detected by the coarseness of their skin, by the size of their extremities, by the ill-development of the breast, and by the assertiveness and aggressiveness of their conversation and schemes.[34]

According to the gynaecologist, a 'normal woman' would not exploit her intellect for her own profit but for the benefit of her descendants:

> It is reasonable to believe from the evidence at our disposal that the application of feminine talents to the competitive work in which men are engaged, which is strictly speaking an evolutionary form of the hunter's craft, is adventitious and injurious to the psychical and physical functions connected with the biological life of women. So long as there are two sexes it is unlikely that women will, without detriment to their own sex-psychology and physical attractions, which are so essential to sexual selection, develop the masterful mind of the male that may attain to the lofty height of genius – a level to which the intellect of no woman has ever yet ascended.[35]

Bell's concerns over women engaging in intellectual activity were reinforced by wider scientific and political preoccupations linked to a decrease in population, the creation of a female labour force as a result of World War One, and the independence of women. By the time Bell was writing, old anxieties that women's higher education might interfere with their childbearing capacity were reinforced by eugenicists, who warned that 'emancipated' women would either limit their families or abstain from childbirth altogether. By the end of the nineteenth century it was increasingly noted that there was a reversal in demographic trends: birth rate statistics had been falling since 1876. By 1901, fertility had dropped by more than 24 per cent, and by 33 per cent by 1914.[36]

Hermaphroditism

Bell was aware that participating in a 'controversy concerning the difference between Man and Woman' – which had social and legal implications – was rarely conducted 'on unbiased and scientific lines'.[37] His studies on hermaphroditism are revealing of his own ambiguous scientific attitude. On the one hand, as the examples above show, Bell took for granted the existence of clear dividing lines between femininity and masculinity, and normal and abnormal women. On the other hand, he based his belief that the human being was fundamentally bisexual on the basis of contemporary scientific observations. Bell considered that regardless of the individual's predominant biological sex, there were always 'latent traits of the opposite sex' in various degrees of development:[38] 'certain men are somewhat effeminate', and many a woman had 'the smallest balance of femininity in her favour'.[39]

While in theory nineteenth-century doctors agreed that sex was determined by the nature of the gonads, in practice, outward appearance continued to be of great importance – particularly in Britain. Every recorded case of hermaphroditism demonstrated that regardless of how much attention was bestowed on the gonads, other 'sex' traits were frequently commented on, not just as an indicator, but as the ultimate determinant of sex.[40] In practice, the function of the gonads in establishing the 'true' sex was not completely clear or fixed. In cases of ambiguous genitals at birth, many physicians insisted that it would not be possible to distinguish between male and female before the onset of puberty, because it was only then that the object of sexual desire and secondary sexual characters became manifest. If a genital examination proved inconclusive, it was necessary to take into account the individual's psychological attitude and sexual inclinations.[41]

Following in this tradition, Bell thought that in cases of unusually mixed sexual characteristics, the 'true sex' could not be established by examining the sex glands because these would be dysfunctional.[42] The findings of new medical technologies seemed to support such claims: by 1915, laparotomies – exploratory surgery – and biopsies – removal of small samples of suspicious tissues – performed on living patients had confirmed the presence of testes in women, ovaries in men, and ovotestes in true hermaphrodites.[43] Two cases of uncertain sex which he published in 1915 – only one year before *The Sex Complex* – prompted Bell to revise the protocols for determining a patient's sex: the first was a case in which a patient had an ovotestis; the second was a case of a very 'womanly' male patient who was shown to have testis and no ovaries – medical science currently refers to this as androgen insensitivity syndrome. On the basis of these case-histories, Bell came to believe that to determine a person's sex according to gonadal anatomy alone was to risk not only the happiness of the patient and his sexual instinct, but social order as well:

> I want [...] to raise the question as to whether we are justified [...] in branding [patients] with a sex which is often foreign not only to their appearance but also to their instincts and social happiness. [...] Both these patients were good-looking and might have been married, just as many other male tubular partial hermaphrodites have been, believing themselves to be women.[44]

Thus Bell wondered whether it was sensible to say that someone was a true hermaphrodite when there was no legal or social way to deal with such a person.[45] Each case of hermaphroditism should be considered 'as a whole' by 'the obvious predominance of characteristics'; undertaking surgery was 'justifiable' as a means 'to establish more completely the dominant sex of the individual, and the one which is most in accordance with the social happiness of the person concerned'.[46] Endeavouring to understand gender beyond the dimension of biological sex, Bell remarked that since it was possible to scientifically demonstrate that the 'psychical and the physical attributes of sex' were not necessarily dependent on the gonads, gynaecologists should evaluate the specificities of each case. Sex should be 'determined by the obvious predominance of characteristics', especially secondary, rather than by focusing solely on the non-functional sex glands, for this was neither scientific nor just.[47]

As Dumorat Dreger has suggested, although Bell's approach to sex diagnosis deliberately moved away from exclusively biological definitions of sex, his position was conservative in two fundamental ways. In the first place, like his predecessors and many of his successors, Bell was interested in maintaining clear, medically sanctioned divisions between the two sexes in each individual case, and in society as a whole. In the second place, Bell held fast to the notion that individuals did indeed have a single true sex, even if neither the sex gland nor the person's physical aspect allowed the physician to form an unambiguous opinion of the patient's true sex.[48] Bell recommended that doctors should not only diagnose a single sex for anomalous bodies, but also eliminate any sexually 'anomalous' bodily parts such ovaries in individuals who were predominantly men, or testis and ovotestis in individuals who appeared to be women. In his analysis of hermaphroditism, where it is quite likely that the word 'gender' was used in medical literature for the first time,[49] Bell concluded:

> Our opinion of the gender [of a given patient] should be adapted to the peculiar circumstances and to our modern knowledge of the complexity of sex, and [...] surgical procedures should in these special cases be carried out to establish more completely the obvious sex of the individual.[50]

In the social order there existed only two true sexes and it was the gynaecologist's prerogative to discern them.

In practice, hermaphrodites posed conceptual and practical problems for physicians who claimed men and women were naturally, fundamentally, and obviously different, and who also claimed that sexual desires – heterosexuality, that is – was the obligatory result of the existence of two distinct sexes, which in turn were the product of human evolution with its inherent sexual differentiation. It was not a coincidence that Bell openly linked hermaphroditism to sexual inversion. The human being, he wrote, was:

> originally bisexual, and in the adult there is always evidence – structural if not functional – of the recessive sex.
>
> In some cases the latent *secondary characteristics* become pronounced in one or more particulars, without any real disturbance of the feature of the predominant sex-characterisation. Such conditions are known as '*Inversions*', and they are common in minor degrees, in which such phenomena as hair on the face, coarse skin and other masculine characteristics are seen in women. [...] Hermaphroditism may with justice be held to be *akin* to inversion.[51]

The notion that same-sex desires and hermaphroditism were in some way related was an old medical idea.[52] Bell agreed with earlier medical writers that individuals who loved their own sex and hermaphrodites were part of the same phenomenon as neither of them conformed to the traditional needs of the sexual reproductive pattern. Therefore, hermaphrodites and sexual inverts unsettled dominant ideas about sexuality in similar ways, and forced physicians such as Bell to recast traditional medical theories.

Female sexual inversion

In Bell's definition, the term (female) 'sexual inversion' described the 'derangements' of the 'normal' feminine secondary characteristics.[53] In the female invert, the 'masculine-producing secretion' altered the metabolism of the 'lime salts'. In other words, the organism produced an excess of calcium that became manifest in masculine sexual secondary characteristics. Normally, the male skeleton was heavier and stronger than the female's because a woman's organism would retain less calcium in the course of her biological development. When those parts of a woman's 'hormonopoietic system', whose function was to store calcium, were 'abnormally active' she would develop 'other masculine characteristics, such as the growth of hair on the face and alterations in the formation of the larynx and breast'.[54] Female inverts did indeed have low, male voices, and were flat-chested.

Interestingly, sexual inversion was conceived as a bodily dysfunction: as an imbalance in internal secretions. Bell's explanation of sexual inversion does not sit comfortably with other accounts of same-sex desires such as the anatomical or psychiatric models described in this book. Bell's endocrinological explanation rested on the idea that the body was regulated by chemicals spread through the entire organism by the blood. To a certain extent this endocrinological model is closer to ancient humorism than to pure anatomy. In explaining sexual inversion, Bell did not refer to abnormal genitals. Nevertheless, the new endocrinological model that rested on the notion that bodily functions were regulated by chemical substances shared a number of similarities with the model that stated bodily functions were regulated by the nervous system. Sexual perversion would be caused by internal secretions rather than by the brain's neurophysiology, but in any event it would not be caused by the sexual instinct. Therefore both explanations are based exclusively on the body. While Bell's endocrinological view of sexual behaviour may raise different interpretations, it is clear that both the anatomical and endocrinological explanations of sexual inversion refuse to conceive of sexual behaviour in psychological terms. As discussed before, although in distinguishing normal from abnormal women Bell borrowed from psychiatric terminology, he did not engage with the descriptions of the individual's feeling and desires. His analysis of sexual inversion is an example of how, well into the twentieth century, British medicine declined – in Davidson's words – a 'psychiatric style of reasoning' in favour of bodily oriented explanations.

Although in approaching sexual inversion Bell did not adopt Continental psychiatric explanations such as Krafft-Ebing's, he nevertheless exploited certain psychiatric themes. For example, he explained that 'if, with the onset of puberty, marriage took place in European countries as it does in some Eastern countries, such indulgences [as masturbation and sexual inversion] would not be practiced in educated women'.[55] Bell indeed listed sexual inversion amongst the 'psychical disturbances' of late puberty.[56] Matrimony, and of course, regular healthy sexual intercourse with a man, could cure masturbation and even sexual inversion. He did warn, however, that if these habits had become established, even matrimony may fail 'to provide a satisfactory substitute'.[57] Bell also believed that masturbation in a woman might be a sign of 'excessive sexuality', but it might be 'an accompaniment to mental instability or insanity other than that of sexual origin; and in such circumstances it is the result of a weakened moral outlook and will-power'.[58]

The habit of masturbating could also be acquired in childhood from nurses or other older women playing with the child's external genitalia; or it might be taught by one girl to another in a school or workroom. Bell added to these comments the Italian lawyer Niceforo's observation – quoted in Ellis – that 'all girls in certain workrooms in Italy were known to masturbate themselves or one other in the heat of the day'.[59] Bell concluded by stating that a 'watchful mother will detect the habit' and, if the patient was 'taken to the physician, the cause of irritation may be removed':[60]

> If masturbation be practised at puberty and afterwards, when orgasm can be experienced, the girl may suffer in health. In one case the patient's distress and remorse at her own evil ways, which she found impossible to check, were such that we excised her clitoris and nymphae. This method of treatment may be adopted with excellent results if the right type of case be selected: the girl who is not suffering with excessive sexuality, but, rather, with the fascination of a bad but pleasant habit, to the detriment of her moral and physical equilibrium.[61]

Excessive sexuality could be manifested in a woman in three ways: 'excessive sexual indulgence with its train of neurasthenic symptoms', 'masturbation or sexual inversion', and 'sexual insanity'.

Bell wrote that sexual inversion was not a common 'practice' amongst married women unless they had indulged in it before matrimony; 'if so, the woman may be comparatively apathetic to the normal act and prefer abnormal satisfaction'. If the patient's overall health were at all affected, Bell stressed, the symptoms were the same as those displayed by a neurasthenic person.[62] This reasoning harks back to the old psychiatric association between masturbation, sexual inversion, and moral insanity. According to Bell, only women who were 'apathetic' or whose sexuality tended to excess, or who masturbated, were sexually inverted.

It must be noted that Bell used the term 'practice' in regards to homosexuality. While only a few decades earlier sexologists such as Ellis considered sexual inversion an innate inclination that could not be limited to episodic sexual practices because it had to do also with an individual's psychological make-up, Bell spoke of sexual inversion as a sexual practice in the late 1910s. Yet according to his own description of female sexual inversion, the latter was something more than a mere sexual practice because female sexual inverts were readily recognisable due to their masculine appearance, which, as shown, was caused by an unbalance of internal secretions. In Bell's gynaecological arguments, women who indulged in same-sex acts were easily detected due to their distinctive masculine aspect, an idea that mirrored the Continent's psychiatric and especially criminal anthropological literature. In fact, to a certain extent, Bell's portrayal of the body of female sexual inverts resembled criminal anthropological descriptions: features such as flat breasts, a heavy skeleton and coarse skin were routinely mentioned in Italian case-histories that followed Lombroso's model. Although Bell omitted the requisite lists of bodily measurements present in case-histories written by criminal anthropologists, his description of the sexual invert did echo the stereotype proposed by some Italian medical writers. Both Bell's choice of language and his remarks about the link between civilisation and women's sexuality also suggest that he was familiar with Continental criminal anthropology and other Continental medical influences.

Unlike Continental theories of sexual inversion, however, Bell's accounts of same-sex desires focused almost exclusively on the body, rather than on both the mind and the body. As a result, Bell did not survey his patients' fantasies, predilections, habits, or childhood experiences. While in Continental literature female inverts had a mannish personality, a masculine lifestyle, and often a masculine body as well, the female invert portrayed by Bell only had a man-like physical appearance and an endocrinological imbalance. In linking female masculinity to same-sex desires, Bell arguably recast concepts of same-sex desires that predated sexual inversion into the new fashion of endocrinology, which was fundamentally a bodily oriented explanation. Further, Bell's refusal to adopt the Continental psychiatric model of sexual inversion illustrates the extent to which this model did not transverse all medical disciplines in the same way. It also shows how, in the medical literature analysed in this book, female same-sex desires were multiple in the sense that as soon as different medical fields started looking at female same-sex desires, this apparently single object was decentralised into a multitude of meanings.

Just as Bell's idea of women as reproductive machines was integral to his larger project of justifying his own profession, so too his idea of sexual inversion as the result of unbalanced internal secretions promoted the gynaecological study of endocrinology. Moreover, his endocrinological formulation of sexual inversion implicitly advanced a possible cure for individuals who might seek, or were forced to seek, treatment in medical hospitals of various kinds. Therefore, Bell competed with psychiatrists and criminal anthropologists for the medical jurisdiction over the diagnosis and treatment of female sexual perversion.

Bell's studies are riddled with ambiguities. His references to tearing aside 'the veil overhanging the mystery of sex' through gynaecology, and to the 'ravages of science', seem to suggest that Bell was aware of both the need to study sexual matters and the limits of his discipline, and more generally science, in approaching the topic.[63] He alluded to the effects of civilisation on sexual behaviour,[64] suggesting that the latter was also a result of historical developments, and yet at the same time he believed that sexual behaviour was exclusively determined by internal secretions. He stated that the removal of ovaries – a common gynaecological practice used to treat most female problems – was useless; yet he treated female masturbation by excising the clitoris and insisted the same method should be applied in similar cases.[65] He thought that a physician had to consider a hermaphrodite's happiness when defining their sex, rather than adhering strictly to anatomical evidence. He nonetheless berated 'modern women' for rejecting the maternal role and expected every female to fit the stereotype of normalcy set by gynaecologists. While he declared that it was senseless to consider the male superior to the female sex, he insisted that men were suited to business whereas reproduction and the propagation of the species were a woman's reason for being. He also believed that all individuals were latently bisexual, but advocated a scientific view whereby the intellectual capabilities of the two sexes were different.

Bell's writings on sexual deviation reveal the extent to which he was deeply committed to Victorian traditions. The language of sexual glands was then new and based on the latest research, but had a number of elements in common with earlier explanations of same-sex desires; thus it was masculine internal secretions rather than degenerative signs that singled out abnormal from normal women. Bell not only invoked a kind of 'sick' woman who was at odds with accepted standards of the physical beauty of the time, but also identified such abnormality with the kind of highly educated, independent women who had decided to cultivate their minds. These women, he claimed, were physically masculine. Perhaps he was thinking of feminists and was anxious about women occupying traditional male roles; certainly he promoted the image of the mother as a healthy woman.

Like Ellis's, Bell's outlook on human sexual behaviour was materialistic and his views on sexual dimorphism were biologically essentialist. Both Ellis and Bell subscribed to the theory of an original bisexuality in human beings. Their positions on sexual inversion, however, were sharply contrasting. Whereas Ellis's psychology of sex had an open approach to female non-procreative sexuality, Bell's gynaecology promoted a female sexuality strongly oriented towards reproduction. Ironically Ellis, who presented sexual inverts as normal individuals, had originally supported Lombroso's criminal anthropology, whereas Bell, who did not credit the influence of criminal anthropology in his gynaecological studies, implicitly supported the physical stereotype put forward by Continental criminal anthropology.

Finally, the contrast with Bell's work is clear evidence of the progressive nature of Ellis's theories, and further confirms that a part of British medicine was reluctant to accept the more radical implications of Ellis's work *Sexual Inversion*. Almost twenty years after the first edition of *Sexual Inversion*, Bell adopted the term 'sexual inversion', but rejected Ellis's conclusions, and so sexual inverts continued to be represented as abnormal and diseased individuals in successive decades. Contemporary British medical writers' failure to embrace Ellis's position does not mean that Ellis had no impact outside medical circles: despite *Sexual Inversion* being banned in England, it was nonetheless able to reach a broad audience and those individuals who recognised their own story in Ellis's case-histories felt that they were accepted, not criminalised or pathologised. In the long term, sexologists such as Alfred Kinsey, William Masters and Virginia Johnson recognised the important role Ellis had played, and forgot Bell's contributions.[66]

9
Concluding Remarks

Medical writers did not have to wait for gender historians and contemporary social scientists to undermine the biological foundation of sex. Before the 1970s, and even before surgical operations allowed changes to a naturally given sex, medical scientists had already interpreted the sexual body as something contingent, a product of organic, social, and psychological factors and, at times, a combination of all of them. Contemporary historians have overlooked the implications of the professionalisation of medicine in the development of knowledge about sexuality: practitioners held different views of the sexual body and had competing interests in treating abnormal sexual behaviours depending on their specialty. If one looks at how different medical disciplines approached the study of abnormal sexual behaviours in the second half of the nineteenth century, we see that medical practitioners looked at various aspects of human sexuality and were aware of a wide range of social, cultural, and psychological influences. When medical writers looked at sexuality, they did not see only a fact of nature; they did not conceive of the human body only as a strictly physiological or biological entity. Instead, they saw physiological, social, and psychological interactions at the centre of medical studies, and medical writers such as Penta and Ellis attempted to combine these elements in their analyses. Despite his publicised insistence on constitutional influences, Lombroso weighed the effect of the environment when analysing sexual perversions. Moreover, even those practitioners who relied heavily on very physical elements to explain same-sex desires did not agree on what ultimately determined sexual perversions; medical writers nominated diverse organic points of origin for same-sex desires, ranging from the ovaries and an enlarged clitoris, to the brain and internal fluids. Adopting such diverse physiological, social, and psychological approaches allowed late nineteenth-century medical writers first to medicalise sexual behaviour, and then to destabilise the biological foundations of sex.

While it is certainly true that practitioners followed local traditions when medicalising sexuality, local discourses also crossed national borders, so that ideas, new technologies, methods of scientific enquiry, and physicians travelled around Europe, between Britain and the Continent, and ultimately between Britain and Italy. British and Italian late nineteenth-century medical writers, especially psychiatrists, shared some of those intellectual foundations that would eventually lead to the emergence of the psychiatric concept of sexual inversion. Both countries defined mental illness as an organic disorder, classed perversions of feelings into categories such as mania or moral insanity, and, to varying degrees, saw mental disorders as the result of hereditary degeneration.

Nonetheless, the paths that led to the emergence of sexual inversion in these two countries were different. Whereas Italian sexologists were eager to embrace and expand the new psychiatric category of sexual inversion, British physicians were wary of addressing the issue, which is not to say they neglected it altogether. This is just one aspect of two distinct, although at times intertwined, narratives. In charting the transformations of these different scientific traditions, the historian cannot help but wonder about the underlying reasons why same-sex desires were formulated in such diverse ways in Italy and Britain.

The historical, professional, and cultural contexts of the Italian and British medical communities provide some answers. Despite its recent unification, the new Italian kingdom was deeply fragmented, as was evident in the division between north and south. The *questione meridionale* and the violent repression of the *Fasci Siciliani* testify that, until the end of the nineteenth century, the political situation remained largely unstable. A significant proportion of the population living both in the south and in the north still had to fight poverty: in rural areas of the north, for example, poor diet continued to cause endemic diseases such as pellagra, which was common until the end of the century. As the preceding chapters have shown, in the last decades of the nineteenth century, Italian psychiatrists in particular were involved in all sorts of pioneering activities, in both practical and theoretical areas. Despite the many problems that afflicted Italy, many intellectuals felt that the years following the unification were the beginning of a new era in which science would be a crucial instrument of civil progress and emancipation. A. De Waterville, one of the editors of the British journal *Brain*, commented: 'The political resurrection of Italy has been speedily followed by a scientific renaissance amongst its people. On every side signs of increasing activity become manifest.'[1] Pioneering scientific enterprise was seen as a critical step to developing the nation and sexology was one of the disciplines that gained the most ground in the later decades of the nineteenth century. At the very moment when German and French psychiatrists – considered the forerunners of sexology by most historians – were turning their attention to sexual inversion, Italian practitioners were going through a phase of enthusiastic commitment to science. The social and cultural role of scientists in building the new nation, and the fact that they were part of the ruling class, rendered them even more enterprising. The lack of government intervention in the management of the insane had meant that although they were left to fend for themselves, Italian psychiatrists had more freedom of action than their British counterparts.

British psychiatry, on the other hand, was a long-standing discipline that had continuously negotiated a strong government presence in its professional activities and had struggled against the legal profession. Throughout the nineteenth century, a number of official enquiries into the management of asylums and the enforcement of laws regulating the profession discouraged practitioners from trying to go against the guidelines set by the authorities. The debate about criminal responsibility epitomised the struggle between the legal and the psychiatric professions, both of which were attempting to claim the right to define criminal responsibility. Maintaining that sexual inversion was an innate condition would have been tantamount to arguing that the law should have no right to punish male same-sex acts. Ellis and Symonds advanced this marginal position, but their theory was not instantly popular with members of the British medical community. Within the British psychiatric community, few individuals were willing to adopt a view of sexuality that might result in professional and even social ostracism. For instance, Savage and Mercier, who did address sexual inversion (if only cursorily), were noticeably prudent in their handling of the topic. The 1857 Obscene Publications Act indirectly prompted the guarded tone that predominates in British psychiatric debates of sexual issues. In principle, physicians had leave to write about sexuality, but in practice there were harsh constraints and standards of decency that had to be scrupulously upheld. Both the reaction of the medical establishment to Ellis's *Sexual Inversion* and the problems he encountered following the publication of his work exemplify the kind of obstacle experts faced when engaging in discussions about sexuality – especially 'deviant' sexuality.

Neither in Britain nor in Italy was the relationship between legal and psychiatric experts smooth. British psychiatrists, however, were less inclined than their Italian colleagues to challenge the law. In late nineteenth-century Britain it was believed that male homosexuality was a wilful vice – rather than an inborn condition or a diseased state – that merited legal punishment. Physicians would have had an arduous battle arguing that sexual inversion was a congenital or even an acquired disease, because this directly undermined the concept of free will, which was a tenet of British law. In Italy, the Roman Catholic Church was equally attached to the principle of free will, but most criminal anthropologists and psychiatrists joined forces to repel religious authority. This is not to say that there were no Catholic physicians in Italy, but until the turn of the century the scientific field dealing with the study of sexuality was predominantly led by positivists. With the rise of Benedetto Croce's neo-idealism in the first decades of the twentieth century, positivism entered a phase of decline, but not without leaving a lasting imprint on scientific thought: in the long term, even Catholics would insist that homosexuality was a mental disease.

In both countries, sexology in general, and the study of same-sex desires in particular, evolved in relation to quite specific professional and political interests, and thus there was nothing monolithic about the development of the research on so-called sexual psychopathologies. The scientific and cultural stances of both Penta and Ellis were equally radical, and both of them were prepared to risk professional ostracism. On the other hand, contemporaneous medical writers such as Lombroso and Bell held conservative views on same-sex desires. But not in absolute terms: when Lombroso spelled out the problem of sexual abuse of children, or when Bell advocated that hermaphrodites should have the right to choose which biological sex they belonged to, they were taking up positions that went against the grain of consensual medical and social assumptions. The ambiguity of much medical analysis of same-sex desires derived not only from single physicians' often contradictory views, but from sexology's inherent ambiguities. Sexual experts compiled endless lists of the physical and psychological traits of sexual perverts, which contributed to singling out deviant sexualities; this, in turn, would widen the breach between normal and pathological. In the long run, the methodological preference for taxonomies consolidated such a distinction, but physicians who classified individuals as types had been trained to regard the pathological as an exaggeration of the normal. In other words, in the nineteenth century, medical science's understanding of the 'normal' depended problematically on its understanding of the 'pathological'.[2] In psychiatry in general, and in sexological writings in particular, the boundary between normal and abnormal was not clearly set, and by the end of the nineteenth century, sexologists such as Ellis and Penta had begun to challenge the notion that same-sex desires were pathological.

Love between women incited a good measure of curiosity and fear in the nineteenth century, and this made the subject attractive to a wide range of medical writers and psychiatrists. Italian and British medical discourses on female same-sex desires overlapped and intertwined, so that they had many elements in common, including the stereotypes or figures used to describe same-sex desires, like the tribade-prostitute. Italian and British descriptions of relationships amongst female prostitutes betray the influence of foreign scholars such as Parent-Duchâtelet and the reliance on common medical sources. The figure of the tribade-prostitute was by no means created by Parent-Duchâtelet: it had been in use for centuries. Thus, the adoption of such an association in both countries shows how older ideas about female same-sex desires continued to have currency, even after the introduction of the concept of sexual inversion.

The possibility of an excessive female sexuality was a source of anxiety for both Italian and British doctors. Representations of female deviancy in Italian criminal anthropological writings single out women's active and excessive sexuality, an issue that was less commonly discussed in relation to male offenders. The fact that, in Lombroso's view, prostitution was the equivalent phenomenon of male criminality illustrates the extent to which women's sexuality was always taken into account when defining female deviancy. In Britain, female same-sex desires were often discussed in contexts such as medical observations of nymphomania, in which the main concern with regards to female perversion was less the choice of object than the sexual excess in which some women engaged. Both countries also associated women's virility and tribades' hypertrophied clitorises with the problem of immoderate sexuality. Women with masculine personalities and bodies blurred the divisions between the two sexes, and in threatening sexual dimorphism they compromised evolution. The progress of civilisation was potentially forestalled by gender confusion, by masculine women whose excessive sexuality itself proved that they were degenerate.

Feminine women are at the other end of the ample spectrum of medical representations of female same-sex desires. The fragility of female chastity and the fear that men were unable to control female sexuality pervades medical discourses about young girls living together. Doctors both in Italy and in Britain believed that female same-sex desires were widespread in 'normal' women-only environments such as schools. British warnings about young women sleeping in the same bed and the Italian figure of the *fiamma* applied not to members of the poorer classes or outsiders of society, but to middle-class women. This apprehension fits in with broader concerns about women's access to education: in the last decades of the nineteenth century, British and Italian societies engaged in intense debates over the question of women's education, which drew attention to schools and colleges. Medical debates focusing on all-girls boarding schools are predictably riddled with anxieties about women living beyond the authority of men and about shifts in feminine social roles. Further, these medical warnings betrayed the fear that same-sex relationships, endemic to boarding schools, might affect physicians' daughters.

On the whole, the aspects in which Italian medical discourses on female same-sex desires differ from British formulations reflect the broader differences between the two scientific communities. Medical writings about female sexual inversion appeared and increasingly proliferated in Italy well before the late 1890s, whereas in Britain the figure of the female invert was virtually unknown prior to the publication of Ellis's study. In Italy, both prominent and marginal medical writers directed a voyeuristic gaze towards female homoerotic relationships, and to female sexuality in general. While they indulged in lengthy descriptions of female same-sex acts, their British counterparts were consistently prudish in their descriptions of love between women and their language choices come across as extremely cautious. While in British medical literature there are no instances of authors describing women masturbating with eggs and crucifixes, or using dildos during intercourse, such detailed accounts are indeed frequent in the writings of

Italian authors – Lombroso himself provided the examples above. British physicians were never so explicit and took great care to maintain a standard of decency; for this reason, they would typically hint at sexual acts to avoid shocking their readers. This difference in approach is evident even in anatomical accounts, so that where British physicians spoke about lesbians' hypertrophied clitorises, Italian doctors described the consistency and the sensitivity to the touch of female genitalia. Even Ellis, for all his radicalism, weighed his words carefully when describing female sexual inverts.

Style was not the only aspect of medical writings to be affected by the British cautious attitude; there were also laws regulating the production and circulation of writings of this kind, as evidenced by the fact that Ellis was eventually forced to publish *Sexual Inversion* with an American publisher. Italian medical literature had a titillating quality that British practitioners avoided at all costs. It was not a coincidence that even though they were translated by respectable medical publishing houses and popularised in medical journals, many British readers considered works such as Krafft-Ebing's *Psychopathia Sexualis* to be scientific pornography.

If there is any one reason that explains the eagerness of Italian positivist physicians to speak openly about sex, it is the silence of organised religion. The Roman Catholic Church had explicitly condemned sexual knowledge, advocating ignorance as the best protection against sexual vice.[3] Italian sexologists inverted the terms in the relationship and insisted there was nothing worse than ignorance in the sexual sphere. No matter how 'distasteful' a subject might be, Italian physicians argued that it was not possible to devise a solution to a problem without exhaustively analysing the topic first. During the nineteenth century, medical discourses explicitly denounced Christian principles as an unnatural denial of human sexuality: sex was part of human, especially male, nature, and Italian writers from Mantegazza to Lombroso made constant references to the value of sexual knowledge and to the harm caused by religious prudery. There was a further reason for Italian physicians to speak openly about sex. Although medical writers such as Mantegazza or Lombroso were prestigious professors employed by universities and other institutions and thus reasonably well-paid, other less established practitioners who could not make ends meet found the study of psychopathologies to be a profitable source of additional income. In the early twentieth century it was common for obscure physicians such as Alberto Orsi to publish erotic novels in the guise of popular expositions of the new medical theories.[4] Sexual subjects sold well in Italy, and titillating works were profitable for both publishers and physicians.

The concept of female sexual inversion incorporated older ideas of same-sex desires. Indeed, medical literature had featured descriptions of gender inversion before the appearance of 'sexual inversion', with the old cultural stereotype of the *virago* being the lesbian *par excellence*. Nonetheless, if there was something new in the study of sexual inversion, it was perhaps how sexual fantasies and desires were scrutinised in case-histories. This diagnostic tool meant that doctors, especially in Italy, increasingly enquired into the invert's childhood, personal and family medical history, favourite pastimes, and irregular ways of life. Yet, case-histories did not only focus on sexual inverts and perverts: in late nineteenth-century medical literature, any mental disorder known to medical science merited a case-history, as did murder and anarchism, which were considered forms of degeneration. Moreover, in psychiatric practice there existed a long tradition of compiling case-histories. The father of Italian psychiatry, Vincenzo Chiarugi, concluded his *Della Pazzia* [*On Insanity*] (1793) with an appendix of one hundred case-studies. While his descriptions of patients' feelings were relatively simple, throughout the nineteenth century case-histories became increasingly sophisticated as the discipline developed a consistent literature. Hence, at the turn of the twentieth century, when Penta explained to his students how to compile a case-history, he covered seventy-seven points ranging from basic data like name and age of the patient, to bodily measurements such as skull size. He also instructed his students to assess 'psychic functions' such as level of awareness, personality, erotic thoughts, impulses, social and moral feelings, dreams, and sleep patterns.[5] Penta's sophisticated survey was the culmination of at least a century of case-based reasoning in psychiatry.[6] Investigations into anomalous desire were already a feature of case-histories published at the beginning of the nineteenth century, and as the century wore on, these investigations became increasingly sophisticated.

The investigation of sexual fantasies was also central to psychoanalysis. At the beginning of the twentieth century, psychoanalysis and sexology were competing against each other to assert their jurisdiction over human sexuality and sexual desires as fields of study.[7] Sigmund Freud acknowledged his debt to sexologists such as Ellis in *Drei Abhandlungen zur Sexualtheorie* [*Three Essays on the Theory of Sexuality*] (1905).[8] In the first of these essays, 'Sexual Aberrations', Freud focuses on the subject of 'inversion', carefully separating it from the phenomenon of inherited degeneration.[9] He noted that most male inverts maintained the 'mental quality of masculinity' and that, overall, their appearance was masculine.[10] Even if he conceded that a better knowledge of the facts might reveal a different picture, the descriptions of female inversion in *Three Essays* are nonetheless striking: 'the active inverts exhibit masculine characteristics, both physical and mental, with peculiar frequency and look for femininity in their sexual objects'.[11] This inconsistency in Freud's analyses of male and female homosexuality is puzzling. It is unclear why descriptions of women inverts continued to rely heavily on their physical appearance, and whether their masculine attitudes influenced Freud's perception of female inverts' bodies. But Freud's observation on female inverts shows that female homosexuality was still linked to bodily anomalies, something that he was less willing to accept for men. To a certain extent, Freud's remark on the masculinity of female inverts harks back to the traditional association of female same-sex desires with virility.

There are many elements in Freud's text that are reminiscent of sexological discourses. For instance, he proposed that the experience of early childhood had a determining effect on homoerotic desires in later life,[12] which recalls the theory of association devised by Binet to explain and analyse inverted sexual feelings. Freud also remarked that the importance of sexual 'abnormalities' lay in facilitating the physician's understanding of 'normal development'.[13] This matched one of the basic premises of positivist Italian science: to understand the normal through the abnormal or, as Livi said in his 1875 article that would become one of the founding works of Italian psychiatry, madness represented a microscope that magnified and exaggerated normal feelings.[14] Freud also spoke of a universal bisexual disposition,[15] an idea that was common amongst sexologists in the 1890s. Finally, Freud argued that even a healthy person might display some form of sexual perversion, a view shared by many late nineteenth-century sexologists.[16]

Unlike several earlier sexologists studied in this book, Freud was not overly keen to interpret physical characteristics as clear signs of (male) sexual inversion. By contrast, even those sexologists who adopted a resolutely psychological approach to sexual inversion could not easily refuse the constitutional approach to sexuality. It is Freud's decidedly psychological turn that makes his characterisation of female homosexuality so striking. Despite challenging some sexological assumptions of his time, Freud was not able to take distance from the idea that female homosexuality was related to masculinity, at both mental and physical levels. This idea predated the concept of female sexual inversion, and was just one of the many older notions of female same-sex desires that continued to inform medical discourses throughout the nineteenth century and beyond.

Despite the fact that Freud refuted the link between homosexuality and degeneration, various twentieth-century homosexual writers have regarded psychoanalysis with suspicion, preferring sexological accounts of same-sex desires instead.[17] Numerous factors explain this lack of popularity amongst homosexuals, but the main objections are directed towards Freud's theory of the Oedipal Complex, and the definition of psychoanalysis as a 'talking cure'. On the one hand, accepting the Oedipal Complex implies accepting that there is a developmental difference between heterosexual and homosexual individuals whereby the latter remain fixed at an early stage of psychosexual development. On the other hand, in adopting the 'talking cure' one tacitly admits that homosexuality needs 'curing'.[18] From this perspective, Freud appears to be closer to physicians such as Lombroso and Bell – who would have been pleased to cure sexual inversion – than to Ellis's research into sex psychology. Rather than 'curing' differences, Ellis's work was predominantly interested in describing sexual behaviour and advocating acceptance, which in turn explains why twentieth-century 'sexual perverts' were so drawn to it.

To say that the medical model of sexual inversion did not replace other forms of understanding same-sex desires is not to contest the fact that at the turn of the nineteenth century psychiatric ideas about same-sex desires underwent a series of conceptual shifts. Nor is it to deny that in countries like Italy a particular model achieved hegemony over others. In the course of this book, I have shown both that the concept of sexual inversion incorporates many of the ideas it allegedly supersedes and that Britain resisted the theoretical implications of sexual inversion. The importance of these two points is that Britain's reluctance to accept sexual inversion and the persistence of older ways of thinking about same-sex desires allows us to understand that medical ideas of sexual inversion were not homogeneous, and in fact assimilate a range of different influences, including sexual fantasies.

These points have further implications that historians need to readdress. Studies of sexual inversion were part of medical science's growing interest in 'sexual aberrations', an area that had been conventionally associated with mental illness. Although medicine had a long-standing tradition of addressing problematic desires, historians interested in nineteenth-century medical representations of same-sex desires have failed to engage with pre-sexological writings, focusing only on writings published over the last decades of the nineteenth century. This gap in the existing research has indirectly consolidated a view of the late nineteenth century as a turning point in the history of sexuality. If present-day history of sexuality emphasises epistemological ruptures, it is also because its empirical research into early nineteenth-century sources is still scarce. Scholars often trace the history of the categories that have meaning in their own time, investigating the past to understand the present. They pose questions relevant to today's society and turn to the past in search of specific answers, rather than allowing the past to tell its own tale. Thus, outdated representations of female same-sex desires have been forgotten in dusty archives.

In the vast amount of literature that has exploited the idea of epistemological ruptures, most scholars employ this concept diachronically. There is another way to think about ruptures, however, and that is synchronically.[19] This book has shown that medical writings of the late nineteenth-century could be interpreted as marked by ruptures between fields: different medical specialties looked at the 'same' object as though it were two different things. One only needs to think about the ways in which in Britain Ellis and Bell looked at sexual inversion. Yet these constructions are also interrelated. Therefore, at the same time, it is very difficult to define precisely ruptures between medical fields. As I have shown in this book, competing medical explanations were not impermeable discourses; the case is, rather, that these explanations inform one another to varying degrees. Medical conceptualisations of sexual inversion, particularly of female same-sex desires, should not be thought of as a coherent whole. Instead, they form a fractured theory in which different elements are highlighted in different contexts.

This book has focused on debates about female same-sex desires in the narrow field of medical discourses, rather than on representations of female homosexuality across a broader range of fields such as law, religion, or literature.[20] Much work is still to be done in exploring how medical discourses about female same-sex desires influenced cultural constructions of lesbianism and how cultural assumptions influenced medical writings. A host of phantasms seems to haunt medical writings, which makes the task of compiling an exhaustive historical account of medical discourses much more complex than it would appear. For instance, it was a common practice for sexological works to quote from and briefly discuss novels such as Denis Diderot's *La Religieuse* [*The Nun*] (1760), Honoré de Balzac's *La Fille aux Yeux d'Or* [*The Girl with the Golden Eyes*] (1835), Théophile Gautier's *Mademoiselle de Maupin* (1835), or Émile Zola's *Nana* (1880). It is impossible to determine to what extent those readings may have influenced Ellis's or Lombroso's own sexological understanding of female homosexuality. In the second half of the nineteenth century there was a renewed interest in ancient cultures, especially Greek antiquity, which may also have influenced sexological representations of same-sex desires. Before co-authoring *Sexual Inversion* with Ellis, Symonds had relied on evidence from classical studies to frame his ideas about homosexuality. Both Lombroso and Penta mentioned classical texts such as Lucian's *Dialogues of the Courtesans* in their discussions of female same-sex desires.

My aim in this book has been to show how love between women was medicalised when sexual inversion emerged as a psychiatric category. I have also traced the roots and successive developments of various British and Italian medical figures of female same-sex desires, showing how the view of these two medical communities at times converged, and at other times diverged. Nevertheless, there is much to be gained from looking at other different models for understanding same-sex desires available to late nineteenth-century intellectuals. After all, physicians worked and wrote within a wider context that informed their thought. Moving beyond medicine, further prospective research might probe representations across different discursive domains and their mutual influences. Before the emergence of sexology, the question of what should be considered appropriate and inappropriate sexual acts had been addressed by medical, religious, and legal discourses. These disperse sexual knowledges are at the base of sexological theories: far from being spontaneously generated in the mind of a sexologist, the concept of sexual inversion has its own multi-rooted genealogies. Rather than emphasising apparent discontinuities, scholars would do well to produce longer narratives and to explore the cultural baggage of the concept of sexual inversion and various representations of same-sex desires.

It is also possible to trace links between late nineteenth-century sexological enquiry into the origin of sexual perversion, and twentieth-century medical research into the field of genetics. The quest for a 'gay gene' allegedly present in the chromosomes, which determines 'symptoms' such as finger length or ear-lobe size, reiterates sexological discourses that locate the source of same-sex desires in the body. When one considers the scientific commitment to measure sexual inclinations in the body, modern-day science differs very little from nineteenth-century sexology. While British gynaecologists and psychiatrists thought that an enlarged clitoris and abnormal uterus size were clear signs of innate same-sex preferences, contemporary scientists have turned to finger length. In 2000, *Nature* published an article declaring that finger length could reveal whether people were gay or straight. A group of scientists from the University of California, Berkeley, decided to interview 720 people in San Francisco about their sexual orientation. They then considered the finger length ratios, the influence of prenatal androgen exposure, and the mother's body memory to conclude that the body or, more precisely, finger length, could reveal an individual's sexual orientation.[21] These scientists were not the first to measure fingers: Lombroso even measured the degenerate's toes. The language has certainly changed since the nineteenth century, but some of the methods employed in today's scientific enquiry remain the same.

The legacy of nineteenth-century sexual science continues today in other forms, such as in contemporary psychiatry. The classification of 'Sexual Disorders' listed in the American Psychiatric Association's *Diagnostic and Statistical Manual of Mental Disorders* (hereafter *DSM-IV*), can be grouped into three categories. The first is sexual dysfunctions, which incorporates desire disorders, aversion disorder, arousal disorder (male and female), orgasmic disorder, sexual pain disorders, and sexual dysfunctions not otherwise specified. The second category comprises paraphilias: exhibitionism, fetishism, frotteurism, paedophilia, sexual masochism, sexual sadism, voyeurism, and paraphilias not otherwise specified. The last category refers to gender identity disorders, which are characterised by 'a strong and persistent cross-gender identification, which is the desire to be, or the insistence that one is of the other sex' and 'evidence of a persistent discomfort about one's assigned sex or a sense of inappropriateness in the gender role of that sex'.[22] Homosexuality does not appear in the list of 'paraphilias' and yet, the idea that there is a 'right' sexuality, that desires can be appropriate or inappropriate, pervades the *DSM-IV*. So, the idea that there are two natural genders remains. A girl's preference for 'boyhood games' and 'little interest in dolls or any form of feminine dress up or role play activity' might be an indication of gender identity disorder.[23] Certainly women are now encouraged to experience sexual pleasure, but pleasure is far from being a central issue in scientific studies. This is the legacy of nineteenth-century sexology: the

DSM-IV is not concerned to find out whether sexual masochists, fetishists, and so on are perfectly happy with their 'paraphilia'.

The medicalisation of sexual behaviour is not confined to homosexuality or paraphilias – twentieth-century psychiatric categories, after all, incorporate many of the groupings that nineteenth-century medical practitioners called sexual perversions. One needs only consider how sexual desire has been transformed into an addiction treatable in support groups within specialised clinics, although one might point out that nineteenth-century nymphomaniacs were confined to asylums. If we move away from sexuality, the invention of attention deficit disorders (ADD) illustrates how common behaviours can become medicalised. We live in a medicalised society and one cannot dismiss the power of medical ideas in shaping our attitudes towards more or less common behaviours.

In the twentieth century sexological research was often put to use in non-medical contexts. Although it has been suggested that twentieth-century psychoanalysis captured the popular imagination and displaced sexology, the lesbian writer Radclyffe Hall modelled the central character in the novel *The Well of Loneliness* (1928) on Ellis's and Hirschfeld's sexological studies, rather than on Freud's.[24] Hall believed that her novel would provide a moral argument against those who posed homosexuality was an example of indecent behaviour.[25] While Hall's book precipitated a trial for obscenity and was banned in England, it has become a classic and a bestseller. It was probably the most popular novel amongst female homosexuals until the 1960s, and has generated endless debates in feminist and lesbian studies.[26] The nineteenth-century image of the sexual invert did not disappear with the advent of psychoanalysis, nor did it curtail the appeal of sexology to self-identified homosexuals.

If in the first half of the twentieth century homosexuals turned to Ellis's work to find social acceptance, science in general still has a strong appeal to the gay and lesbian community. At a time when science is valued, biological and genetic evidence that sexual inclination is pre-determined is often believed to bring social acceptance. Since Ellis's time, sexual subcultures or, as we might call them today, sexual minorities, have applauded science. Since the end of the nineteenth century there have been examples of prominent campaigners for social acceptance of homosexuality, such as Hirschfeld, who were also medically trained and who used science to advance their political cause. In the 1990s a number of scientists and self-identified gay people became prominent within and outside the international scientific community for their scientific studies on sexual orientation. The geneticist Dean Hamer, for example, attempted to link male homosexuality to a gene that some men inherit from their mothers.[27] The neuroscientist Simon LeVay researched the differences between heterosexual and homosexual men's brains.[28] As with Hamer's research, LeVay's findings were widely reported in the media as proof that sexual orientation is linked to anatomy, despite the fact that LeVay cautioned against misrepresenting his results, stressing that he did not find homosexuality was genetic.[29]

The scientific study of sexuality is never an ideologically neutral pursuit and can imply different things in different cultural contexts. Thus, science can serve many political purposes. The Italian fascist regime used sexological research to build a theoretical apparatus designed to identify and round up homosexuals. It is well known that Lombroso's disciple, Salvatore Ottolenghi, founded the *Polizia scientifica* [Scientific Police] at the beginning of the twentieth century, and that criminal anthropology supplied the fascist regime with interpretative and practical tools to suppress various forms of deviance. During this period Italian police also investigated homosexuality amongst female prostitutes.[30] Long after the fall of fascism, the language of sexological research is still in force in Italy: in March 2007, the proposal of a bill that would grant rights such as the recognition of civil partnerships for homosexual couples generated a political debate of considerable proportions. In this context, the Catholic senator Paola Binetti, who used to belong to the centre-left coalition, asserted that homosexuals were 'deviations' from 'the norm inscribed into the morphological, genetic, endocrine, and personality codes'.[31] A few days later, the Italian psychiatric association replied that they had not endorsed such a position in decades, that they did not support the senator, and that her statements were not founded upon science. In the ensuing weeks, Italian politicians, journalists, Catholic spokespeople and scientists argued about the nature of homosexuality. Surprisingly, many Catholics used science to justify their claims that homosexuality is a form of abnormality; homosexuality, they said, is a condition requiring treatment. Even the Vatican has recently embraced science in its battle against homosexuality. In April 2010, the Vatican's Secretary of State, Cardinal Tarcisio Bertone, perhaps the most important person in the Vatican hierarchy after the Pope, suggested that paedophilia is linked to homosexuality. He remarked that 'many psychologists and psychiatrists have demonstrated [...] that there is a relation between homosexuality and paedophilia. That is true. That is the problem. [...] It's a pathology that relates to all social categories, and to a lesser extent priests.'[32] Invoking the authority of psychiatry and associating homosexuality with paedophilia, Bertone aimed to establish a link between homosexuality and mental pathology. Psychologists and psychiatrists promptly disowned such statements.[33]

Alternatively championed and spurned by self-identified homosexuals, scientific discourse has become the cardinal reference point in popular debates. Nineteenth-century medical discourses greatly contributed to shape modern ideas of homosexuality. While some might argue that the concept of sexual identity itself has limitations from the outset by defining a person, it has nonetheless been embraced by minority groups with long-lasting effects.[34] This does not mean that I am underestimating the individual's sexual fluidity, or that I assume all individuals who find pleasure in same-sex activities share an identity, a natural essence, or a psychology. Nor do I think that it is impossible to resist the medicalisation of sexuality. I am merely suggesting that, regardless of whether attempts to understand human sexuality conform to scientific standards, it is clear that scientific theories are rife with political implications. The study of medical concepts and their uneven and overlapping trajectories remains therefore a privileged means of bringing such implications to the fore.

Notes

Note on Terminology

1. E. Willard, *Sexology as the Philosophy of Life* (1867), 11–16.
2. C. Waters, 'Sexology', in H. G. Cocks and M. Houlbrook (eds.), *Palgrave Advances in the Modern History of Sexuality* (2006), 42–3.

1 Introduction: Female Sexual Inversion and other Medical Embodiments of Female Same-Sex Desires in Italy and Britain, circa 1870–1920

1. C. F. Michéa, 'Des deviations de l'appétit vénérien', *Union Medicale*, July 1849, vol. 3, 338–9; J. L. Casper, 'Ueber Nothzucht und Päderastie und deren Ermittelung Seitens des Gerichtesarztes', *Vierteljahrschrift für gerichtliche öffentliche Medizin*, 1852, vol. 1, 21–78. On the importance of these texts, see for example G. Hekma, 'A History of Sexology: Social and Historical Aspects of Sexuality', in J. Bremmer (ed.), *From Sappho to De Sade: Moments in the History of Sexuality* (1989), 173–93; H. Oosterhuis, *Stepchildren of Nature: Krafft-Ebing, Psychiatry and the Making of Sexual Identity* (2000), 39; I. Crozier, 'Introduction', in Havelock Ellis and John Addington Symonds, *Sexual Inversion: A Critical Edition*, ed. I. Crozier (2008), 18–19. The novelist Karl Maria Kertbeny coined the term 'homosexuality' between 1869 and 1870. In the early nineteenth century works on same-sex desires were published now and then, but most historians have argued that these early works did not stimulate any specific growth in the field of sexology. For early works on same-sex desires, see P. Gutmann, 'On the Way to Scientia Sexualis: "On the relation of the sexual system to the psyche in general and to cretinism in particular" (1826) by Joseph Häussler', *History of Psychiatry*, 2006, vol. 17, 45–53. For the growth of sexology, homosexuality and Westphal's text, see also G. Hekma, 'A Female Soul in a Male Body: Sexual Inversion as Gender Inversion in Nineteenth Century Sexology', in G. Herdt (ed.), *Third Sex, Third Gender* (1994), 213–39. The impact of homosexuals such as Karl Heinrich Ulrichs on the development of sexology has also been noted: see V. L. Bullough, 'The Physician and Research into Human Sexual Behavior in Nineteenth-Century Germany', *Bulletin for the History of Medicine*, 1989, vol. 63, 247–67; J. Hutter, 'The Social Construction of Homosexuals in the Nineteenth Century: The Shift from the Sin to the Influence of Medicine on Criminalizing Sodomy in Germany', *Journal of Homosexuality*, 1993, vol. 24, 73–93.
2. W. Griesinger, 'Vortrag zur Eröffnung der psychiatrischen Clinik', *Archiv für Psychiatrie und Nervenkrankheiten*, 1868, vol. 1, 651.
3. Ibid.
4. C. Westphal, 'Contrary Sexual Feeling: Symptom and a Neuropathic (Psychopathic) Condition', trans. M. A. Lombardi-Nash, in *Sodomites and Urnings: Homosexual Representations in Classic German Journals* (2006), 87–120. The

original article was C. Westphal, 'Die conträre Sexualempfindung: Symptom eines neuropathischen (psychopathischen) Zustandes', *Archiv für Psychiatrie und Nervenkrankheiten*, 1869–70, vol. 2, 73–108. The importance of this case and its effects on psychiatric writing about homosexuality is discussed in I. Crozier, 'Pillow Talk: Credibility, Trust and the Sexological Case History', *History of Science*, 2008, vol. 46, 383–6. Westphal also presented a case-study, although the main characteristic of this case was the man's cross-dressing.

5. Westphal, 'Contrary Sexual Feeling', 88–96.

6. R. von Krafft-Ebing, 'Über gewisse Anomalies des Geschlechtstriebs und die klinisch-forensich Verwertung derselben als eines wahrscheinlich funktionellen Degenerationszeichens des centralen Nervensystems', *Archiv für Psychiatrie und Nervenkrankheiten*, 1877, vol. 7, 291–312.

7. On Tamassia, see C. Beccalossi, 'The Origin of Italian Sexological Studies: Female Sexual Inversion ca. 1870–1900', *Journal of the History of Sexuality*, 2009, vol. 18, 109–11. On Charcot and Magnan, see V. A. Rosario, *The Erotic Imagination: French Histories of Perversity* (1997), 69, 83–9. Some of the early American cases include: M. Wise, 'Case of Sexual Perversion', *Alienist and Neurologist*, 1883, vol. 4, 87–91; J. C. Shaw and G. N. Ferris, 'Perverted Sexual Instinct (Contrare Sexualempfindung: Westphal; Inversione dell'instinto sessuale: Arrigo Tamassia; Inversion du sens génital: Charcot et Magnan)', *Journal of Nervous and Mental Disease*, 1883, vol. 10, 185–204; J. C. Kiernan, 'Perverted Sexual Instinct', *Chicago Medical Journal and Examiner*, 1884, vol. 48, 263–5. On these American writings, see B. Hansen, 'American Physicians' Earliest Writings about Homosexuals, 1880–1900', *The Milbank Quarterly*, 1989, vol. 67, suppl. 1, 99–108. On English articles, see Chapter 4.

8. For an overview, see Ellis and Symonds, *Sexual Inversion. A Critical Edition*, ed. Crozier, 115–23.

9. M. Foucault, *The Will to Knowledge, The History of Sexuality*, vol. 1, trans. R. Hurley (1998), 43.

10. M. McIntosh, 'The Homosexual Role', *Social Problems*, 1968, vol. 16, 182–92.

11. V. L. Bullough, 'Homosexuality and the Medical Model', *Journal of Homosexuality*, 1974, vol. 1, 99–110; J. Weeks, 'Movements of Affirmations: Sexual Meanings and Homosexual Identities', *Radical History Review*, 1979, vol. 20, 164–79; J. Weeks, *Sex, Politics and Society: The Regulation of Sexuality since 1800* (1981).

12. A. I. Davidson, *The Emergence of Sexuality: Historical Epistemology and the Formation of Concepts* (2001). This book contains a series of essays published since the late 1980s, see especially A. I. Davidson, 'Sex and the Emergence of Sexuality', *Critical Inquiry*, 1987, vol. 14, 16–48 and A. I. Davidson, 'Closing up the Corpses: Diseases of Sexuality and the Emergence of the Psychiatric Style of Reasoning', in G. Boolos (ed.), *Meaning and Method: Essays in Honor of Hilary Putnam* (1990), 295–326.

13. Davidson, *The Emergence of Sexuality*, 3.

14. Ibid.

15. Ibid.

16. Ibid., 35–6.

17. Ibid., 16–22.

18. Ibid., 11. Davidson also specified that his 'three-stage structural partition [in modes of thinking about sexual perversion] does not precisely coincide with historical chronology', because the three styles of reasoning were often combined, sometimes even in a single medical text. More specifically, the shift from brain to sexual instinct aetiologies of sexual perversion cannot be separated by a fixed, datable dividing line. Despite this remark, Davidson is more concerned with

epistemic ruptures throughout time rather than pinpointing the implications of the coexistence of 'different styles of reasoning' within medicine. Nor has he explained how different 'styles of reasoning' manage to coexist. Ibid., 3. See also C. Beccalossi, 'Female Same-Sex Desires: Conceptualizing a Disease in Competing Medical Fields in Nineteenth-Century Europe', *Journal of the History of Medicine and Allied Sciences* (forthcoming 2012).

19. See for example: S. Brady, *Masculinity and Male Homosexuality in Britain, 1861–1913* (2005); H. G. Cocks, *Nameless Offences: Homosexual Desire in the Nineteenth Century* (2003); M. Cook, *London and the Culture of Homosexuality, 1885–1914* (2003).

20. R. A. Nye, 'The History of Sexuality in Context: National Sexological Tradition', *Science in Context*, 1991, vol. 4, 387–406.

21. E. K. Sedgwick, *Epistemology of the Closet* (1990), see especially at 40–8.

22. More recently, medievalists and early modern historians have taken issue with the dichotomy between pre- and post-nineteenth-century notions of same-sex desires, which is correlative with the so-called 'acts paradigm' and the 'identity paradigm'. M. D. Jordan, *The Invention of Sodomy in Christian Theology* (1997); T. van der Meer, 'Sodomy and its Discontents: Discourse, Desire, and the Rise of a Same-sex Proto-something in the Early Modern Dutch Republic', *Historical Reflections*, 2007, vol. 33, 41–67; K. Borris and G. Rousseau (eds.), *The Sciences of Homosexuality in Early Modern Europe* (2008).

23. M. D. Halperin, *One Hundred Years of Homosexuality* (1990); M. D. Halperin, *How to Do the History of Homosexuality* (2002).

24. Halperin, *How to Do the History of Homosexuality*, 105–6.

25. Ibid., 109–10. For a search of types in lesbian history, see V. Traub 'The Present Future of Lesbian Historiography', in G. E. Haggerty and M. McGarry (eds.), *A Companion to Lesbian, Gay, Bisexual, Transgender, and Queer Studies* (2007), 124–45.

26. Scholars working within science and technology studies have shown that medicine is not a homogeneous and unified field. See for example: A. Mol, *The Body Multiple: Ontology in Medical Practice* (2002).

27. Nye, 'The History of Sexuality in Context'.

28. Ibid., 387–406; A. Copley, *Sexual Moralities in France* (1989).

29. Oosterhuis, *Stepchildren of Nature*.

30. J. D. Steakley, ' "Per scientiam ad justitiam": Magnus Hirschfeld and the Sexual Politics of Innate Homosexuality', in V. A. Rosario (ed.), *Science and Homosexualities* (1997), 133.

31. L. Hall, 'Hauling Down the Double Standard: Feminism, Social Purity and Sexual Science in Late Nineteenth-Century Britain', *Gender & History*, 2004, vol. 16, 36–56.

32. I. Crozier, 'Nineteenth Century British Psychiatric Writing on Homosexuality before Havelock Ellis: The Missing Story', *Journal of the History of Medicine and Allied Sciences*, 2008, vol. 63, 65–102.

33. More work has been done in the American context, see for example: L. Duggan, *Sapphic Slashers: Sex, Violence and American Modernity* (2000); J. Terry, *An American Obsession: Science, Medicine and Homosexuality in Modern Society* (1999).

34. For an overview, see N. Milletti, 'Analoghe sconcezze. Tribadi, saffiste, invertite e omosessuali: categorie e sistemi sesso/genere nella rivista di antropologia criminale fondata da Cesare Lombroso (1880–1949)', *DWF*, 1994, vol. 4, 50–122.

35. Recently, Valerie Traub has urged historians interested in the history of female same-sex desires to situate their work in a comparative and transnational

perspective. Traub, 'The Present Future of Lesbian Historiography'. For a rare example of comparative history in the field, see S. Marcus, 'Comparative Sapphism', in M. Cohen and C. Dever (eds.), *The Literary Channel: The Inter-National Invention of the Novel* (2002), 251–85. For a discussion of how German sexological ideas found their way into English culture, see H. Bauer, *English Literary Sexology: Translations of Inversion, 1860–1930* (2009).

36. On sexology, see E. Brecher, *The Sex Researchers* (1969); R. Pearsall, *The Worm in the Bud* (1983) [1969]. On psychoanalysis, see E. Glover, 'Victorian Ideas of Sex', in BBC (ed.), *Ideas and Beliefs of the Victorians* (1949), 362–4.

37. Foucault, *The History of Sexuality*; M. Foucault, *Abnormal* (2003) [1999].

38. See for example M. Vicinus (ed.), *Suffer and Be Still: Women in the Victorian Age* (1972). For women's history in Italy and Britain, see A. Rossi-Doria (ed.), *A che punto è la storia delle donne in Italia* (2003); S. Alexander, *Becoming a Woman* (1994). On psychoanalysis and women's history, see Alexander at 225–30.

39. See for example L. Faderman, *Surpassing the Love of Men: Friendships between Women from the Renaissance to the Present* (1991) [1981]; S. Jeffreys, *The Spinster and Her Enemies: Feminism and Sexuality* (1985).

40. C. Smith-Rosenberg, 'The Female World of Love and Ritual', *Signs*, 1975, vol. 1, 1–29; Faderman, *Surpassing the Love of Men*; L. Faderman, *Odd Lovers and Twilight Girls: A History of Lesbian Life in Twentieth Century America* (1991); J. D'Emilio and E. B. Freedman, 'Problems Encountered in Writing the History of Sexuality: Sources, Theory and Interpretation', *Journal of Sex Research*, 1990, vol. 27, 481–95.

41. L. Doan, *Fashioning Sapphism: The Origin of the Modern Lesbian Culture* (2001). See also, M. Vicinus, 'They Wonder to Which Sex I Belong: The Historical Roots of the Modern Lesbian Identity', in D. Altman, C. Vance, and M. Vicinus (eds.), *Homosexuality, Which Homosexuality?* (1989), 485.

42. Ellis and Symonds, *Sexual Inversion, a Critical Edition*, ed. Crozier, 161.

43. R. von Krafft-Ebing, *Psychopathia Sexualis with Special Reference to Contrary Sexual Instinct: A Medico Legal Study*, trans. C. G. Chaddock [of the seventh German edition, 1892] (1893), 185–319.

44. Ellis and Symonds, *Sexual Inversion, a Critical Edition*, ed. Crozier, 124–80. In addition, Ellis and Symonds published three American cases (a man and two women) provided by the American psychiatrist G. Franck Lydston. Ibid., 187–98.

45. See for example Rosario, *The Erotic Imagination*, 108; A. Oram and A. Turnbull (eds.), *The Lesbian History Sourcebook: Love and Sex between Women in Britain from 1780 to 1970* (2001), 93.

46. T. Castle, *The Apparitional Lesbian: Female Homosexuality and Modern Culture* (1993).

47. On early modern literary studies see, for example, E. S. Wahl, *Invisible Relations: Representations of Female Intimacy in the Age of Enlightenment* (1999); V. Traub, *The Renaissance of Lesbianism in Early Modern England* (2002). On literary studies see H. Bauer, 'Theorizing Sexual Inversion: Sexology, Discipline and Gender at the Fin de Siècle, *Journal of the History of Sexuality*, 2009, vol. 18, 84–102; J. Halberstam, *Female Masculinity* (1998).

2 Sexuality in *Post-Risorgimento* Italy and Victorian Britain

1. See Chapter 6.

2. M. A. Raffalovich, *L'uranismo, inversione sessuale congenita* (1896); P. Näcke, (review) 'M. A. Raffalovich, *Uranisme et unisexualité*', *APS*, 1896, 168;

G. C. Ferrari, (review) 'M. A. Raffalovich, *L'uranismo. Il processo Oscar Wilde*, Bocca, Torino, 1896; A. V. Schrenck Notzing, *La terapia suggestiva delle psicopatie sessuali*, Bocca, Torino, 1897; A. Niceforo, *Il gergo nei normali, nei degenerati e nei criminali*, Bocca, Torino, 1897', *RSF*, 1896, vol. 22, 888–9.

3. M. A. Raffalovich, 'Gli studi sulle psicopatie sessuali in Inghilterra', *APS*, 1896, 176–181.

4. J. Pemble, *The Mediterranean Passion: Victorians and Edwardians in the South* (1987), 154–69.

5. R. Aldrich, *The Seduction of the Mediterranean: Writing, Art, and Homosexual Fantasy* (1993).

6. See the correspondence between Ellis and Symonds: J. A. Symonds, *The Letters of John Addington Symonds, vol. 3, 1885–1893* (1969), ed. H. M. Schuller and R. L. Peters, 710; H. Ellis, BL ADD 70524, Letter to J. A. Symonds, 21 Dec. 1892, 3 Jan. 1893.

7. Pemble, *The Mediterranean Passion*, 142–3; P. Grosskurth (ed.), *The Memoirs of John Addington Symonds* (1984), 271–83; Aldrich, *The Seduction of the Mediterranean*.

8. Pemble, *The Mediterranean Passion*, 100–1; Aldrich, *The Seduction of the Mediterranean*.

9. J. Money, *Capri: Island of Pleasure* (1986); C. Gargano, *Capri pagana* (2007).

10. M. Vicinus, *Intimate Friends* (2004), 31–55.

11. C. P. Brand, *Italy and the English Romantics: The Italian Fashion in Early Nineteenth-century England* (1957); J. Buzard, *The Beaten Track: European Tourism, Literature and the Way to 'Culture', 1800–1918* (1993).

12. Ibid., ix.

13. Ibid.; M. O'Connor, *The Romance of Italy and the English Political Imagination* (1998); M. Pfister and R. Hertel (eds.), *Performing National Identity: Anglo-Italian Cultural Transactions* (2008).

14. E. M. Forster, *Where Angels Fear to Tread* [1905] (1976); E. M. Forster, *A Room with a View* [1908] (2000).

15. G. M. Trevelyan, *Garibaldi's Defence of the Roman Republic* (1907); G. M. Trevelyan, *Garibaldi and the Thousand* (1909); G. M. Trevelyan, *Garibaldi and the Making of Italy: June–November 1860* (1911).

16. J. A. Davis, 'Introduction: Italy's Difficult Modernization', in J. A. Davis (ed.), *Italy in the Nineteenth Century, 1796–1900* (2000), 1–24.

17. C. Duggan, 'Gran Bretagna e Italia nel Risorgimento', in A. M. Banti and P. Ginsborg (eds.), *Storia d'Italia. Il Risorgimento* (2007), 777–96.

18. M. Isabella, *Risorgimento in Exile: Italian Émigrés and the Liberal International in the Post-Napoleonic Era* (2009).

19. The *Statuto Albertino* was the constitution granted by King Charles Albert of Piedmont-Sardinia in 1848, which subsequently became the constitution of the new kingdom of Italy.

20. For an overview, see L. Riall, *Risorgimento: The History of Italy from Napoleon to Nation-State* (2009). On religion, see D. I. Kertzer, 'Religion and Society, 1789–1892', in Davis (ed.), *Italy in the Nineteenth Century*, 181–205.

21. Kertzer, 'Religion and Society, 1789–1892'.

22. F. Giacanelli, 'Appunti per una storia della psichiatria italiana', in K. Dörner, *Il borghese e il folle* (1975), xxix–xxx.

23. Ibid., v–xxxii.

24. P. Martucci, *Le piaghe d'Italia* (2002).

25. Like many of his followers, Lombroso joined the Italian Socialist Party after it was founded in 1892 because the liberal government's inability to improve the lot of the poor left him disillusioned.
26. G. Cosmacini, *Storia della medicina e della sanità in Italia* (1987), 394–5.
27. B. Wanrooij, 'The History of Sexuality in Italy (1860–1945)', in P. Wilson (ed.), *Gender, Family and Sexuality: The Private Sphere in Italy, 1860–1945* (2004), 177; B. Wanrooij, *La storia del pudore* (1990).
28. H. G. Cocks, 'Religion and Spirituality', in H. G. Cocks and M. Houlbrook (eds.), *The Modern History of Sexuality* (2006), 164.
29. L. Colley, *Britons: Forging the Nation, 1707–1837* (1992).
30. J. Wolffe, *Great Deaths: Grieving, Religion, and Nationhood in Victorian and Edwardian Britain* (2000), 5.
31. P. E. Stepansky, 'A Footnote to the History of Homosexuality in Britain: Havelock Ellis and the Bedborough Trial of 1898', *Essays in the History of Psychiatry* (1980), 72–102.
32. Cocks, *Nameless Offences*, 78; Brady, *Masculinity*, especially chapter 5.
33. J. Antonelli, *Medicina pastoralis in usum confessarium et curiarum ecclesisticarum* (1905), vol. 2, 203–5; A. Gemelli, *Non Moechaberis. Disquisitiones Medicale in Usum Confessarioum. Quaestiones Theologiae Medico-Pastoralis* (1911), especially chapter 4.
34. B. Wanrooij, 'La passione svelata: sessualità, crimine ed educazione in Italia tra Ottocento e Novecento', *Sanità, scienza e storia*, 1987, vol. 5, 397; A. Gemelli, *La tua vita sessuale. Lettera ad uno studente universitario* (1946) [1941], 11–17.
35. Wanrooij, 'The History of Sexuality in Italy', 178.
36. L. Benadusi, *Il nemico dell'uomo nuovo* (2005), 103; G. Dall'Orto 'La "tolleranza repressiva" dell'omosessualità', in Arci Gay Nazionale (ed.), *Omosessuali e stato* (1988), 37–57. I will return to this point later, when discussing the implications of the 1889 Zanardelli Penal Code for male same-sex practices.
37. *Querela di parte* is the option for the victim of a crime, or his or her guardian where the victim is underage, to bring legal action against an offender. This is to say that the legal system is not compelled to proceed.
38. A. Buttafuoco, *Le mariuccine* (1985), 110.
39. C. Lombroso, 'Delitti di libidine', *AP*, 1883, vol. 4, 338.
40. Cocks, 'Religion and Spirituality', 163–5. For the Tractarians see also Lytton Strachey's description of Cardinal Manning, in his *Eminent Victorians* (1918).
41. F. Knight, '"Male and Female He Created Them": Men, Women and the Question of Gender', in J. Wolffe (ed.), *Religion in Victorian Britain* (1997), vol. 5, 26.
42. N. F. Cott, 'Passionlessness: An Interpretation of Victorian Sexual Ideology, 1790–1850', *Signs*, 1978, vol. 4, 219–36; K. Harris, *Sexual Ideology and Religion* (1984).
43. Mason stresses that understanding this notion of anti-sensualism is essential to a proper understanding of Victorian middle-class society. M. Mason, *The Making of Victorian Sexuality* (1994).
44. Bénédict Augustin Morel was born in Vienna, but grew up in France where he received his medical training.
45. I. A. Dowbiggin, *Inheriting Madness: Professionalisation and Psychiatric Knowledge in Nineteenth-Century France* (1991), 116–43.
46. D. H. Tuke, 'Degeneration', in D. H. Tuke (ed.), *A Dictionary of Psychological Medicine* (1892), vol. 1, 332.

47. P. Penta, 'L'origine e la patogenesi della inversione sessuale, secondo Krafft-Ebing e altri autori', *APS*, 1896, 54. For more on Penta and degeneration, see Chapter 6.
48. J. E. Chamberlain and S. L. Gilman (eds.), *Degeneration: The Dark Side of Progress* (1985); D. Pick, *Faces of Degeneration* (1989); M. Gervasoni, ' "Cultura della degenerazione" tra socialismo e criminologia alla fine dell'Ottocento In Italia', *Studi Storici*, 1997, vol. 3, 1087–1119.
49. Pick, *Faces of Degeneration*.
50. Ibid., 114–15.
51. Ibid., 176, 184–5. See also M. Neve, 'The Influence of Degenerationist Categories in Nineteenth-Century Psychiatry, with Special Reference to Great Britain', in Y. Kawakita, S. Sakai, and Y. Otsuka (eds.), *History of Psychiatric Diagnosis* (1997), 141–63.
52. S. L. Gilman, *Difference and Pathology* (1985), 191–2.
53. J. J. Howard, 'The Civil Code of 1865 and the Origins of the Feminist Movement in Italy', in B. Boyd Caroli, R. F. Harney, and L. F. Tomasi (eds.), *The Italian Immigrant Woman in North America* (1978), 14–24; P. Willson, 'Introduction: Gender and the Private Sphere in Liberal and Fascist Italy', in P. Willson (ed.), *Gender, Family and Sexuality* (2004), 8.
54. E. Sarogni, *La donna italiana, 1861–2000. Il lungo cammino verso i diritti* (2004), 1–97.
55. Howard, 'The Civil Code of 1865', 12–24.
56. Sarogni, *La donna italiana*, 97.
57. Of course they could only do this if they met the patrimony and literacy criteria, as with men.
58. Howard, 'The Civil Code of 1865', 14–16.
59. L. A. Jackson, 'The Regularisation of Sexuality in the Age of Empire', in C. Beccalossi and I. Crozier (eds.), *A Cultural History of Sexuality in the Age of Empire* (2010), 87.
60. J. Butler, 'An Appeal to the People of England' (first published 1870), reproduced in S. Jeffreys (ed.), *The Sexuality Debates* (1987), 126.
61. L. Stone, *Road to Divorce* (1990), 368–90.
62. L. A. Hall, *Sex, Gender and Social Change in Britain since 1880* (2000), 10–12.
63. Gilman, *Difference and Pathology*, 76–108.
64. J. R. Walkowitz, *Prostitution and Victorian Society* (1980); M. Gibson, *Prostitution and the State of Italy* (1986).
65. Hall, *Sex, Gender and Social Change*, 22.
66. A. J. B. Parent-Duchâtelet, *De la prostitution* (1835–6); W. Acton, *Prostitution* [1857] (1870).
67. G. Gattei, 'La sifilide: medici e poliziotti intorno alla "Venere politica" ', in F. Della Peruta (ed.), *Storia d'Italia, Annali 7* (1984), 755–75.
68. Ibid.
69. The Act of 1864 stated that women found to be infected could be interred in locked hospitals for up to three months, which was gradually extended until the 1869 Act increased the time to a year.
70. L. Mahood, *The Magdalenes: Prostitution in the Nineteenth Century* (1990), 1–2, 63–7; R. Davidson, *Dangerous Liaisons* (2000), 26.
71. A. Summers, ' "The Constitution Violate": The Female Body and the Female Subject in the Campaigns of Josephine Butler', *History Workshop Journal*, 1999, vol. 48, 1–15.
72. B. Caine, *English Feminism, 1780–1920* (1997), 123.

73. Howard, 'The Civil Code of 1865', 16.
74. Ibid.; M. A. Manacorda, 'Istruzione ed emancipazione della donna nel Risorgimento', in S. Soldani (ed.), *L'educazione delle donne* (1989), 1–33.
75. Ibid., 7–8.
76. A. Buttafuoco, 'In servitù regine', in Soldani (ed.), *L'educazione delle donne*, 363–91, especially at 365–8.
77. A. Buttafuoco, *Cronache femminili* (1988).
78. M. De Giorgio, *Le italiane dall'Unità ad oggi* (1992), 26.
79. In my view Mary Gibson has overestimated the role of Italian feminists in challenging sexual moralities. M. Gibson, *Born to Crime* (2002), 55–60.
80. De Giorgio, *Le italiane dall'Unità ad oggi*, 21–3.
81. V. P. Babini, F. Minuz, and A. Tagliavini, *La donna nelle scienze dell'uomo* (1986); V. P. Babini, 'Un altro genere. La costruzione scientifica della "natura femminile"', in A. Burgio (ed.), *Nel nome della razza. Il razzismo nella storia d'Italia, 1870–1945* (1999), 475–89; A. Rossi-Doria, 'Antisemitismo e antifemminismo nella cultura positivistica', in Burgio (ed.) *Nel nome della razza* (1999), 455–73.
82. A similar analysis is provided by Babini.
83. In 1832, Pope Gregorio XVI established sodomy should be punished with a life sentence: '*delitto consumato contro natura*'. See Regolamento Gregoriano, art. 178. Quoted in E. Oliari, *L'omo delinquente* (2006), 17.
84. Art. 425 'Codice Penale per il Regno di Sardegna, 20/11/1859 – Libro II, Titolo VII, Dei Reati contro il Buon Costume'. Quoted in ibid.
85. Benadusi, *Il nemico dell'uomo nuovo*, 98–9.
86. Giovanni Carmignani (1768–1847) was Professor of Penal Law at the University of Pisa and remained an authority in his field throughout the nineteenth century.
87. Benadusi, *Il nemico dell'uomo nuovo*, 103.
88. Ibid.
89. Stepansky, 'A Footnote to the History of Homosexuality in Britain', 96–7. Incidentally, Havelock Ellis translated *La Terre* from the French.
90. Camera dei Deputati, *Progetto del Codice penale per il Regno d'Italia e disegno di legge che ne autorizza la pubblicazione* (1887), vol. 1, 213–14 (22 novembre 1887); Senato del Regno, *Relazione della commissione speciale che autorizza il Re a pubblicare il Codice penale per il Regno d'Italia*, s.i.t. (1888), 183–5. On the English legal tradition of preventing discussion about (male) same-sex acts, see L. J. Moran, *The Homosexual(ity) of Law* (1996), ch. 3. Of course, one might argue that although the Italian and British goverments used opposing methods for dealing with male same-sex acts, both goverments aimed to promote ignorance. Yet, it is clear that the Italian goverment did not ban public discussion of the subject.
91. Symonds, *The Letters of John Addington Symonds*, vol. 3, 507.
92. The Buggery Act of 1533 remained the basis for English legal treatment of homosexuality until 1967. Brady, *Masculinity*, 27. Homosexuality was not decriminalised in Scotland until 1980.
93. Cocks, *Nameless Offences*, 17–18.
94. F. B. Smith, 'Labouchère's Amendment to the Criminal Law Amendment Act', *Historical Studies*, 1976, vol. 17, 159–69.
95. This was similar to the situation of German homosexuals, who were judged under the Prussian legal code after the unification. J. Steakley, *The Homosexual Emancipation Movement in Germany* (1975).
96. Pemble, *The Mediterranean Passion*, 159.

97. L. Bland, *Banishing the Beast* (1995), 289; Cook, *London and the Culture of Homosexuality*.
98. M. S. Foldy, *The Trials of Oscar Wilde* (1997), 89–90; D. Pritchard, *Oscar Wilde* (2001), 149.
99. Oliari, *L'omo delinquente*.
100. In Germany, the Eulenburg affair achieved greater notoriety. Prominent members of Kaiser Wilhelm II's cabinet and entourage between 1906 and 1909 were accused of homosexuality, which at the time was a legally punishable offence in Germany. Several trials followed.
101. Oliari, *L'omo delinquente*, 78–90. Baron Wilhelm von Gloeden was another German enamoured of southern Italy who was involved in sexual scandals. He moved to Taormina (Sicily) where he worked as a photographer specialising in male nudes. Alongside this activity he organised meetings between the subjects of his pictures and men from Germany and England who wrote to von Gloeden wanting to experience some of the beauties of Italy. The police discovered his trade, and in 1908 he was sentenced to one month in prison and fined for procuring male minors for prostitution. His trial and sentence stirred up a debate between local Sicilian newspapers defending von Gloeden's art (and the economic benefits of his art for Taormina), and socialist and Catholic newspapers associating homosexuality with prostitution, paedophilia, and 'foreign vices'.
102. Oliari, *L'omo delinquente*; A. Oram, 'Cross-Dressing and Transgender', in Cocks and Houlbrook (eds.), *The Modern History of Sexuality*, 263–6.

3 Italy: The Fashionable Psychiatric Disorder of Sexual Inversion and other Medical Embodiments of Same-Sex Desires

1. For an overview of the Italian primary sources on sexual inversion, see G. Dall'Orto, *Leggere Omosessuale* (1984); Milletti, 'Analoghe sconcezze'.
2. For endorsement of the view that nineteenth-century medical writers did not look at female-same sex desires, see for example Rosario, *The Erotic Imagination*, 108–9; Castle, *The Apparitional Lesbian*.
3. F. De Peri, 'Il medico e il folle: introduzione psichiatrica, sapere scientifico e pensiero medico fra Otto e Novecento', in F. Della Peruta (ed.), *Storia d'Italia. Annali 7* (1984), 1059–1140. For historical accounts of the situation of Italian asylums as described by British physicians during the nineteenth century, see J. H. Davidson, 'Remarks on Some of the Large Asylums of Italy', *JMS*, 1874, vol. 20, 410–15; H. C. Burnett, *Hospitals and Asylums of the World* (1891), 58–9, 111–33, 463–77; F. Needham, 'A Visit to Some Foreign Asylums', *JMS*, 1892, vol. 38, 222–7.
4. Andrea Verga (1811–95), alienist and one of the editors of the *Gazzetta medica*, was supervisor of the asylum at Senavra from 1848 to 1852, and subsequently Director of the Ospedale Maggiore in Milan.
5. Giacanelli, 'Appunti per una storia', xiv–xx.
6. Ibid.
7. Serafino Biffi and Cesare Castiglioni were well-known alienists working in the Milan area.
8. De Peri, 'Il medico e il folle', 1081–2.
9. The term *'freniatria'* comes from the Greek *'fren'*, brain. It was preferred to *'psyche'*, which referred to the soul and had metaphysical connotations.

10. The sense in which these Italian psychiatrists were positivists should be understood by analogy to the 'scientific naturalism' claimed by Victorian psychiatrists in Britain, rather than as a direct legacy of Comte.
11. A. Tagliavini, 'La scienza psichiatrica', in V. P. Babini, M. Cotti, F. Minuz, and A. Tagliavini, *Tra sapere e potere* (1982), 98–9; V. P. Babini, *La questione dei frenastenici* (1996); A. Tagliavini, 'Aspects of the History of Psychiatry in Italy in the Second Half of the Nineteenth Century', in W. F. Bynum, R. Porter, and M. Shepherd (eds.), *The Anatomy of Madness* (1985), vol. 2, 179; De Peri, 'Il medico e il folle', 1086.
12. Tagliavini, 'La scienza psichiatrica', 85–6.
13. Carlo Livi (1823–77) first worked in Siena as Professor of Legal Medicine and director of the city asylum. From 1873 he was Director of the Reggio Emilia asylum, which became a famous psychiatric research centre. Amongst his studies was *Frenologia forense* (1865–8).
14. Tagliavini, 'Aspects of the History of Psychiatry', 178–83.
15. Enrico Morselli (1852–1929) was Director of the Macerata asylum, and later ran the mental clinic of Turin and Genoa. He founded journals such as *Rivista sperimentale di freniatria*, *Rivista di filosofia scientifica*, and *Rivista di patologia nervosa e mentale*. Some of his works are well-known: *Manuale di semejotica delle malattie mentali* (1885–9) and *La psicanalisi* (1944). On Morselli, see P. Guarnieri, *Individualità difformi* (1986). Augusto Tamburini (1848–1919) became Director of the Reggio Emilia asylum in 1877, and from 1905 was Professor of Psychiatry in Rome. His work *Sulla genesi delle allucinazioni* (1880) became well known in Italy and abroad.
16. The *Rivista sperimentale di freniatria* amalgamated with the *Archivio italiano per la malattie mentali* in 1892.
17. C. Livi, 'Del metodo sperimentale in freniatria e medicina legale', *RSF*, 1875, vol. 1, 4.
18. Ibid., 5–6. All translations from Italian to English are mine, unless specified.
19. '[L]'uomo fisico'.
20. Livi, 'Del metodo sperimentale', 7.
21. On how borders between fields of knowledge are created, demarcated, and contested, see T. F. Gieryn, *Cultural Boundaries of Science: Credibility on the Line* (1999).
22. It should be mentioned that Lombroso had already been criticised for his less than scientific method. De Peri, 'Il medico e il folle', 1184–8; D. Frigessi, *Cesare Lombroso* (2003), 170–1.
23. In 1882 it became *Archivio di psichiatria, scienze penali ed antropologia criminale* and in 1904 the name changed again to *Archivio di psichiatria, neuropatologia, antropologia criminale e medicine legale*.
24. Ibid., 180; Giacanelli, 'Appunti per una storia', xxix–xxx. Giacanelli has also argued that it is possible to see the development of Italian psychiatry as an expression of the rise of an Italian bourgeois rank that performed managing functions in the new nation. Giacanelli, 'Appunti per una storia', xiv.
25. Tagliavini, 'Aspects of the History of Psychiatry', 178.
26. Ibid., 175–96.
27. C. Lombroso, 'Prefazione', in E. Fornasari di Verce, *La criminalità* (1894), xx.
28. G. Monaco, *Manuale di medicina legale* (1887), 35.
29. P. Zacchia, *Quaestionum medico-legalium* (1621–35), L. IV, T. II, Q. V.

30. S. Laura, *Trattato di medicina legale* (1874), 421–2.
31. The noun 'tribade' has a Greek origin. Both French and English texts record it from the sixteenth century onwards. In Italy it was a classic term used in works of literature and science to name female same-sex acts. Literally, it meant a woman who rubs, who would enjoy 'tribady', or tribadism (rubbing clitoris on clitoris, pubic bone, leg) with another woman. E. Donoghue, *Passions between Women* (1993), 5.
32. A. Filippi, *Manuale di medicina legale* (n.d.), 736–7. Filippi wrote that women broke the child's hymen to make the child's first sexual intercourse easier. In legal medical treatises published before 1880 – before physicians drew attention to 'sexual inversion' – treatises are concerned with 'pederasty' to various extents. There are many such works in the ancient collection of the Carlo Livi library of San Lazzaro, which was one of the most important asylums of the time: F. Bonucci, *Medicina legale delle alienazioni mentali* (1863); F. Gandolfi, *Fondamenti di medicina forense analitica* (1862–3); H. Maudsley, *La responsabilità nelle malattie mentali* (1875) [1874]; A. Filippi, *Manuale di afrodisiologia civile* (1878); F. Freschi, *Manuale teorico pratico di medicina legale* (1855). Aside from Laura's treatise, only the Italian translation of G. [sic] L. Casper's *Manuale pratico di medicina legale* (1859) [1852] briefly examined 'tribadism'. Casper was much quoted by physicians until the 1890s.
33. C. Lombroso, (review) 'Contribuzione alla casuistica della inversione dell'istinto sessuale di G. Cantarano', *AP*, 1884, vol. 5, 133–4. Arrigo Tamassia (1848–1917) was Lombroso's assistant during his medical training in Pavia.
34. A. Tamassia, 'La pazzie morale', *RSF*, 1877, vol. 3, 169.
35. See Maudsley, *La responsabilità nelle malattie mentali*; Tamassia, 'La pazzie morale', 160–2.
36. G. Bock Berti, 'Sulla "formazione" medico-legale di Arrigo Tamassia (1849–1917)', *Rivista di storia della medicina*, 1992, vol. 23, 37–43.
37. A. Tamassia, 'Sull'inversione dell'istinto sessuale', *RSF*, 1878, vol. 4, 97–8.
38. Ibid., 98.
39. Krafft-Ebing, 'Über gewisse Anomalies des Geschlechtstriebs'.
40. Tamassia, 'Sull'inversione dell'istinto sessuale', 110–11.
41. Ibid., 99.
42. Ibid.
43. '[N]europatico' and 'psicopatico'.
44. '[C]ongenito perverimento dell'istinto sessuale'.
45. '[V]era forma di alienazione mentale'.
46. Tamassia, 'Sull'inversione dell'istinto sessuale', 101–2, 112–13.
47. Krafft-Ebing had also noted that Westphal did not settle on whether sexual inversion was a symptom of a neuropathic condition or a proper disease in itself. Krafft-Ebing, *Psychopathia Sexualis with Special Reference to Contrary Sexual Instinct*, 224–5.
48. Tamassia, 'Sull'inversione dell'istinto sessuale', 115.
49. Ibid.
50. Ibid., 112–13.
51. Ibid., 116.
52. '[U]n profondo stato psicopatico [. . .] grave degenerazione funzionale'. Ibid.
53. '[N]ervoso'.
54. '[M]ezza isterica'.

55. '[M]ezzo idiota'.
56. '[E]ccentrico'.
57. '[R]istrettezza di mente'. Ibid., 103–10. At the time, this would have meant lacking mental agility.
58. '[C]oscienza del proprio pervertimento'.
59. Tamassia, 'Sull'inversione dell'istinto sessuale', 110, 116.
60. Ibid., 110.
61. Rosario, *The Erotic Imagination*, 70.
62. Other than Tamassia's, there were a few well-known case histories: G. Cantarano, 'Contribuzione alla casuistica della inversione dell'istinto sessuale', *La psichiatria, la neurologia, e le scienze affini*, 1883, vol. 1, 201–16; C. Lombroso, 'L'amore nei pazzi', *AP*, 1881; vol. 2, 1–32 ; C. Lombroso, 'Del tribadismo nei manicomi', *AP*, 1885, vol. 4, 218–21; A. Zuccarelli, *Inversione congenita dell'istinto sessuale in due donne* (1888). See J. M. Charcot and V. Magnan, 'Inversion du sens génital', *Archives de neurologie*, 1882, vol. 3, 53–60; concluded in vol. 4, 296–322; H. Ellis and J. A. Symonds, *Studies in the Psychology of Sex: Sexual Inversion*, vol. 1 (1897), 27–8; R. von Krafft-Ebing, *L'Inversione sessuale nell'uomo e nella donna* (1897) [1894], 83–109.
63. The *Società di freniatria italiana* accepted the first official classification written by Verga in 1874. This did not list same-sex desires as a mental disorder. In 1902 the Italian psychiatric association approved the second official classification of mental disorders written by Sante De Sanctis (1862–1935). Sexual inversion, along with other 'sexual psychopathies', '*frenestesie*' [insanities], and moral insanity, was grouped amongst the 'congenital psychoses'. L. Bianchi, *Trattato di psichiatria* (1904), 396–7.
64. Conolly Norman, 'Sexual Perversion', in Tuke (ed.), *A Dictionary of Psychological Medicine* (1892), 1156–7.
65. H. Ellis, *A Note on the Bedborough Trial* (1898), 12. Ellls's *Sexual Inversion* was not, strictly speaking, a psychiatry manual. On Ellis, see Chapter 7.
66. Morselli, *Manuale di semejotica delle malattie mentali*. On Morselli, see P. Guarnieri, 'Between Soma and Psyche: Morselli and Psychiatry in Late-Nineteenth-Century Italy', in W. F. Bynum, R. Porter, and M. Shepherd (eds.), *The Anatomy of Madness* (1988), vol. 3, 102–4.
67. Morselli, *Manuale di semejotica*, 213.
68. Ibid., 213–14.
69. On Lombroso's theory of atavism, see Chapter 5. On Ernst Haeckel's evolution theory, see S. G. Gould, *Ontogeny and Phylogeny* (1977), 76–85.
70. On sexology and phylogeny, see F. J. Sulloway, *Freud, Biologist of the Mind* (1992) [1979], 290–6. These developments will be further illustrated in Chapters 5 and 6.
71. Morselli, *Manuale di semejotica*, 668–81.
72. Ibid., 213–15.
73. Ibid., 214.
74. Ibid., 666–83. See also E. Tanzi, *Trattato delle malattie mentali* (1905), 247–8.
75. V. P. Babini, 'Il lato femminile della criminalità', in Babini, Minuz, and Tagliavini, *La donna nelle scienze dell'uomo*, 32–3.
76. Morselli referred to both 'homosexuality' and 'sexual inversion'. Morselli, *Manuale di semejotica*, 680–3.
77. Tanzi, *Trattato delle malattie mentali*, 246–7; Bianchi, *Trattato di psichiatria*, 396.

78. Leonardo Bianchi (1848–1927) was a neurologist and psychiatrist who taught at the universities of Palermo and Naples. In 1882 he founded the Istituto psichiatrico di Napoli [Neapolitan Psychiatric Institute] and, in 1891, the journal *Annali di neurologia* [*Neurological Annals*]. He was a Member of Parliament and Minister of Education in 1905. He wrote *Trattato di semiotica delle malattie del sistema nervoso* (1889) and *Trattato di psichiatria* (1904), two well-known textbooks in Italy and abroad.

79. In 1897, a new section titled 'Progress in Psychiatry' was introduced in the *Journal of Mental Science* and Bianchi was in charge of writing the section on Italian psychiatry. L. Bianchi, 'The Function of the Frontal Lobes', *Brain*, 1895, vol. 18, 497–522. For British appreciation of Bianchi's work, see for example: W. L. Andriezen, 'On the Basis and Possibilities of a Scientific Psychology and Classification in Mental Disease', *JMS*, 1899, vol. 45, 272; T. Williamson, 'On the Symptomatology of Gross Lesions (Tumours and Abscesses) Involving the Prae-Frontal Region of the Brain', *Brain*, 1896, 19, 346–65; W. F. Robertson, (review) 'Trattato di Psichiatria by Prof. L. Bianchi', *JMS*, 1902, vol. 48, 558; W. F. Robertson, (review) 'Trattato di psichiatria ad uso dei medici e degli studenti by L. Bianchi', *JMS*, 1905, vol. 51, 411–13.

80. Bianchi, *Trattato di psichiatria*, 398–401.

81. Ibid., 619–27.

82. Ibid.

83. Eugenio Tanzi (1856–1934) worked at the Reggio Emilia asylum, and later as Professor of Psychiatry in Cagliari (1893) and Florence (1895). In 1896 he founded *Rivista di patologia nervosa e mentale*.

84. Tanzi, *Trattato delle malattie mentali*, 618.

85. Ibid., 49–56, 251–60.

86. Ibid., 623–4.

87. Ibid., 619–27.

88. E. Tanzi, *Psichiatria forense* (1911), 247–54.

89. Ibid.

90. Like Tanzi, Penta moved away from degeneration and adopted Binet's explanations of sexual perversion. See Chapter 6.

91. S. De Sanctis, *Trattato di medicina sociale* (1911), 94–5.

92. Valeria P. Babini, *Liberi Tutti. Manicomi e psichiatri in Italia: una storia del Novecento* (2009), 28.

93. The French psychologist Alfred Binet's (1857–1911) approach was opposed to what most psychologists of his day held. He was interested in the workings of the normal rather than the pathological mind. He wanted to find a way to objectively measure the ability to think and reason, without regard to education in any particular field.

94. Sulloway, *Freud*, 290–6. Darwin had thought that every female and every male possessed the secondary sexual characteristics of the other sex in latent form, which could become manifest under special conditions. C. Sengoopta, *Otto Weininger: Sex, Science and Self in Imperial Vienna* (2000), 85–90.

95. J. Chavalier, *Una Maladie de la personalité. L'inversion sexuelle* (1893), 408–19.

96. R. von Krafft-Ebing, 'Zur Erklärung der conträren Sexualempfindung', *Jahrbücher für Psychiatries und Neurologie*,1895, vol. 13, 1–16.

97. A. Binet, 'Le fétichisme dans l'amour', *Revue Philosophique*, 1887, vol. 24, 153. See also R. A. Nye, 'The Medical Origins of Sexual Fetishism', in E. Apter and W. Pietz (eds.), *Fetishism as Cultural Discourse* (1993), 13–30.

98. Binet, 'Le fétichisme dans l'amour', 166–7.
99. For more on atavism, see Chapter 5.
100. Lombroso himself – who paid extensive attention to the topic of sexual deviance – confirmed it was the third published Italian case-history to discuss the inversion of sexual instinct: Lombroso, (review) 'Contribuzione', 133–4. I will pay attention to the second case-history published in 1881 by Lombroso in Chapter 5.
101. Ellis and Symonds, *Sexual Inversion* (1897), 27–8, 102–3; Krafft-Ebing, *L'Inversione sessuale nell'uomo e nella donna*, 83–109.
102. On the importance of case-histories in sexology, see I. Crozier, 'Philosophy in the English Boudoir: Havelock Ellis, "Love and Pain", in the Sexological Discourses on Algophilia', *Journal of the History of Sexuality*, 2004, vol. 13, 294–5; for the importance in science of 'reasoning in cases': J. Forrester, 'If p, then what? Thinking in Cases', *History of the Human Sciences*, 1996, vol. 9, 1–25.
103. Cantarano, 'Contribuzione alla casuistica della inversione dell'istinto sessuale', 201.
104. Ibid., 201–16.
105. Ibid., 201–2.
106. C. Beccalossi, 'Nineteenth-Century European Psychiatry on Same-Sex Desires: Pathology, Abnormality, Normality and the Blurring of Boundaries', *Psychology & Sexuality*, 2010, vol. 1, 226–38.
107. Cantarano, 'Contribuzione', 201–2.
108. G. Cantarano, 'Inversione e pervertimenti dell'istinto sessuale', *La psichiatria la neurologia, e le scienze affini*, 1890, vol. 8, 280.
109. Cantarano, 'Contribuzione', 204–7.
110. Ibid., 213.
111. Ibid., 203, 212–13.
112. Ibid., 208.
113. Ibid., 209.
114. Ibid.
115. Ibid.
116. Ibid., 205, 210.
117. Ibid., 205.
118. Charcot had the same view on the topic. J. M. Charcot and V. Magnan, 'Inversione del senso genitale e altre perversioni genitali', in J. M. Charcot, *La donna dell'isteria* (1989) [1882], 22–3.
119. Gock was a German physician and Wise was American. H. Gock, 'Beitrag zur Kenntniss der conträren Sexualempfindung', *Arch. Psychiatr. Nervenkr.*, 1875, vol. 5, 564–74; Wise, 'Case of Sexual Perversion'.
120. Cantarano, 'Contribuzione', 214.
121. Lombroso, 'Del tribadismo nei manicomi' and Zuccarelli, *Inversione congenita dell'istinto sessuale in due donne* are the most significal case-histories on women of the period. On Lombroso, see Chapter 5.
122. Babini, Minuz, and Tagliavini, *La donna nella scienze dell'uomo*; V. P. Babini, 'L'infanticida tra letteratura medica e letteratuta giuridica', in P. Rossi (ed.), *L'età del positivismo* (1986), 453–74 ; A. Cavalli Pasini, 'Ruolo e figura femminili nella pubblicistica e nella letteratura popolare', in Rossi (ed.), *L'età del positivismo*, 405–38; F. Minuz, 'La norma del femminile nell'antropologia', in Rossi (ed.), *L'età del positivismo*, 439–52; A. Tagliavini, 'La "mente femminile" ', in Rossi (ed.), *L'età del positivismo*, 475–91.

123. Babini, Minuz, and Tagliavini, *La donna nella scienze dell'uomo*.
124. F. Tonini, *Igiene e fisiologia del matrimonio* (1862), vol. 2, 198–211.
125. Ibid., 190–1.
126. Ibid., 191.
127. Ibid., 194–5, 367, 371, 387.
128. Ibid., 194. Italics are mine.
129. L. Gowing, 'Lesbians and Their Like in Early Modern Europe, 1500–1800', in R. Aldrich (ed.), *Gay Life and Culture* (2006), 126–143.
130. On the contribution of Paolo Mantegazza (1831–1910) to sexual morality, see L. Tasca, 'Il "senatore erotico" ', in B. Wanrooij (ed.), *La mediazione matrimoniale in Europa fra Otto e Novecento* (2004), 295–322.
131. P. Mantegazza, *Gli amori degli uomini* (1886), English translation by S. Putman with an introduction by Victor Robinson: P. Mantegazza, *The Sexual Relations of Mankind* (1935). Mantegazza also wrote *Fisiologia del piacere* (1854); *Fisiologia dell'amore* (1872); *Igiene dell'amore* (1877). Regardless of the titles, these were mostly philosophical works on different aspects of love rather than physiological treatises.
132. Tasca, 'Il "senatore erotico" ', 316–17.
133. In *Gli amori degli uomini*, Mantegazza continued: 'It is true that with a more reasonable and scientific morality, such as that which is to come, hygiene and ethics ought to go perfectly in accord; but up to the present time, the two are very frequently at daggers drawn and in contradiction to each other, a certain proof either that hygiene is ignorant or that morality is false.' Mantegazza, *The Sexual Relations of Mankind*, 78–9; Mantegazza, *Gli amori*, vol. 1, 131–2.
134. Mantegazza, *Gli amori*, 154.
135. Krafft-Ebing, *L'inversione sessuale*, 103–4; on Britain and Mantegazza see Chapter 7. Outside medical circles Richard Burton reported Mantegazza's theory: R. F. Burton, 'Terminal Essay. Section D: Pederasty', in *A Plain and Literal Translation of the Arabian Nights Entertainments, Now Intituled the Book of the Thousand Nights and a Night with Introduction Explanatory Notes on the Manners and Customs of Moslem Men and Terminal Essay upon the History of the Nights* (1886), vol. 10, 208–9.
136. Mantegazza, *Gli amori*, 148.
137. Ibid., 314–15.
138. Ibid., 133–4.
139. Ibid., 136.
140. Ibid. The original reads: *[Tribadismo] è 'una pratica [usata] da una femmina che fornita di clitoride eccezionalmente lungo può simulare l'amplesso con altra femmina'*. Mantegazza is therefore implying that sexual orgasm between two women is a simulacrum of heterosexual intercourse.
141. See for example Krafft-Ebing, *L'inversione sessuale nell'uomo e nella donna*, 86–96. His reviews show that Mantegazza was a careful reader of medical research on sexual matters, see P. Mantegazza, (review) 'A. Moll, "Les perversions de l'instinct génital" ', *Archivio per l'antropologia e l'etnologia*, 1893, vol. 23, 464–5; P. Mantegazza, (review) 'P. Penta, "I pervertimenti sessuali nell'uomo" ', *Archivio per l'antropologia e l'etnologia*, 1893, vol. 23, 476; P. Mantegazza, (review) 'P. Penta, "L'origine e la patologenesi delle inversioni sessuali" ', *Archivio per l'antropologia e l'etnologia*, 1897, vol. 27, 433; P. Mantegazza, (review) 'A. Niceforo, "Le psicopatie sessuali acquisite e i reati sessuali" ', *Archivio per l'antropologia e l'etnologia*, 1897, vol. 27, 440.

142. P. Mantegazza, *Fisiologia della donna* (1893) [1891], 264.
143. J. R. Walkowitz, 'Dangerous Sexualities,' in G. Duby and M. Perrot (eds.), *A History of Women* (1993), vol. 4, 395. In 1836, Parent-Duchâtelet published a study on Parisian prostitution, *De la prostitution dans la ville de Paris*. This research was a demographic study of 12,000 prostitutes who had been observed over a 15-year period, between 1816 and 1831. On Parent-Duchâtelet and the tribade-prostitute, see Beccalossi, 'Female Same-Sex Desires: Conceptualizing a Disease in Competing Medical Fields in Nineteenth-Century Europe'.
144. See for example C. Lombroso and G. Ferrero, *La donna delinquente* (1893) (see Chapter 5).
145. On the prostitute as a figure perceived as an embodiment of sexuality, disease, and passion, see S. L. Gilman, *Sexuality: An Illustrated History* (1989), 296–307.
146. C. Andronico, 'Prostitute e delinquenti', *AP*, 1882, vol. 3, 143.
147. Ibid., 143–6.
148. Krafft-Ebing also read Parent-Duchâtelet's work on the tribade-prostitute; see Krafft-Ebing, *Psychopathia Sexualis*, 429.
149. Scipio Sighele (1868–1913) is considered one of the founders of mass psychology. He taught criminal law at the universities of Rome and Pisa, and courses on criminal sociology and mass psychology at the Institut des Hautes Études of Brussels University. His books were read not only by specialists, but also by the general public.
150. S. Sighele, 'La coppia criminale', *AP*, 1892, vol. 13, 530–2. This article was then incorporated in Sighele's book, S. Sighele, *La coppia criminale* (1893). Amongst others, Sighele quoted Parent-Duchâtelet; see for example *La coppia criminale*, 167.
151. The *souteneur* was a stable lover and protector of the prostitutes. He could be also the manager of the brothel.
152. Sighele, 'La coppia criminale', 532–3.
153. A. J.-B. Parent-Duchâtelet, *Prostitution in the City of Paris*, trans. R. Ridley-Smith (2008), 72–80, 99–100; A. Corbin, *Women for Hire: Prostitution and Sexuality in France after 1850* (1990), 7.
154. Sighele, 'La coppia criminale', 533–4.
155. G. B. Moraglia, 'Nuove ricerche su criminali, prostitute e psicopatiche', *AP*, 1895, vol. 16, 309.
156. A. Niceforo, *Le psicopatie sessuali acquisite e i reati sessuali* (1897), 87–9, 93.
157. Ibid., 93.
158. Ibid., 91.
159. I. Callari, 'Prostituzione e prostituta in Sicilia', *AP*, 1903, vol. 24, 197–203.
160. Una tribade, *Tribadismo. Saffismo-Clitorismo* (1914).
161. F. Stura, *Le miserie di Venere* (1904), 119–20. For similar narratives, see A. Orsi, *La donna nuda* (1905).
162. S. Soldani (ed.), *Educazione delle donne* (1989), xv, xvi. In Italy at the turn of the century, teaching was the only respectable profession for lower-middle-class women. In 1901 there were about 62,600 female teachers, most of whom were appointed in the last decade of the nineteenth century. This phenomenon was also the result of the government's decision to encourage mass education after the unification of Italy. On the role of *Risorgimento* in female education, see S. Soldani, 'Il libro e la matassa', in Soldani (ed.), *Educazione delle donne*, 87–126.
163. De Giorgio, *Le italiane dall'Unità ad oggi*, 118–30.
164. '[A]mor disordinato', see the 1842 edition: A. de' Liquori, *La vera sposa di Gesù Cristo* (1842) [1760], 220.

165. Ibid., 221. The Italian is *'amicizie avvelenate'*.
166. G. Frassinetti, *La monaca in casa* (1862) [1859], 136. Italics are mine.
167. P. Penta and A. d'Urso,'Sopra un caso d'inversione sessuale in donna epilettica', *APS*, 1896, 33–39 and Niceforo, *Le psicopatie sessuali acquisite*, 25–40, 43–67; Niceforo, *Il gergo nei normali, nei degenerati, nei criminali*, 37– 43; Moraglia, 'Nuove ricerche su criminali, prostitute e psicopatiche', 510–15.
168. In the second edition of *Sexual Inversion*, Ellis added a section based on Marchesini and Obici's work, see Appendix B, 'The School-Friendships of Girls,' in H. Ellis, *Studies in the Psychology of Sex: Sexual Inversion* (1901), vol. 2, 243–57; H. Ellis, *Studies in the Psychology of Sex: Sexual Inversion* (1924), vol. 2, 368–84. Ellis reported that the *'fiamma* phenomenon' in Britain was called 'rave' or 'spoon'. On Ellis and school friendships, see Chapter 7. For similar accounts of female same-sex desires in schools, see also Chapter 6.
169. G. Marchesini and G. Obici, *Le 'amicizie' di collegio* (1903) [1898]. This work became very popular in Italy and was moderately successful abroad.
170. Ibid., 22–4.
171. Ibid., 116.
172. Ibid., 115–16.
173. Ibid., 240–3.
174. Ibid., 8–9.
175. Ibid., 245–67.
176. Ibid., 298–307, 311–34.
177. Ibid., 195–8.
178. E. Morselli, 'Introduzione', in ibid., xvi–xviii.
179. Ibid., xxxii–xxiv.
180. Ibid., xl–xli. See section three of this chapter: 'The establishment of sexual inversion as a mental illness in psychiatric manuals'.

4 Britain: Oblique Discourses Surrounding 'Lesbic Love'

1. L. Hall, 'Malthusian Mutations: The Changing Politics and Moral Meaning of Birth Control in Britain', in B. Dolan (ed.), *Malthus, Medicine and Morality: Malthusianism after 1798* (2000), 142.
2. R. Porter and L. Hall, *The Facts of Life* (1995), 155.
3. C. Darwin, *The Descent of Man, and Selection in Relation to Sex* (1981) [1871], part 2.
4. On the British reluctance to take up Continental sexology, see L. Hall, '"The English Have Not Hot Water Bottles": The Morganatic Marriage between Sexology and Medicine since William Acton', in R. Porter and M. Teich (eds.), *Sexual Knowledge, Sexual Science* (1994), 350–66; Brady, *Masculinity*. On the psychiatric approach to sexological writings on male homosexuality before Ellis, see Crozier, 'Nineteenth-Century British Psychiatric Writing on Homosexuality'.
5. This assumption has been supported by a number of scholars, see for example Castle, *The Apparitional Lesbian*, 4–6; Oram and Turnbull (eds.), *The Lesbian History Sourcebook*, 93.
6. K. Jones, 'The Culture of the Mental Hospital', in G. E. Berrios and H. Freeman (eds.), *150 Years of British Psychiatry* (1991), 17–19.
7. Berrios and Freeman (eds.), *150 Years of British Psychiatry*; A. Scull, C. MacKenzie, and N. Hervey, *Masters of Bedlam* (1996).
8. K. Jones, 'Law and Mental Health: Sticks or Carrots', in Berrios and Freeman (eds.), *150 Years of British Psychiatry*, 95.

9. A. Scull, *Museums of Madness* (1975); E. Renvoize, 'The Association of Medical Officers of Asylums and Hospitals for the Insane, the Medico-Psychological Association, and their Presidents', in Berrios and Freeman (eds.), *150 Years of British Psychiatry*, 29–78.

10. Renvoize, 'The Association of Medical Officers', 34–41. The Medico-Psychological Association became the Royal College of Psychiatrists in 1971.

11. Forbes Benignus Winslow (1810–74) graduated as a Doctor of Medicine from the University of Aberdeen in 1849; he became a Fellow of the Royal College of Physicians of Edinburgh the following year and a member of the London College in 1859. He was President of the Medical Society of London and became President of the Association of Medical Officers of Asylums and Hospitals for the Insane in 1857. Renvoize, 'The Association of Medical Officers', 44–5.

12. John Charles Bucknill (1817–97) qualified in 1840 in both medicine and surgery at the University of London. In 1844 he was appointed medical superintendent at the Devon County Asylum at Exminster. He played a key role in the development of psychiatry in Britain. Renvoize, 'The Association of Medical Officers', 45–6 and Scull, MacKenzie, and Hervey, *Masters of Bedlam*, 187–225.

13. Scull, MacKenzie, and Hervey, *Masters of Bedlam*, 244.

14. P. Koehler, 'The Evolution of British Neurology in Comparison with Other Countries', in F. C. Rose (ed.), *A Short History of Neurology* (1999), 62–3.

15. Daniel Hack Tuke (1827–95) graduated in medicine at Heidelberg University in 1853. He became visiting physician at the York Dispensary and at the York Retreat. He also lectured in psychology at the York School of Medicine. Renvoize, 'The Association of Medical Officers', 50–2; Scull, MacKenzie, and Hervey, *Masters of Bedlam*, 188.

16. Scull, MacKenzie, and Hervey, *Masters of Bedlam*, 204.

17. Ibid., 246–7.

18. Berrios and Freeman (eds.), *150 Years of British Psychiatry*.

19. D. Garland, 'Of Crimes and Criminals: The Development of Criminology in Britain', in M. Maguire, R. Morgan, and R. Reiner (eds.), *The Oxford Handbook of Criminology* (1997) [1994], 35.

20. H. Maudsley, 'Criminal Responsibility in Relation to Insanity', *JMS*, 1895, vol. 51, 662. Maudsley did not mention Lombroso explicitly, but it seems clear he was talking about him.

21. Garland, 'Of Crimes and Criminals', 35.

22. Ibid., 12–36.

23. T. S. Clouston (ed.), 'David Skae, M.D., F.R.C.S.E. (by the late), The Morisonian Lectures on Insanity', *JMS*, 1873–4, vol. 19, 340–55, 491–507. On David Skae (1814–73), see F. Fish, 'David Skae, M.D., F.R.C.S.E. Founder of the Edinburgh School of Psychiatry', *Medical History*, 1965, vol. 9, 36–53. In the 1880s, Clouston (1840–1915) was a medical superintendent of the Royal Edinburgh asylum.

24. Clouston (ed.), 'David Skae'.

25. T. S. Clouston, *Clinical Lectures on Mental Diseases* (1883).

26. Clouston (ed.), 'David Skae', 348–51.

27. Ibid., 348. Italics are mine and are used to underline the forms of insanity relating to sexual matters. In 1883, Clouston wrote that Skae's classification, following Morel, looked at mental disease using a 'clinical method'. This method endeavoured to take account of causes, of the relationship the different varieties of the disease have to physiology, and of the activities of the body. In other

words 'it regards the whole *natural history* of the disease'. Clouston, *Clinical Lectures*, 20.

28. Clouston, 'David Skae', 348–51.
29. Ibid., 499.
30. Ibid. Italics are in the text.
31. Ibid.
32. Ibid., 500–1. Italics are mine.
33. Ibid., 351.
34. On Prichard's concept of moral insanity see H. F. Augstein, 'J. C. Prichard's Concept of Moral Insanity: A Medical Theory of the Corruption of Human Nature', *Medical History*, 1996, vol. 40, 311–43.
35. Clouston, 'David Skae', 351. Skae considered nymphomania, erotomania, kleptomania, and so forth diseases, and not symptoms.
36. For a list of all states and mental disorders, see the division of chapters in various editions of Clouston's *Clinical Lectures*.
37. T. S. Clouston, *Clinical Lectures* (1883), 24.
38. Ibid., 329.
39. Ibid., 330.
40. Ibid., 329–30.
41. Anon., 'Aberrations of the Sexual Instinct', *Medical Times and Gazette*, 1867, 142. On this article, see Brady, *Masculinity*, 129–30; Crozier, 'Nineteenth-Century British Psychiatric Writing', 75–7.
42. Anon., 'Aberrations of the Sexual Instinct', 143. On phrenology and homosexuality, see M. Lynch, ' "Here is adhesiveness": From Friendship to Homosexuality', *Victorian Studies*, 1985, vol. 29, 67–96.
43. Anon., 'Aberrations of the Sexual Instinct', 143.
44. Ibid.
45. Ibid.
46. Ibid., 145–6.
47. Ibid., 146.
48. For the introduction of sexual inversion in Britain, see discussion about Krueg later in this chapter.
49. See Chapter 7.
50. See Beccalossi, 'Nineteenth-Century European Psychiatry on Same-Sex Desires'.
51. On various forms of partial insanity, see J. Goldstein, *Console and Classify: The French Psychiatric Profession in the Nineteenth Century* (2001), 152–96. It should be noted that in the late nineteenth century the two categories of psychopathy and psychopathology were often conflated when approaching the study of human sexuality.
52. J. C. Bucknill and D. H. Tuke, *A Manual of Psychological Medicine Containing the Lunacy Laws, Aetiology, Statistics, Description, Diagnosis, Pathology, and Treatment of Insanity with an Appendix of Cases* (1879) [1858], 454–7.
53. Ibid., 487.
54. At the time, George Savage (1842–1921) was a medical superintendent of Bethlehem Royal Hospital, Lecturer of Mental Insanity at Guy's Hospital, and Joint Editor of the *JMS*. For the link between sexuality and moral insanity, see G. Savage, 'Moral Insanity', *JMS*, 1881, vol. 27, 147–55, especially at 152. For sexual perversion and insanity, see G. Savage, *Insanity and Allied Neuroses* (1886) [1882].

55. Savage, *Insanity and Allied Neuroses*, 24–5.
56. Ibid., 273.
57. Ibid., 59.
58. Ibid., 63.
59. Ibid., 473.
60. Charles Mercier (1852–1919) was a psychiatrist who wrote extensively on forensic medicine. During the 1890s, Mercier was a lecturer in Insanity at the Westminster Medical School, and at the Medical School for Women. About Mercier, see Scull, MacKenzie, and Hervey, *Masters of Bedlam*, 254, 352; and his work *Sanity and Insanity* (1890); in relation to same-sex desires, see also Crozier, 'Nineteenth-Century British Psychiatric Writing about Homosexuality', 84–5.
61. Mercier, *Sanity and Insanity*, 149.
62. Ibid., 152
63. Ibid., 150–3. Italian physicians called this phenomenon atavism.
64. Ibid., xix.
65. Ibid, 153–4.
66. Ibid., 154. Mercier dealt briefly with homosexuality in *Criminal Responsibility* (1905), 145, 201; C. Mercier, *Crime and Criminals* (1918), 198–9.
67. See Chapter 5 on Lombroso.
68. See the first section in this chapter.
69. J. Krueg, 'Perverted Sexual Impulses', *Brain*, 1881, vol. 4, 368–76. Krueg worked at a private lunatic asylum, Ober Döbling, Vienna.
70. Crozier, 'Nineteenth-Century British Psychiatric Writing on Homosexuality', 77–80.
71. Krueg, 'Perverted Sexual Impulses', 368–9.
72. Ibid., 370.
73. Ibid., 371–2.
74. Ibid., 375.
75. Ibid., 370.
76. Ibid., 375.
77. On Tamassia's response to Westphal, see Chapter 3.
78. Krueg, 'Perverted Sexual Impulses', 374.
79. Westphal, 'Die Conträre Sexualempfindung'; Gock, 'Beitrag zur Kenntniss der conträren Sexualempfindung'.
80. Krueg, 'Perverted Sexual Impulses', 375.
81. See for example Cantarano in Chapter 3 and Ellis in Chapter 7.
82. G. Savage, 'Case of Sexual Perversion in Man', *JMS*, 1884, vol. 30, 390–1; Dr Urquhart, 'Case of Sexual Perversion', *JMS*, 1891, vol. 37, 94–6; W. Sullivan, 'Notes on a Case of Acute Insanity with Sexual Perversion', *JMS*, 1893, vol. 39, 225–6; G. Savage and C. Mercier, 'Insanity of Conduct', *JMS*, 1896, vol. 42, 1–17. It should be noted that none of these psychiatrists mentioned the term 'sexual inversion' openly in these articles. Instead, they employed the more general term 'sexual perversion'. However, the patients' symptoms centred around same-sex desires, while they showed traditionally feminine psychological characteristics and were interpreted as suffering from a mental pathology. It is therefore possible to talk about sexual inversion. Moreover, Urquhart's case-history referred to Krafft-Ebing's work, so the indirect reference to 'sexual inversion' is clear. Ivan Crozier has also interpreted these case histories as sexual inversion, see Crozier, 'Nineteenth-Century British Psychiatric Writing about Homosexuality'.

83. Ellis's work *Sexual Inversion* was first published in Germany in 1896, and in Britain in 1897. See Chapter 7.
84. C. Norman, 'Sexual Perversion', in Tuke (ed.), *A Dictionary of Psychological Medicine*, 1156–7. Norman also explained the most important Continental sexological theories available despite his reluctance to enter into details. Conolly Norman (1853–1908) was medical superintendent of the Richmond Asylum, Dublin, and had been President of the Medico-Psychological Association. In 1874 he received the licences of the King's and Queen's College of Physicians and the Royal College of Surgeons of Ireland.
85. Medical reactions to Ellis's publication of *Sexual Inversion* are further evidence of British medicine's reluctance to accept sexual inversion as a psychiatric category. See Chapter 7.
86. Savage and Mercier, 'Insanity of Conduct'; Crozier, 'Nineteenth-Century British Psychiatric Writing on Homosexuality', 89.
87. Savage and Mercier, 'Insanity of Conduct', 14; Crozier, 'Nineteenth-Century British Psychiatric Writing on Homosexuality', 90.
88. Ellis's case, discussed in more detail in Chapter 7, illustrates this point.
89. See Chapter 3, and my discussion of Parent-Duchâtelet and the tribade-prostitute: Beccalossi, 'Female Same-Sex Desires: Conceptualizing a Disease in Competing Medical Fields in Nineteenth-Century Europe'.
90. For the early nineteenth-century medical view, see F. Churchill, *Outlines of the Principal Diseases of Females* (1835), specifically the chapter on 'Diseases of External Organs of Generations'; T. Laycock, *A Treatise on the Nervous Diseases of Women* (1840), 176–98.
91. For an early nineteenth-century medical overview on nymphomania, see M. Magendie, (1836–7) 'Lectures of the Physiology of the Nervous System, delivered in 1836, in the College of France', *The Lancet*, vol. 2, 24 June 1837, 463–6. For an overview of nymphomania as a medical disease in the nineteenth century, see C. Groneman, 'Nymphomania: The Historical Construction of Female Sexuality', *Signs*, 1994, vol. 19, 337–67.
92. The term 'nymphomania' comes from the Greek term *nymphae*. The term 'nymphae' was also used in medicine to indicate female genitalia, in particular the labia.
93. Bucknill and Tuke, *A Manual of Psychological Medicine*, 282.
94. Ibid., 282–4. For an overview of erotomania see G. E. Berrios and N. Kennedy, 'Erotomania: A Conceptual History', *History of Psychiatry*, 2002, vol. 13, 381–400.
95. Bucknill and Tuke, *A Manual of Psychological Medicine*, 783.
96. Ibid.
97. Ibid., 783–4.
98. The *Dictionary* was re-edited several times; this case was reported under the heading nymphomania until 1879.
99. See Skae's analysis above.
100. G. Bouchereau, 'Nymphomania', in Tuke (ed.), *A Dictionary of Psychological Medicine*, vol. 2, 863.
101. Ibid., 864.
102. Ibid., 865. For a brief discussion on the introduction of words such as 'lesbianism', 'lesbic', and 'lesbian' to the English language, see Donoghue, *Passion between Women*, 2–8.

103. Ibid.
104. A. S. Taylor, *The Principles and Practice of Medical Jurisprudence* (1905) (5th edn., ed. Fred J. Smith), vol. 2, 316, 321. Alfred Taylor (1806–80), Professor of Medical Jurisprudence at Guy's Hospital Medical School.
105. Ibid., 316.
106. Ibid., 321.
107. Ibid.
108. The second (American) edition of Ellis's *Sexual Inversion* was published in 1901.
109. A. J. B. Parent-Duchâtelet, *On Prostitution in the City of Paris* (1840) [1835–6]. It was the second edition and an abridged version of the original work. Nevertheless, same-sex acts amongst female prostitutes were extensively analysed, see at 31–9. Compare with the original French edition: A. J. B. Parent-Duchâtelet, *De la prostitution*, 162–72.
110. Walkowitz, *Prostitution and Victorian Society*, 33–47.
111. Anon., 'On Prostitution in the City of Paris: Considered under the Heads of Public Hygiene, Morals, and Internal Police', *The Lancet*, 1836–7, vol. 28, 16–23; 41–8; 106–11; 157–63; 755–63.
112. Ibid., 47.
113. M. Ryan, *Prostitution in London* (1839), 178–9.
114. Porter and Hall, *The Facts of Life*, 148–50, 152–3.
115. On Drysdale's work and homosexuality, see Brady, *Masculinity*, 125–8. In 1882, Drysdale's book reached its 21st edition.
116. G. Drysdale, *The Elements of Social Science* (1882) [1854], 14–44.
117. Ibid., 44.
118. Ibid.
119. Ibid., 166–7.
120. Ibid., 73.
121. Ibid., 168. Constantly indulging in venereal pleasures meant that 'the mind becomes effeminate, and the nerves lose their tone; the power of thought becomes impaired, cloyed as it were by sweetness. Nature never meant that we should be absorbed in one set of feelings, not steeped in sexual indulgences, as some of the southern nations are.' Ibid., 187.
122. Ibid., 168.
123. Brady, *Masculinity*, 125–8.
124. Drysdale, *The Elements*, 247.
125. Ibid., 248.
126. Parent-Duchâtelet, *Prostitution in the City of Paris* (2008), 72–7.
127. Drysdale, *The Elements*, 248.
128. Parent-Duchâtelet, *Prostitution in the City of Paris* (2008), 77.
129. Lombroso and Ferrero, *La donna delinquente*, 414–15.
130. H. Ellis, *The Criminal* (1901), 107. Ellis and Symonds, *Sexual Inversion* (1897), 100–3. Still, at the beginning of the twentieth century, Abraham Flexner, a well-known American reformer of medical education dealing with homosexuality in female prostitutes in Europe, was still relying on Parent-Duchâtelet. A. Flexner, *Prostitution in Europe* (1914), 31–3, 109, 196.
131. Brady, *Masculinity*, 119–55.
132. Fleetwood Churchill was a Fellow, and later President, of the King and Queen's College of Physicians in Ireland, and Professor of Midwifery.
133. F. Churchill, *On the Diseases of Women* (1874) [1850], 67. This book went through several editions. On enlarged clitoris and female same-sex acts, see below.

134. Parent-Duchâtelet, *Prostitution in the City of Paris* (2008), 99–100.
135. On forensic medicine and male same-sex practices, see: I. Crozier 'All the Appearances Were Perfectly Natural: The Anus of the Sodomite in Nineteenth-Century Medical Discourse', in C. E. Forth and I. Crozier (eds.), *Body Parts* (2005), 65–84.
136. For the growth of gynaecology in England, see O. Moscucci, *The Science of Woman: Gynaecology and Gender in England, 1800–1929* (1990).
137. For an earlier example of the medical association between an enlarged clitoris and female disease, see J. Blundell, 'Lectures on the Diseases of Women and Children', *The Lancet*, 1828–9, 5 Sept. vol. 2, 707. On the association between a hypertrophied clitoris and tribadism in the early modern period, see Gowing, 'Lesbians and Their Like', 126–8; Traub, *The Renaissance of Lesbianism in Early Modern England*, 188–228.
138. Heywood Smith was a member of the Royal College of Physicians, a physician working at the Hospital for Women, and at the British Lying-in Hospital (London). H. Smith, *Practical Gynaecology* (1877), 116–17.
139. K. Park, 'The Rediscovery of the Clitoris: French Medicine and the Tribade, 1570–1620', in D. and C. Mazzio (eds.), *The Body Parts: Fantasies of Corporeality in Early Modern Europe* (1997), 171–94; Traub, *The Renaissance of Lesbianism in Early Modern England*, 188–228; Wahl, *Invisible Relations*, 17–42.
140. Lawson Tait was Professor of Gynaecology in the Queen's College, Birmingham, and a surgeon at the Birmingham and Midland Hospital for Women.
141. L. Tait, *Diseases of Women and Surgery* (1889), vol. 1, 43.
142. Ibid., 44.
143. Ibid.
144. Ibid., 46–7.
145. Ibid., 47.
146. Ibid., 49–50.
147. Ibid., 57–8.
148. Ibid.
149. A pioneer of surgical gynaecology, since the 1870s Barnes had been a member of the Royal College of Physicians. He had been an examiner in Obstetrics and the Diseases of Women at the University of London and at the Royal College of Surgeons; he had also lectured on Obstetrics and the Diseases of Women at St Thomas's Hospital. In 1884 he founded the British Gynaecological Society. A month after this meeting, on 13 November 1890, same-sex desires were discussed at another meeting of the Medico-Psychological Association, but this meeting seemed inconclusive on the subject of female same-sex desires. The Secretary, Dr Urquhart, gave a paper on male sexual inverts. Dr Ireland, who discussed Urquhart's paper, noted that same-sex desires existed in women, but was not clear what stance should be taken towards them: 'There are women who have a depraved taste for women. It has been so through the ages, if we are to trust classical authorities. What is the lesbian passion? Some of the finest odes of Sappho signalize this unnatural love.' Urquhart, 'Case of Sexual Perversion'; Anon., 'Discussion at the Quarterly Meeting of the Medico-Psychological Association, November 13, 1890', *JMS*, 1891, vol. 37, 200.
150. R. Barnes, 'On the Correlations of the Sexual Functions and Mental Disorders of Women', *British Gynaecological Journal*, 1890, vol. 4, 390–1.
151. Ibid., 395–8.
152. Ibid., 407–8.

153. Ibid., 408.
154. G. H. Savage, 'Case of Malformation of Genitalia in Insanity', *JMS*, 1878, vol. 24, 459–61. I would like to thank Ivan Crozier for drawing my attention to this case.
155. Ibid., 459.
156. Ibid.
157. Ibid., 460.
158. Ibid., 461.
159. Savage, 'Case of Sexual Perversion in Man', 390–1.
160. Barnes, 'On the Correlations of the Sexual Functions and Mental Disorders of Women', 421–7.
161. Ibid., 430.
162. C. H. F. Routh, 'The Conservative Treatment of Disease of the Uterine Appendages', *British Gynaecological Journal*, 1894, vol. 10, 58–61.
163. Routh described nymphomania in a similar way to other gynaecologists who described same-sex acts: see C. H. F. Routh, 'On the Etiology and Diagnosis, considered specially from a Medico-legal Point of View, of those Cases of Nymphomania which lead Women to make False Charge against their Medical Attendants', *BMJ*, 1887, vol. 2, 485–511.
164. Porter and Hall, *The Facts of Life*, 144–54; L. Hall, 'Forbidden by God, Despised by Men: Masturbation, Medical Warnings, Moral Panic, and Manhood in Great Britain, 1850–1950', in J. C. Fout (ed.), *Forbidden History* (1990), 293–315. On the scientific and cultural attitudes towards masturbation in the nineteenth and twentieth centuries, see T. W. Laqueur, *Solitary Sex* (2003), 247–420.
165. Hall, 'Forbidden by God', 302.
166. I do not necessarily agree with Brady. Physicians occasionally observed that masturbation was a cause of male same-sex acts, as some of the cases analysed by Crozier show. Brady, *Masculinity*, chapter 5; Crozier, 'Nineteenth-Century British Psychiatric Writing on Homosexuality'.
167. There was no agreement as to whether girls masturbated more or less than boys. James Copland, for instance, thought that girls masturbated as much as boys. J. Copland, *Dictionary of Practical Medicine* (1844–58) [1838], vol. 1, 41–3. Clouston thought that boys masturbated more than girls and also linked male masturbation to homosexuality. T. S. Clouston, *Clinical Lectures on Mental Diseases* (1898) [1883], 532–35.
168. W. Acton, *The Functions and Disorders of the Reproductive Organs, in Childhood, Youth, Adult Age, and Advanced Life, Considered in the Physiological, Social, and Moral Relations* (1857).
169. See for example Tuke (ed.), *A Dictionary of Psychological Medicine*, vol. 2, 864.
170. J. Crichton-Browne, 'Psychical Diseases of Early Life', *JMS*, 1860, vol. 6, 284–320.
171. K. Rowold, ' "The Academic Woman": Minds, Bodies and Education in Britain and Germany, 1860–1914' (1996), 80.
172. Copland, *Dictionary of Practical Medicine*, vol. 1, 42.
173. Ibid.
174. Ibid., 282. James Copland (1791–1870), physician, began his career as medical officer to the settlements of the Royal African Company. He became editor of the London Medical Repository in 1822.
175. Laycock, *Treatise on the Nervous Diseases of Women*, 141. Thomas Laycock (1812–76) was educated at University College London, but commenced his medical training while studying in Paris. He became a member of the Royal College of Surgeons and practised at the York County Hospital.

176. Savage, *Insanity and Allied Neuroses*, 59, 64. Savage also made connections between children's insanity and masturbation, see the second section above in this chapter.
177. Tait, *Diseases of Women and Surgery*, vol. 1, 60.
178. Ibid., 61.
179. Ibid., 61–2.
180. Ibid., 62.
181. H. M. Jones, *Practical Manual of Diseases of Women* (1884), 107. Throughout his career Henry Macnaughton Jones had worked in Ireland, Scotland, and England; at the beginning of the twentieth century he became President of the British Gynaecological Society.
182. H. M. Jones, *Points of Practical Interest in Gynaecology* (1901), 83–4.
183. F. W. S. Browne, *Sexual Variety and Variability among Women* (1915), 6. On Stella Browne, see L. A. Hall, 'Feminist Reconfigurations of Heterosexuality in the 1920s', in L. Bland and L. Doan (eds.), *Sexology in Culture* (1998), 135–49.
184. Strachey, *Eminent Victorians*.
185. Only medical texts could deal freely with sexual matters. The selling of Émile Zola's novels, for example, was forbidden due to the Obscene Publications Act. Stepansky, 'A Footnote to the History of Homosexuality in Britain', 96–7.
186. N. Davie, *Tracing the Criminal: The Rise of Scientific Criminology in Britain, 1860–1918* (2005).

5 Cesare Lombroso and Italian Criminal Anthropology

1. G. Canguilhem, *Il normale e il patologico* (1998) [1966].
2. Pick, *Faces of Degeneration*, 112.
3. On how his experiences in southern Italy influenced Lombroso's thought, see ibid., 114–15.
4. G. Colombo, *La scienza infelice* (1975), 41–53.
5. Pick, *Faces of Degeneration*, 111–13.
6. For an account of this moment, see ibid., 122.
7. Ibid., 125–6.
8. On Haeckel, see Chapter 3.
9. The term 'born criminal' was originally coined by the criminologist Enrico Ferri in 1880. R. Villa, *Il deviante e i suoi segni: Lombroso e la nascita dell'antropologia criminale* (1985), 169, 188.
10. Ibid., 144–9.
11. Davie, *Tracing the Criminal*, 22.
12. J. B. Thomson, 'The Psychology of Criminals', *JMS*, 1870, vol. 16, 321–50.
13. Lombroso, *L'uomo delinquente* (1876), vol. 2, 34–5; Lombroso, *L'uomo delinquente* (1889), vi, lix.
14. D. Nicolson, 'The Morbid Psychology of Criminals', *JMS*, 1873, vol. 19, 222–31, 398–409, 1874, vol. 20, 20–37, 167–85, 527–51, 1875, vol. 21, 18–31, 225–50. For an overview, see H. Ellis, 'The Study of the Criminal', *JMS*, 1890, vol. 36, 1–15.
15. Quoted in Ellis, 'The Study of the Criminal', 5. On Maudsley's influence in Italian sexology see Chapter 3 and Tamassia.
16. N. H. Rafter, *Creating Born Criminals* (1997), 81.
17. Pick, *Faces of Degeneration*; Davie, *Tracing the Criminal*.

18. Davie, *Tracing the Criminal*, 127ff.
19. Tuke wrote that Lombroso came from Leibniz (Austria), which suggests that perhaps Tuke did not know much about Lombroso.
20. T. S. Clouston, 'The Developmental Aspects of Criminal Anthropology', *Journal of the Anthropological Institute of Great Britain and Ireland*, 1894, vol. 23, 215–25; Ellis, *The Criminal* (1901), ix–x; H. Ellis, 'The Progress of Criminology' in his *Views and Reviews: A Selection of Uncollected Articles, 1884–1932* (1932), 308.
21. D. Pick, (1988) ' "Terrors of the night": *Dracula* and "Degeneration" in the Late Nineteenth Century', *Critical Quarterly*, 1988, vol. 30, 71–87.
22. J. M. Guy (ed.), *The Complete Works of Oscar Wilde* (2007), vol. 4, 79; Oscar Wilde wrote to the Home Secretary on 2 July 1896, reprinted in M. Jay and M. Neve (eds.), *1900: A Fin-de-siècle Reader* (1999).
23. In other words, neither male and female offenders, nor savage and black people, displayed secondary sexual characteristics to match their biological sex. In the context of evolutionary theory this was a sign of atavism. C. Lombroso, *L'uomo delinquente* (1876), 199.
24. Ibid., 129–32.
25. Ibid., 200.
26. J. R. Gasquet, 'Italian Psychological Literature', *JMS*, 1884, vol. 29, 587–8.
27. Krafft-Ebing, 'Über gewisse Anomalies des Geschlechtstriebs und die klinisch-forensich Verwertung derselben als eines wahrscheinlich funktionellen Degenerationszeichens des centralen Nervensystems'. In this article the Austrian psychiatrist had organised sexual abnormalities into three categories: 'wrong time' of sexual activity (childhood or old age); 'wrong amount' of sexual activity (complete absence or excess of sexual drive); 'wrong aim of the sexual act' ('contrary sexual feeling', cannibalism, and necrophilia). This first attempt to classify sexual perversions was later systematically developed in *Psychopathia Sexualis* (1931) [1886]. Late nineteenth-century sexual taxonomies increasingly identified non-procreative sexualities with specific groups of men.
28. Lombroso, 'L'amore nei pazzi', 24, 26–7.
29. These case-histories soon became well known to the members of the international medical community. They were some of the earliest published cases of 'sexual inversion'.
30. Lombroso, 'L'amore nei pazzi', 27.
31. J. L. Casper, *Handbook for the Practice of Forensic Medicine* (1865) [1852], vol. 4, 289. Casper's text has been considered a turning point in medico-legal discourses about same-sex desires because he associated pederasty with a 'congenital psychic condition'. See for instance Ellis and Symonds, *Sexual Inversion*, 26.
32. Ibid., 30. On the study of pathology as a means to understand the normal in Lombroso's medical lectures, see Bulfaretti, *Cesare Lombroso*, 124.
33. Lombroso, 'L'amore nei pazzi', 31–2.
34. Lombroso, 'Delitti di libidine', 340.
35. Ibid., 169–78.
36. Ibid., 338.
37. Ibid., 342.
38. Ibid.
39. Ibid., 345–6.
40. Ibid., 332–4. Lombroso used the words '*rei nati*', '*rei nati di libidine*', and '*stupratori nati*'.
41. Ibid., 335.

42. Ibid., 337.
43. Gibson, *Born to Crime*, 22–6.
    Lombroso published an article on moral insanity in 1884 where he noted in passing that Krafft-Ebing's observations on sexual inversion were linked to theories of moral insanity: C. Lombroso, 'Pazzia morale e delinquente nato', *AP*, 1884, vol. 5, 17–22.
44. Frigessi, *Cesare Lombroso*, 105–6.
45. C. Lombroso, *L'uomo delinquente* (1884), 453.
46. '[G]rande cultura e ingegno', ibid.
47. The feminised tastes of upper-class men were a stereotype of bourgeois culture. Given the perception of aristocrats as depraved, corrupt, and idle, it was assumed pederasty was an eminently aristocratic pastime.
48. Lombroso, *L'uomo delinquente* (1884), 453–4.
49. See Chapter 2 and the Zanardelli Penal Code.
50. Krafft-Ebing, *L'inversione sessuale*, 83–109.
51. C. Lombroso, *L'uomo delinquente* (1889), vol. 1, 36–7, 294, 300, 452–3.
52. Ibid., 120, 593.
53. Ibid., 595.
54. Ibid., vol. 2, 223–46.
55. Ibid., 235. Italics in the original.
56. Ibid., 223.
57. Ibid., 224.
58. For an overview of how early nineteenth-century psychiatric categories such as monomania were associated to same-sex desires, see Beccalossi, 'Nineteenth-Century European Psychiatry on Same-Sex Desires'.
59. Lombroso, *L'uomo delinquente* (1889), vol. 2, 235–6.
60. For example John Addington Symonds, see Chapter 7.
61. Krafft-Ebing, *L'inversione sessuale*, 83–109.
62. C. Lombroso, 'Verzeni e Agnoletti', *Rivista di discipline carcerarie in relazione con l'antropologia, col diritto penale, colla statistica*, 1873, vol. 3, 193–213. Vincenzo Verzeni, the 'Vampire of Bergamo', was an Italian rural version of Jack the Ripper. In 1873 he was accused of murdering two young women, removing their genitals and sucking their blood. The case generated a huge controversy in Italian psychiatric circles, and Verzeni was examined by eleven famous medical and legal experts. For a detailed discussion of this case, see Chapter 6.
63. See Chapter 3.
64. See for example Lombroso, 'L'amore nei pazzi', 1–32; Lombroso, *L'uomo delinquente* (1889), vol. 2, 235–40.
65. R. von Krafft-Ebing, *Le psicopatie sessuali* (1889) [1886]. Lombroso wrote the introduction to this book.
66. Ibid., xi.
67. C. Lombroso, 'Du parallélisme entre l'homosexualité et la criminalité innée', *AP*, 1906, vol. 27, 378–81. It was not the first time a scientist gave a paper on homosexuality. Italian medical literature records that in 1896 the Viennese physician Luzenberger had attended the Ninth Conference of the *Società di freniatria* at Florence, see A. Luzenberger, 'Sul meccanismo dei pervertimenti sessuali e la loro terapia', *APS*, 1896, 265–71. In 1901 a Professor of Criminal Anthropology at Amsterdam University called Aletrino gave a paper on the topic of homosexuality to the Fifth International Congress of Criminal Anthropology in Amsterdam. See A. Aletrino, 'La situation sociale de l'uraniste', *La scuola positiva*, 1901, vol. 11, 481–96.

68. Lombroso, 'Du parallélisme', 378–81.
69. Lombroso, 'Del tribadismo nei manicomi', 219. A *sifilicomio* was a specialised hospital in which prostitutes were locked up and treated when suffering from syphilis. It was similar to the British lock hospital.
70. This study was conducted some time before 1876, because that was the year Lombroso moved to Turin.
71. Lombroso, 'Del tribadismo nei manicomi', 218–19.
72. Ibid., 219.
73. In the same year, however, Lombroso published a case-study on 'paradoxical nymphomania'. Because the nymphomaniac patient did not present abnormal genitalia, he concluded that sexual perversion had cerebral origins. This deduction coincided with Magnan's findings of that same year, which he had introduced in a paper delivered as to the Académie Impériale de Médecine. C. Lombroso, 'Ninfomania paradossa', AP, 1885, vol. 6, 362–9.
74. Italics are mine.
75. Lombroso, 'Del tribadismo nei manicomi', 219–21.
76. Lombroso, 'L'amore nei pazzi', 32.
77. Ibid., 29–30, 32.
78. For gynaecological methods, see Moscucci, *The Science of Woman*.
79. On the use of surgical operations to treat not only self-eroticism, but tribadism, see L. Martineau, *Le deformazioni vulvari ed anali* (1896) [1884], 103.
80. C. Lombroso, 'Amori anomali e precoci nei pazzi', AP, 1883, vol. 4, 23–5. These two children were not identified as tribades because they masturbated, even though one of the children 'corrupted' ['*corruppe*'] her four-year-old sister.
81. D. G. Horn, *The Criminal Body* (2003), 52. The English version omitted all parts dealing with sexual psychopathologies.
82. Lombroso and Ferrero, *La donna delinquente*, 57.
83. Babini, Minuz, and Tagliavini, *La donna nelle scienze dell'uomo*.
84. Lombroso and Ferrero, *La donna delinquente*, 56.
85. Ibid., 57.
86. Ibid. Italics are in the text.
87. Horn, *The Criminal Body*, 98–9.
88. Lombroso, *L'uomo delinquente* (1876), 142. In the 1876 edition of *L'uomo delinquente* Lombroso mentioned that old prostitutes often tattooed on their own body the name of their tribade lovers, an observation he had already made in an article originally published in 1874, see below note 105.
89. Lombroso and Ferrero, *La donna delinquente*, 541–3.
90. Ibid., 467.
91. Ibid., 571–3.
92. Ibid., 359.
93. Ibid., 284–334.
94. Ibid., 388–9.
95. Ibid. 177.
96. Ibid., 186, 365, 369, 532, 594–5, 613–14. In *La donna delinquente*, the parts regarding the evolution of the two sexes, women's psychology, and women's intelligence were written by Ferrero, whereas the chapters dealing with criminal anthropology were written by Lombroso. Broadly speaking, the first part of the book on the normal woman was Ferrero's, the rest Lombroso's. Therefore the parts on female sexual deviancy mainly belong to Lombroso.
97. Ibid., 365, 369.

98. Ibid., 532.
99. Ibid., 594–5, 613–14.
100. Ibid., 186. If one is to infer that the elements of the comparison reflect on each other beyond the terms of the comparison, it should be noted that the animals listed were traditionally associated with dullness.
101. Ibid., 417–18.
102. Ibid., 417.
103. Lombroso relied heavily on Parent-Duchâtelet for this association, and on Scipio Sighele to a lesser extent.
104. [Le] '*più degradate*'.
105. C. Lombroso, 'Sul tatuaggio in Italia fra i delinquenti. Studio medico legale', *Archivio per l'antropologia e l'etnologia*, 1874, vol. 4, 393–4, 400. A prostitute would be considered old at merely twenty-five years of age.
106. Lombroso and Ferrero, *La donna delinquente*, 411.
107. Ibid.
108. Ibid., 412–13.
109. Ibid., 413–14.
110. See Chapter 4.
111. Lombroso and Ferrero, *La donna delinquente*, 411–15
112. See Chapter 3.
113. For the reception of this case in newspapers, see Oliari, *L'omo delinquente*, 61–9.
114. C. Lombroso, 'La psicologia di una uxoricida tribade', *AP*, 1903, vol. 24, 6–7.
115. Ibid., 7–10.
116. Babini, Minuz, and Tagliavini, *La donna nelle scienze dell'uomo*.
117. Gibson, *Born to Crime*, 60.
118. This point is also made by Horn, *The Criminal Body*, 140.
119. Gibson, *Born to Crime*, 60–1.
120. Babini, 'Un altro genere'.
121. Ibid.
122. R. Macrelli, *L'indegna schiavitù* (1981), 137.
123. M. Montessori, *La morale sessuale nell'Educazione* (1912), 272–81. I would like to thank Valeria Babini for drawing my attention to this document.
124. F. Mort, *Dangerous Sexualities: Medico-Moral Politics in England since 1830* (1987), 60–1.
125. Although Rossi-Doria has studied Lombroso's theorisation of female sexuality in some depth, she has nevertheless failed to identify any association between lesbianism and feminism in his writings. Rossi-Doria 'Antisemitismo e antifemminismo'.
126. H. Kurella, 'Osservazioni sul significato biologico della bisessualità', *AP*, 1896, vol. 17, 424–5. Kurella was a Lombroso follower.
127. For some examples of doctors who popularised Bloch and Weininger, see Pellegrini, *Sessuologia*, 739–42; L. Mongeri, *Patologia speciale delle malattie mentali* (1907), 28; R. Fronda, 'L'omosessualità nella donna', *Il manicomio*, 1912, vol. 17, 123–5; V. Masserotti, *Nel regno di Ulrichs* (1913), 27–36; A. La Cara, *La base organica dei pervertimenti sessuali* (1924) [1902], 105–14.
128. R. Pellegrini, *Sessualogia* (1954), 742; I. Bloch, *La vita sessuale dei nostri tempi* (1910) [1907].
129. See for example, C. Lombroso, 'Due geni nevrotici femminili', *AP*, 1891, vol. 12, 478–84; S. Venturi, *Le degenerazioni psico-sessuali* (1892), 290–8; Lombroso and

Ferrero, *La donna delinquente*, 160–1; P. Celesia, 'Sull'inversione sessuale', *AP*, 1900, vol. 21, 218–20.
130. C. Lombroso, 'Introduzione', in R. von Krafft-Ebing, *Le psicopatie sessuali*.
131. Ibid., xxxi.
132. V. P. Babini and L. Lama, *Una 'donna nuova'. Il femminismo scientifico di Maria Montessori* (2000).
133. On the relationship between Lombroso, his daughters, and feminism, see D. Dolza, *Essere figlie di Lombroso* (1990); V. P. Babini, 'In the Name of Father. Gina and Cesare Lombroso', in V. P. Babini and R. Simili (eds.), *More than Pupils: Italian Women in Science at the Turn of the 20th Century* (2007), 75–106.
134. Lombroso states these points both in '*Delitti di libidine*' and in his introduction to Krafft-Ebing's *Psychopathia Sexualis*.
135. M. Gibson, 'Cesare Lombroso and the Italian Criminology', in P. Becker and R. F. Wetzell, *Criminals and Their Scientists* (2006), 150.
136. V. P. Babini, 'In the Name of Father'.

6   Pasquale Penta, 'First Class Sexologist'

1. M. Hirschfeld, *Homosexuality of Men and Women* (2000) [1914], 431–2; 655, 571; Ellis, *A Note on the Bedborough Trial*; Bloch, *La vita sessuale*, 569; A. Épaulard, *Vampirisme, nécrophilie, nécrosadisme, nécrophagie* (1901), 41, 77–8. I am grateful to Lisa Downing for drawing my attention to Alexis Épaulard.
2. P. Näcke, 'Penta als einer der besten Kenner und Förderer der Sexualwissenschaft', *Zeitschrift für Sexualwissenschaft*, 1908, vol. 1, 75. Paul Näcke graduated in medicine from Würzburg University in 1873. From 1880 he occupied various posts in a number of different asylums. From 1901, Näcke began writing a series of articles in the main French and German journals of criminal anthropology. These articles display Näcke's startling familiarity with the homosexual scene centred on public urinals in Paris, Berlin, and other cities. Näcke denied that homosexuality represented a degenerative phenomenon, and argued instead that men and women were basically bisexual, and that their subsequent development was almost a matter of chance. E. Shorter, *A Historical Dictionary of Psychiatry* (2005), 129.
3. Näcke, 'Penta als einer der besten Kenner', 75.
4. Oosterhuis, *Stepchildren of Nature*, 67.
5. Wanrooij refers briefly to Penta's *APS*. Wanrooij, 'The History of Sexuality in Italy', 177.
6. Neapolitan intellectuals of the time were strongly influenced by neo-Hegelian positions, which had inhibited the spread of positivism in the city.
7. See the list of collaborators of *AP*, in particular for the year 1896, when Lombroso thanked Penta and other assiduous collaborators. Anon., 'Collaboratori', *AP*, 1896, vol. 17, first page.
8. On Bianchi, see Chapter 3.
9. Historians have often assumed that Lombroso was the first Professor of Criminal Anthropology, see for example D. G. Horn, 'Making Criminologists: Tools, Techniques, and the Production of Scientific Authority', in P. Becker and R. F. Wetzell (eds.), *Criminals and Their Scientists* (2006), 318. Horn says that there was no chair of Criminal Anthropology before 1905, and 'this only an *ad personam* appointment to honor Lombroso at Turin'. L. Bianchi, 'Introduzione' (1904); Anon., 'Necrologio di Pasquale Penta', *AP*, 1904, vol. 25, 774.
10. Krafft-Ebing, *Psychopathia Sexualis* (1931), 95–9.

11. For an account of the trial, see M. Tanteri and E. Trimboli, 'Il processo Verzeni', in L. Condò, M. Cotugno, C. Giardiello, G. Malerba, F. Ottaviani, and L. Saletti (eds.), *Modelli, giudizi e pregiudizi* (1999–2000).
12. P. Penta, *I pervertimenti sessuali* (1893).
13. Lombroso, 'L'amore nei pazzi', 11.
14. See Tanteri and Trimboli, 'Il processo Verzeni', 32–3.
15. Ibid.; Lombroso, *L'uomo delinquente* (1889), vol. 1, 587, 589, vol. 2, 24, 74. Lombroso had previously described the case in 'Verzeni e Agnoletti'; Lombroso, 'L'amore nei pazzi', 11–12.
16. Penta, *I pervertimenti sessuali*, 54.
17. Ibid., 61.
18. P. Penta, *Vincenzo Verzeni (Lo strangolatore delle donne). Esame psichiatrico, La tribuna giudiziaria* (1890), vol. 4 (extract), 15.
19. Ibid., 3–11.
20. On Binet, see Chapter 3.
21. Penta, *I pervertimenti sessuali*, 122–46. In 1881, the American Shobal Vail Clevenger argued that the origin of the sexual impulse was hunger. S. V. Clevenger, 'Hunger: The Primitive Desire', *Science*, 2, 1881 (15 January), vol. 2, 14; For similar ideas in Britain see Patrick Geddes and J. A. Thomson, *Evolution of Sex* (1889), 279–81.
22. Ibid., 216–50.
23. Penta, 'L'origine e la patogenesi della inversione sessuale', 64.
24. Penta, *I pervertimenti sessuali*, 149–58.
25. Penta, *Vincenzo Verzeni*, 19.
26. Penta, *I pervertimenti sessuali*, 301–6.
27. Oosterhuis, *Stepchildren of Nature*, 39–43; P. Guarnieri, 'Alienists on Trial: Conflict and Convergence between Psychiatry and Law (1876–1913)', *History of Science*, 1991, vol. 29, 393–410.
28. Ibid.
29. Zuccarelli was one of the first psychiatrists to draw attention to female sexual inversion. In 1888 he published two long clinical cases of tribades, see Zuccarelli, *Inversione congenita*.
30. See Chapter 3.
31. On Näcke and Raffalovich and homosexuality, see Shorter, *A Historical Dictionary of Psychiatry*, 129; Rosario, *The Erotic Imagination*, 97–108.
32. Anon., 'Primo congresso di medicina legale', *RMPF*, 1898 vol. 1, 333.
33. L. Bianchi, 'Prefazione', in P. Penta, *La simulazione della pazzia* (1905) [1904] without page numbers.
34. See for example P. Garnier, *L'onanismo a due* (1907) [1883]; Krafft-Ebing, *L'inversione sessuale*; Martineau, *Le deformazioni vulvari*; A. Moll, *Uranismo e prostituzione mascolina* (1897) [1893]; P. Moreau de Tours, *Le aberrazioni del senso genesico* (1897) [1880]; A. Tardieu, *I delitti di libidine* (1898) [1858].
35. See for example Raffalovich, *L'uranismo, inversione sessuale congenita*; Schrenck-Notzing, *La terapia suggestiva delle psicopatie sessuali*; Niceforo, *Il gergo nei normali*; R. von Krafft-Ebing, *Trattato di psicopatologia forense* (1897) [1875]; A. Forel, *La questione sessuale* (1907) [1905]; La Cara, *La base organica dei pervertimenti sessuali*. See the advertisement for the series 'Biblioteca antropologica-giuridica' in *AP*, 1896, vol. 17, 343. Lombroso's *L'uomo delinquente* was published by Bocca.
36. F. Capaccini, 'Programma', *APS*, 1896, iii–iv.
37. Ibid., iv.

38. Raffalovich, *L'uranismo*, 11–12; La Cara, *La base organica dei pervertimenti sessuali*, 1–4.
39. See H. Ellis, 'Note sulle facoltà artistiche degli invertiti', *APS*, 1896, 243–5.
40. On female homosexuality, see below.
41. A. Cristiani, 'Autopederastia in un alienato, affetto da follia periodica', *APS*, 1896, 183–7; S. A. Neri, 'Pervertito, necrofiliaco, pederasta, masochista, etc.', *APS*, 1896, 109–10.
42. See for example P. Penta, 'Dei pervertimenti sessuali', *APS*, 1896, 1–7, 17–20.
43. Näcke, 'Penta als einer der besten Kenner'.
44. Raffalovich, 'Gli studi sulle psicopatie sessuali in Inghilterra', *APS*, 178–9; Raffalovich, *L'uranismo, inversione sessuale congenita*. For an account of Raffalovich's article, see Chapter 2.
45. P. Penta, (review) 'E. S. Talbot and Havelock Ellis, "A Case of Developmental Degenerative Insanity, with Sexual Inversion, Melancholia following Removal of Testicles, Attempted Murder and Suicide, *Journal of Mental Science*, April 1896', *APS*, 1896, 191–4.
46. P. Penta, (review) 'Havelock Ellis, "A Note on the Treatment of Sexual Inversion" in *Alienist and Neurologist*, 1896', *APS*, 1896, 261–2.
47. Ellis, 'Note sulle facoltà artistiche degli invertiti', 243.
48. Ibid., 244.
49. Penta, 'L'origine e la patogenesi della inversione sessuale', 54; Pick, *Faces of Degeneration*, 8. See also Chapter 5.
50. Penta, 'L'origine e la patogenesi della inversione sessuale', 57.
51. P. Penta, *Dei pervertimenti sessuali* (1896), 5–9.
52. On Binet's theory of fetishism, see Chapter 3.
53. Fratelli Capaccini, Letter, 28 April 1896, attached to *APS*.
54. Advertisement of *Biblioteca delle psicopatie sessuali*, attached to *APS*.
55. '[*I*]*mpudico*', L. Bianchi, 'Introduzione'.
56. Näcke, 'Penta', 78.
57. P. Penta, (review) 'Jahrbuch für sexuelle Zwischenstufen mit besenderer Berucksochtigung der Homosexualität', *RMPF*, 1903, vol. 2, 518.
58. P. Penta, (review) 'R. von Krafft-Ebing, *Psychopathia Sexualis*, 1898', *RMPF*, vol. 1, 300–1.
59. Editorial Board, 'Programma', *RMPF*, 1898, vol. 1, 1–3.
60. Ibid., 2.
61. Ibid.
62. Ibid., 2–3.
63. Ibid., 1–3.
64. *RMPF*, vol. 1, inside cover.
65. P. Penta, 'Sopra un caso di inversione sessuale', *RMPF*, 1898, vol. 1, 109–15.
66. Ibid., 112.
67. Ibid., 114–15.
68. Ibid., 115.
69. P. Penta, 'Un pedofilo fellatore', *RMPF*, 1903, vol. 6, 477–518.
70. Ibid., 482.
71. P. Penta, *Lezioni di psichiatria dettate nell'anno scolastico 1899–1900* (1900), vol. 1 (manuscript).
72. Ibid., 34.
73. Ibid., 42–4.

74. Ibid., 44.
75. For a rejection of Freud's theories in Italy, see Morselli, *La psicanalisi*; M. David, *La psicoanalisi nella cultura italiana* (1966).
76. Morselli, *La psicanalisi*, vol. 1, 25, 74–115.
77. David, *La psicoanalisi*.
78. Penta, *Lezioni di psichiatria*, vol. 1, 102–14.
79. Ibid., 115–19.
80. Ibid., 123.
81. Ibid., 112–23.
82. Penta, (review) 'Jahrbuch für sexuelle Zwischenstufen', 518.
83. Ibid.
84. Ibid., 524.
85. Ibid.
86. Ibid.
87. Oosterhuis, *Stepchildren of Nature*, 49–50.
88. Penta, *I pervertimenti sessuali*, 200.
89. T. Kuhn, *The Essential Tension* (1977), 168.
90. Penta, 'Sopra un caso di inversione sessuale'.
91. Penta, 'Un pedofilo fellatore'.
92. See the analysis of same-sex practices in studies such as Penta, *I pervertimenti sessuali dell'uomo*; P. Penta, 'Parere medico-legale sulle condizioni psichiche del prete Pietro Paolo Potenza, accusato di duplice assassinio', *RMPF*, 1903, vol. 6, 101–11, 253–62, 289–96, 325–49, 361–79; Penta, 'Un pedofilo fellatore'.
93. Penta, *Lezioni di psichiatria*, 53.
94. Ibid., 52.
95. Ibid., 51–3.
96. Lypemania was defined as a form of mania characterised by sadness.
97. Penta, 'Dei pervertimenti sessuali', 15–16.
98. Ibid., 17.
99. Ibid., 17–18.
100. '[A]mori rudimentali', ibid., 18.
101. Ibid., 19.
102. Penta and d'Urso, 'Sopra un caso d'inversione sessuale in donna epilettica'.
103. Ibid.
104. Ibid.
105. P. Penta, (review) 'W. L. Howard, "Sexual Perversion" in *Alienist and Neurologist*', *APS*, 123–5.
106. Penta and d'Urso 'Sopra un caso d'inversione sessuale', 37.
107. See also G. Penta, 'Un altro caso di inversione sessuale in donna', *APS*, 1896, 94–6. The author of the article signed as Giuseppe Penta, but it is probable that it was written by Pasquale Penta, since the author recalls some of Pasquale Penta's earlier work.
108. P. Penta, 'In tema di pervertimenti sessuali', *RMPF*, 1900, vol. 3, 69. Penta failed to mention that Ellis had published some letters by women in *Sexual Inversion*, although Ellis's subjects did not approach sexual matters as openly as the woman Penta presented. Oosterhuis has paid attention to Krafft-Ebing's use of inverts' letters. Oosterhuis, *Stepchildren of Nature*.
109. Penta, 'In tema di pervertimenti umani', 69.
110. Ibid., 70.
111. Ibid.

112. Ibid.
113. Ibid., 70–1.
114. Ibid., 72.
115. Ibid., 71–2.
116. Ibid., 73.
117. Ibid., 69–89.
118. F. Saporito, 'Sulla delinquenza militare', *RMPF*, 1903, vol. 6, 44–68 and 125–84. In the 1892 edition (11th edition) of *Amori degli uomini*, Mantegazza published a letter of a man who loved other men and who had written to him to explain how people like him suffered because of society. Mantegazza, *Gli amori degli uomini*, vol. 1, 278–89. The role of medical texts in contributing to create a 'homosexual identity' has been hugely debated. For recent contrasting views on the topic, see Oosterhuis, *Stepchildren of Nature*; Brady, *Masculinity*.
119. Such as the case of Hirschfeld.

7   Havelock Ellis and Sex Psychology

1. Some of the case-studies were known to Ellis personally, but most of them were supplied by Symonds and Edward Carpenter. The Americans James G. Kiernan and Dr Lydston also assisted Ellis at various points. An American female physician referred to as 'Dr. K.' also helped Ellis to obtain case-histories and supplied an appendix. Ellis also gave thanks to Lombroso. Ellis and Symonds, *Sexual Inversion*, xv.
2. This argument is also advanced in P. Robinson, *The Modernisation of Sex* (1976); J. Weeks, *Sexuality and its Discontents* (1985).
3. Faderman, *Surpassing the Love of Men*, 239–53. By taking words such as 'morbid' out of context, Faderman concludes that Ellis's work regards lesbianism as a pathology. See also Jeffreys, *The Spinster and Her Enemies*, 102–46. Ellis, however, argued that homosexuality was not a disease.
4. P. Grosskurth, *Havelock Ellis* (1980); C. Nottingham, *The Pursuit of Serenity* (1999).
5. A number of historians have paid attention to Symonds's letters to Ellis, but not as many have studied Ellis's letters to Symonds. See for example Grosskurth, *Havelock Ellis*; Nottingham, *The Pursuit of Serenity*.
6. Crozier, 'Introduction', in Ellis and Symonds, *Sexual Inversion: A Critical Edition*, 34.
7. The Contemporary Science Series aimed to popularise scientific works amongst the general public.
8. H. Ellis, BL ADD 70524, Letter to J. A. Symonds, 1 July 1892; Nottingham, *The Pursuit of Serenity*, 140.
9. H. Ellis, BL ADD 70524, Letter to J. A. Symonds, 18 June 1892.
10. Ibid. In *The Criminal* (1890) Ellis had briefly associated femininity with 'pederasts', and masculinity with female criminals. He also observed that tattooing was a common practice amongst 'pederasts' and 'tribades'. Both ideas were undoubtedly drawn from Lombroso. Ellis, *The Criminal*, 53, 106.
11. For an insightful and detailed analysis of Ellis's and Symonds's correspondence, see Crozier, 'Introduction'.
12. H. Ellis, BL ADD 70536, Letter to E. Carpenter, 17 Dec. 1892.
13. Symonds, *Letters*, Letter 1984 to H. Ellis, 694.
14. H. Ellis, BL ADD 70536, Letter to Edward Carpenter, 17 Dec. 1892.

15. Symonds, *Letters*, Letter 1836 to the Rev. A. Galton, 10 Oct. 1890, 506. About Symonds's homosexuality, see Grosskurth (ed.), *The Memoirs of John Addington Symonds*.
16. See Chapter 4.
17. J. A. Symonds, *A Problem in Modern Ethics* (1896), 10–74.
18. Symonds, *Letters*, Letter 1984 to H. Ellis, 693–4.
19. Formal standards of respectability dominated amongst ordinary middle-class readers and in their magazines. The editorial policy of such periodicals sometimes determined their reticence about sex. Neither Ellis nor anyone else could publish an article on sex for the British general interest press prior to 1920. D. Rapp, 'The Early Discovery of Freud by the British General Educated Public, 1912–1919', *Social History of Medicine*, 1990, vol. 3, 229–30.
20. H. Ellis, BL ADD 70524, Letter to J. A. Symonds, 21 Dec. 1892; Crozier, 'Philosophy in the English Boudoir', 246.
21. Symonds, *A Problem in Modern Ethics*, 128.
22. H. Ellis, BL ADD 70524, Letter to J. A. Symonds, 1 July 1892.
23. Symonds, *Letters*, Letter 1996 to H. Ellis, 7 July 1892, 709–10.
24. In *A Problem in Modern Ethics*, Symonds had defined homosexuality as a 'sport of nature in her attempt to differentiate the sexes'; Symonds, *A Problem in Modern Ethics*, 129. In biology, the term 'sport' indicates an organism that shows a marked change from the normal type or parent stock, typically as a result of mutation. Symonds, *Letters*, Letter 2036 to H. Ellis, 29 Sept. 1892, 754–5.
25. Ibid., Letter 2062 to H. Ellis, 1 Dec. 1892, 787.
26. Ibid., Letter 2036 to H. Ellis, 29 Sept. 1892, 754–5.
27. Ibid., Letter 2062 to H. Ellis, 1 Dec. 1892, 787–8.
28. H. Ellis, BL ADD 70524, Letter to J. A. Symonds, 21 Dec. 1892.
29. Ibid.
30. Ibid.
31. Ibid. Teratology is the study of malformations or deviations from the normal type organisms. It is concerned with anatomy and classification of malformed foetuses.
32. H. Ellis, BL ADD 70524, Letter to J. A. Symonds, 3 Jan. 1893.
33. Ellis and Symonds, *Sexual Inversion*, 134.
34. Ibid., 134–5.
35. Grosskurth, *Havelock Ellis*, 189–90. On the artistic talent of homosexuals, see Ellis, 'Nota sulle facoltà artistiche degli invertiti'.
36. Ellis and Symonds, *Sexual Inversion*, xiv–xv.
37. In 1901, Krafft-Ebing detached homosexuality from a state of psychoneuropathic degeneration. Sulloway, *Freud*, 296.
38. H. Ellis, BL ADD 70524, Letter to J. A. Symonds, 19 Feb. 1893.
39. Nottingham, *The Pursuit of Serenity*, 143–4.
40. See Chapter 6.
41. Symonds, *Letters*, Letter 2070 to E. Carpenter, 29 Dec. 1892, 797.
42. In Venice Symonds had a relationship with a gondolier called Angelo Fusato. In his *Memoirs*, Symonds described Fusato's life and that of other Venetian gondolier male prostitutes. J. A. Symonds, *The Memoirs* (1984), 271–83.
43. According to John Beker, who was a prison doctor at Portsmouth, Ellis had brought criminal anthropology into prominence in Britain. J. Baker, 'Some Points Connected with Criminals', *JMS*, 1892, vol. 38, 364. On the

influence of Lombroso's criminal anthropology in Britain see Davie, *Tracing the Criminal*.

44. Ellis, *The Criminal*, 39.
45. H. Ellis, BL ADD 70524, Letter to J. A. Symonds, 1 July 1892.
46. By the time of the Medical Congress in Rome, Ellis had some scientific standing as editor of the Contemporary Science Series, author of *The Criminal*, and of the recently published *Man and Woman*. Ellis made another important contact while in Rome, Dr Hans Kurella, who would later translate his *Studies* into German. H. Ellis, *My Life* (1940), 283; Grosskurth, *Havelock Ellis*, 168.
47. Preface to the third edition of *The Criminal*, see Ellis, *The Criminal* (1901), ix–x.
48. H. Ellis, BL ADD 70524, Letter to J. A. Symonds, 10 July 1891. Ellis was referring to C. Lombroso, 'Anomalie psichiche in Michelangelo e Virgilio', *AP*, 1890, vol. 11, 331–3.
49. Symonds, *Letters*, Letter 1984 to H. Ellis, 20 June 1892, 694.
50. Symonds, *A Problem in Modern Ethics*, 30.
51. Ibid., 66.
52. Ibid., 65–6.
53. Ibid., 61–8. Throughout his career, Symonds endeavoured to discover an enduring Hellenism. J. Pemble, 'Art, Disease, and Mountains', in J. Pemble (ed.), *John Addington Symonds* (2000), 1–21.
54. H. Ellis, BL ADD 70524, Letters to J. A. Symonds, 18 Jan. 1893; 9 Feb. 1893.
55. Symonds, *Letters*, Letter 2070 to E. Carpenter, 29 Dec. 1892, 797–9, 810–11. In the last decade of the nineteenth century, scientists were speculating about the possible rejuvenating effects of the secretions of sexual glands. For example, Charles-Édouard Brown-Séquard, a professor at Harvard, experimented on himself with injections obtained from the testicles of a pig and a dog, believing that they could rejuvenate him. See Sengoopta, 'The Modern Ovary', 441. Symonds had read Venturi carefully as is evident from his comments on Venturi's theory about the relationship between masturbation and the sexual impulse: Venturi believed that a person first began to experience proper sexual pleasure through masturbation, which in turn awakened the desire for reciprocal sexual relations. For Symonds's comments on this see Crozier, 'Introduction', 48.
56. Symonds, *Letters*, Letter 2070 to E. Carpenter, 29 Dec. 1892, 797–9, 810–11.
57. Mantegazza, *Gli amori degli uomini*. On Mantegazza, see Chapter 3.
58. Symonds, *A Problem in Modern Ethics*, 11–12.
59. Ibid.
60. Ibid., 81–2.
61. See Chapter 3.
62. Symonds, *Letters*, Letter 1996 to H. Ellis, 7 July 1892, 710.
63. Ibid., Letter 2036 to H. Ellis, 29 Sept. 1892, 755.
64. H. Ellis, BL ADD 70524, Letter to J. A. Symonds, 21 Dec. 1892.
65. Ellis and Symonds, *Sexual Inversion*, 22–4.
66. Ibid.
67. Burton, *A Plain and Literal Translation of the Arabian Nights*, 205–54. The essay on pederasty appeared only in 1885 and 1886. The 1886 edition was restricted to private subscribers of the publisher, Burton Club.
68. Symonds knew Burton personally: he reported in his letters that he was planning to go to Trieste with him in 1890. Symonds, *Letters*, Letter 1831 to Henry Graham Dakyns, 24 Sept. 1890, 500.
69. Burton, *A Plain and Literal Translation of the Arabian Nights*, 207–8. Burton also expounded Mantegazza's anatomical theory about pederasty and, unlike

Symonds, also reported some observations by Mantegazza according to which there are tribades amongst women who 'can procure no pleasure except by foreign objects introduced a posteriori', 208–9.

70. Symonds, *A Problem in Modern Ethics*, 78–81.
71. See the third section below.
72. Marchesini and Obici, *Le 'amicizie' di collegio*.
73. The criminal anthropologist Dr Hans Kurella (1858–1916) strongly supported Lombroso and was editor of the journal *Centralblatt für Nervenheilkunde und Psychiatrie*. Amongst other things, in 1910, he published *Cesare Lombroso als Mensch und Forscher*, translated into English by M. Eden Paul, see H. Kurella, *Cesare Lombroso, a Modern Man of Science* (1911) [1910].
74. Anon., (review) ' "Das Conträre Geschlechtsgefühl", von Havelock Ellis und J. A. Symonds', *JMS*, 1897, vol. 43, 565–9.
75. H. Ellis, 'The Study of Sexual Inversion', *The Medico-Legal Journal*, 1894, vol. 12, 148–57; H. Ellis, 'Sexual Inversion in Women', *Alienist and Neurologist*, 1895, vol. 16, 141–58; H. Ellis, 'Sexual Inversion in Men', *Alienist and Neurologist*, 1896, vol. 17, 115–50; Ellis, 'Nota sulle facoltà artistiche degli invertiti'.
76. Tuke had also been a close friend of Symonds's father. One of Tuke's sons was a well-known painter called Henry Scott Tuke, who was homosexual. In the atmosphere that prevailed after the Wilde trials, Tuke's assessment that it would be impossible to limit Ellis's book to physicians was probably realistic. Grosskurth, *Havelock Ellis*, 180.
77. In March 1897, Wilde had only just been released from prison.
78. The Legitimation League was founded in 1893. It claimed to exist 'for the purpose of entering a protest against current ironbound marriage customs, which tend to crush individuality'. One of the major objects of the League's campaign was to facilitate 'divorce by mutual consent'; it endorsed sexual enlightenment for children, and treated as 'open questions' the entire gamut of contemporary sexual institutions such as 'prostitution, celibacy, marriage with an accompaniment of adultery', and alternative sexual arrangements 'such as polygamy, polyandry, promiscuity, and matriarchalism'. Stepansky, 'A Footnote to the History of Homosexuality', 80.
79. Anon., 'Charge of Publishing and Selling Obscene Literature', *BMJ*, 1898, vol. 2, 1466; Ellis, *A Note on the Bedborough Trial*, 6; J. Weeks, *Making Sexual History* (2000), 27–8.
80. Stepansky, 'A Footnote to the History of Homosexuality', 82.
81. Anon., 'Charge of Publishing and Selling Obscene Literature'; Ellis, *A Note on the Bedborough Trial*, 9–10.
82. For the reception of the Bedborough trial in medical journals, magazines and newspapers, see Stepansky, 'A Footnote to the History of Homosexuality'.
83. Ellis, *A Note on the Bedborough Trial*, 16–21.
84. Ibid., 12.
85. Ibid., 18. Kiernan was the Secretary of the Chicago Academy of Medicine.
86. Ibid., 19.
87. On Penta, see Chapter 6. Penta promptly reviewed Ellis's articles published in *Alienist and Neurologist*: see P. Penta, (review) 'Havelock Ellis, "A Note on the Treatment of Sexual Inversion" '.
88. Ellis, *A Note on the Bedborough Trial*, 20.
89. Ibid., 14.
90. Ibid., 8.

91. Anon., 'Charge of Publishing and Selling Obscene Literature'.
92. Editors, 'Editorial article', *The Lancet*, 1898, vol. 2, 19 Nov. 1345.
93. Ibid., 1344.
94. Ibid. Ellis replied to this editorial, and in an attempt to defend himself he explained that none of the medical publishers whom he had approached cared to publish a book on sexual inversion. He wrote that Ulrichs's views did not have any scientific value and, as such, he had 'explicitly rejected them. [...] So far as standpoint is concerned, my own position is not materially different from that of Krafft-Ebing, though I am by no means in sympathy with the Viennese Professor's treatment of these subjects. Certainly inversion when it arises on a congenital basis is "natural", but only in the same sense that all pathological anomalies are natural – colour-blindness, for instance, or hypospadias. I am not prepared to admit that congenital inversion is natural in any other sense.' H. Ellis, 'The Question of Indecent Literature', *The Lancet*, 1898, vol. 2, 26 Nov., 1431. The German lawyer Karl Heinrich Ulrichs (1825–95) is considered one of the first modern theorists of homosexuality. He was a tireless campaigner against injustice towards homosexuals and directed early sexologists to the subject of homosexuality. Ulrichs outlined the so-called third sex theory, which he summed up in the Latin phrase *anima muliebris virili corpore inclusa* [a female soul confined in a male body]. According to Ulrichs, homosexuality is an inborn condition rather than a sin, a disease, or a crime.
95. Stepansky, 'A Footnote to the History of Homosexuality', 91.
96. Anon., ' "Das Conträre Geschlechtsgefül" ', 568.
97. Anon., (review) '*Uranisme et Unisexualité* by M. A. Raffalovich, 1896', *JMS*, 1897, vol. 43, 573.
98. Editors, 'The Bedborough Case', *JMS*, 1899, vol. 45, 122–3.
99. Ibid., 123.
100. Ibid., 122.
101. See Chapter 4 for more on their writings on same-sex desires.
102. Lecturer on Psychological Medicine at St Thomas's Hospital, formerly President of the Medico-Psychological Association, and joint editor of the *JMS*.
103. Goodall was a Medical Superintendent of the Joint Counties Asylums, Carmarthen, and joint editor of the *JMS*. On Savage, see Chapter 4.
104. Anon., 'Sexual Psychology and Pathology', *BMJ*, 1902, vol. 1, 339.
105. Ibid., 340.
106. Ibid.
107. Stepansky, 'A Footnote to the History of Homosexuality', 97.
108. Anon., ' "Das Conträre Geschlechtsgefül" ', 565–9; Editors, 'The Bedborough Case', *JMS*, 1899, 122–3; Editors, 'Editorial article', *The Lancet*, 1898, vol. 2, 19 Nov. 1344–5.
109. See Chapter 6.
110. H. Ellis, BL ADD 70524, Letter to J. A. Symonds, 3 Jan. 1893. Symonds added a couple of pages regarding female homosexuality in the 1901 edition of *A Problem in Greek Ethics*, published after his correspondence with Ellis. J. A. Symonds, *A Problem in Greek Ethics* (1901), 70–1.
111. Krafft-Ebing, *Psychopathia Sexualis* (1893), 185–319.
112. H. Ellis, *Studies in the Psychology of Sex: Sexual Inversion*, vol. 2 (1921), 223–44. Ellis and Symonds also included three additional American cases (a man and two women) provided by the American psychiatrist G. Franck Lydston. Ellis and Symonds, *Sexual Inversion*, 187–98.

113. Ellis, *My Life*, 263–70, 324–31, 370–3; H. Ellis, BL ADD 70524, Letter to J. A. Symonds, 1 July 1892.
114. Faderman, *Surpassing the Love of Men*, 239–53; M. Jackson, *The Real Facts of Life* (1994), 106–28; Jeffreys, *The Spinster and Her Enemies*, 102–46.
115. H. Ellis, *Man and Woman* (1934) [1894], 442–60.
116. H. Ellis, *Man and Woman* (1899) [1894], 1–17.
117. Ibid.
118. Ibid.
119. Ibid., 385–6.
120. Ibid., 387. Characteristics associated with masculinity and femininity were not static essences. They developed in a dynamic process in the course of a lifetime. Although manifest from birth, sexual differentiation only came into full being during puberty. Bodily changes were paralleled by a mental transformation. Female puberty, more so than male 'sexual development', was fraught with dangers caused by the appearance of menstruation. This meant that a girl could not bear the same educational strain as a boy. Injury during puberty generally meant that development into proper femininity was irrevocably arrested. Rowold, '*The Academic Woman*', 114–15.
121. Ellis and Symonds, *Sexual Inversion*, 132.
122. See Chapter 3.
123. Ellis and Symonds, *Sexual Inversion*, 132–3.
124. Crozier, 'Philosophy in the English Boudoir', 293.
125. Ellis, *Man and Woman* (1899).
126. Faderman, *Surpassing the Love of Men*, 239–53; Jeffreys, *The Spinster and Her Enemies*, 105–46; Terry, *An American Obsession*, 50; Weeks, *Making Sexual History*, 16–52.
127. See for example Ellis and Symonds, *Sexual Inversion*, 49–55, 63–4.
128. Ibid., 87–8.
129. Crozier, 'Philosophy in the English Boudoir'.
130. Ibid., 294–8.
131. Ellis and Symonds, *Sexual Inversion*, 88.
132. Ibid., 88–91.
133. Ibid., 91.
134. Ibid., 93–4. According to Grosskurth, this case is based on Ellis's wife, Edith. Grosskurth, *Havelock Ellis*, 188.
135. Ellis pointed this out with regard to the 'very homosexual' Vernon Lee whom he met once. Ellis wanted to ask her to provide a case-history. H. Ellis, BL ADD 70524, Letters to A. J. Symonds, 3 Jan. 1893; 9 Feb. 1893.
136. Ellis and Symonds, *Sexual Inversion*, 96.
137. Emphasis in the original. E. Carpenter, *Love's Coming-of-Age* (1948) [1896], 130. This work was originally published in 1896 without the section 'The Intermediate Sex', to which I refer.
138. It should be noted that most researchers seem to refer only to this edition. The second edition was published in 1901, and the third in 1915. On Alice Mitchell, see Duggan, *Sapphic Slashers*.
139. Ellis, *Studies in the Psychology of Sex: Sexual Inversion*, vol. 2 (1921), 196.
140. Renée Vivien was the pseudonym of Pauline Tarn (1877–1909). Her father was of Scottish descent and her mother an American lady from Honolulu. As a child she was taken to Paris, and was raised in France before becoming notorious as a poet and homosexual. Her most important volumes of poetry are *Études et*

Preludes (1901), *Cendres et Poussières* (1902), and *Evocations* (1903). Ellis, *Studies in the Psychology of Sex: Sexual Inversion*, vol. 2 (1921), 200.

141. Ibid., 201.
142. Faderman, *Surpassing the Love of Men*, 242–9; Jeffreys, *The Spinster and Her Enemies*, 105–13; J. Terry, 'Anxious Slippages between "Us" and "Them": A Brief History of the Scientific Search for Homosexual Bodies', in J. Terry and J. Urla, *Deviant Bodies* (1995), 134.
143. L. Doan, 'Acts of Female Indecency: Sexology Intervention in Legislating Lesbianism', in L. Bland and L. Doan (eds.), *Sexology in Culture* (1998), 209.
144. Ellis and Symonds, *Sexual Inversion*, 100.
145. Ibid.
146. Ibid.
147. Ibid.
148. J. Oppenheim, *Shattered Nerves* (1991), 74–91.
149. Rowold, '*The Academic Woman*', 80.
150. The British Society for the Study of Psychology of Sex published the papers given at the Society; they are kept in the BL: *The British Society for the Study of Psychology*, Cup 364a.
151. Ibid.
152. H. Ellis, *The Erotic Rights of Women and the Objects of Marriage* (1918).
153. R. M. MacIver and C. H. Page, *Society* (1964) [1950], 243; K. Davis, *Human Society* (1969) [1948], 187.
154. Nottingham, *The Pursuit of Serenity*, 8–9, 150.
155. Ibid., 150–3. In the broad spectrum of feminism this tension of difference versus equality persists as an insoluble dilemma. Two theories in particular deal with this contradiction: the philosophy of difference, more common on the Continent (S. De Beauvoir, L. Irigaray, J. Kristeva), and liberal feminism, more common in the USA (C. MacKinnon).
156. Ellis and Symonds, *Sexual Inversion*, 84.
157. Ibid., 82–8.

8   William Blair-Bell and Gynaecology

1. Bell died in 1936. On Bell, see J. Peel, *William Blair-Bell* (1986).
2. C. Lawrence, *Medicine in the Making of Modern Britain* (1994), 60; C. Lawrence, 'Incommunicable Knowledge: Science, Technology and the Clinical Art in Britain, 1850–1914', *Journal of Contemporary History*, 1985, vol. 20, 503–20.
3. Moscucci, *The Science of Woman*, 108.
4. G. Pomata, 'Menstruating Men: Similarity and Difference of the Sexes in Early Modern Medicine', in V. Finucci and K. Brownlee (eds.), *Generation and Degeneration: Tropes of Reproduction in Literature and History from Antiquity through Early Modern Europe* (2001), 146.
5. In 1844 the French physician, Achille Chéreau (1817–85), dismissed van Helmont's dictum that woman was what she was because of her uterus. According to Chéreau, the ovaries determined women's nature and were responsible for menstruation, and female disorders such as hysteria were typically caused by ovarian dysfunctions.
6. According to the 'ovular theory' of menstruation, it was the spontaneous release of the egg that caused menstruation, and the onset of the menses coincided not only with the fertile period, but also with the peak of sexual desire in women.

In the course of the nineteenth century it became a virtually undisputed tenet of gynaecological theory that the ovaries were the essential difference from which all others derived. The ovaries were thought to link women to the world of instincts and automatic behaviour. Moscucci, *The Science of Woman*, 34.

7. See V. C. Medvei *The History of Clinical Endocrinology* (1993), 5.
8. Moscucci, *The Science of Woman*, 34 and N. Oudshoorn, *Beyond the Natural Body* (1994), 205. On the development of endocrinology, see Medvei, *The History of Clinical Endocrinology*; C. Sengoopta, *The Most Secret Quintessence of Life* (2006).
9. Medvei discusses this entire literature in detail in V. C. Medvei, *A History of Endocrinology* (1982), chs. 17–19.
10. W. Blair-Bell, *The Sex Complex* (1916). The dictum was written on the inside cover.
11. C. Sengoopta, 'The Modern Ovary: Constructions, Meanings, Uses', *History of Science*, 2000, vol. 38, 451–2.
12. W. Blair-Bell, 'The Correlation of Function: With Special Reference to the Organs of Internal Secretions and the Reproductive System', *BMJ*, 1920, 787–91; Peel, *William Blair-Bell*, 18–19.
13. Bell, *The Sex Complex* (1916), 101.
14. Ibid., 18.
15. W. Blair-Bell, *The Sex Complex* (1920) [1916], 1–5. In the second edition Bell revised and expanded his work.
16. Ibid., 20.
17. Ibid., 15.
18. Bell, *The Sex Complex* (1916), 119.
19. Ibid., 107.
20. Ibid., 106–7.
21. Ibid., 108.
22. Ibid., 111.
23. Ibid., 113–14.
24. Ibid., 118–19, 120.
25. Ibid., 118.
26. Also known as ovarian therapy, this was the precursor to modern hormone replacement therapies. The emergence of organotherapy as a treatment for supposed ovarian insufficiency arose from the fascination with 'internal secretions' of both men and women. Although the thyroid hormone was isolated in 1891, the existence of such 'internal secretions' in ovaries and testes remained speculative until the 1920s. Despite the tentative nature of their understanding at the time, some scientists eagerly promoted the therapeutic use of various extracts from animals.
27. Bell, *The Sex Complex* (1916), 179; 203–4.
28. W. Blair-Bell, 'Disorders of Functions', in T. Watts Eden and C. Lockyer (eds.), *The New System of Gynaecology* (1917), vol. 1, 401.
29. Bell, *The Sex Complex* (1916), 119.
30. Acromegaly or hyperpituitarism is characterised by an enlargement of the bones of hands, feet, and face, often accompanied by headache, muscle pain, and emotional disturbances. It is caused by the anterior pituitary gland's overproduction of growth hormones.
31. Bell, *The Sex Complex* (1920), 167–8. This recalls Tonini's case in Chapter 3.
32. Bell, *The Sex Complex* (1916), 120.
33. Ibid.
34. Ibid., 207.
35. Ibid., 114–15.

36. Rowold, 'The Academic Woman', 171. Also called 'applied biology' or 'applied human genetics', eugenics was born from other sciences such as statistics, biology, genetics, anthropology, and psychiatry, and was developed to solve social issues, mainly the perceived problem of degeneracy. See for example D. B. Paul, *Controlling Human Heredity* (1995); T. M. Porter, *The Rise of Statistical Thinking* (1986). On birth control in England, see H. Cook, *The Long Sexual Revolution* (2004).
37. Blair-Bell, *The Sex Complex* (1916), 1.
38. Ibid., 13.
39. W. Blair-Bell, 'Hermaphroditism', *Liverpool Medico-Chirurgical Journal*, 1915, no. 35, 289.
40. A. D. Dreger, *Hermaphrodites and the Medical Invention of Sex* (1998), 138–66.
41. Moscucci, *The Science of Woman*, 16.
42. Dreger, *Hermaphrodites and the Medical Invention of Sex*, 138–66.
43. Biopsies became standard practice in the 1910s in cases where a person's sex was in doubt.
44. Bell, 'Hermaphroditism', 291.
45. Bell, *The Sex Complex* (1920), 151–2.
46. Ibid., 152.
47. Ibid., 151–2.
48. Dreger, *Hermaphrodites and the Medical Invention of Sex*, 165–6; Bell, 'Hermaphroditism', 277.
49. Dreger, *Hermaphrodites and the Medical Invention of Sex*, 166.
50. Bell, 'Hermaphroditism', 292.
51. Bell, *The Sex Complex* (1920), 134–5. Italics are mine.
52. Park, 'The Rediscovery of the Clitoris'; Wahl, *Invisible Relations*, 17–42.
53. Bell, *The Sex Complex* (1920), 165.
54. Ibid., 165–6.
55. Ibid., 211.
56. Bell, 'Disorders of Functions', 301.
57. Bell, *The Sex Complex* (1920), 211.
58. Bell, 'Disorders of Functions', 300.
59. Ibid.
60. Ibid.
61. Ibid., 300–1.
62. Ibid., 403.
63. Bell, *The Sex Complex* (1916), 2.
64. Ibid.
65. See for example Bell, 'Disorders of Functions', 300.
66. On Alfred Kinsey's, William Masters's and Virginia Johnson's acknowledgement of Ellis's work see Crozier, 'Introduction'.

9   Concluding Remarks

1. A. De Watteville, (review) ' "Rivista di filosofia scientifica" by Morselli, and "L'origine dei fenomeni psichici" by G. Sergi', *JMS*, 1885, vol. 8, 105.
2. Canguilhem, *Il normale e il patologico*.
3. Wanrooij, 'La passione svelata', 396.
4. Orsi, *La donna nuda*; A. Orsi, *Lussuria e castità* (1907); Stura, *Le miserie di Venere*. Orsi's *La donna nuda* was re-edited five times between 1905 and 1916. A further example of this genre is the anonymous book signed by 'a tribade' published in

1914, which exploited medical theories to tell the titillating story of a woman who indulged in same-sex pleasures. *Una tribade*, *Tribadismo, saffismo, clitorismo*.

5. Penta, *Lezioni di psichiatria*, 6–10.
6. On the importance of case-based reasoning in psychoanalysis and related disciplines, see Forrester, 'If *p*, then what?'
7. On the uptake of psychoanalysis in Italy and Britain, see respectively David, *La psicoanalisi nella cultura italiana*; G. Richards, 'Putting Britain on the Couch: The Popularisation of Psychoanalysis in Britain, 1918–1940', *Science in Context*, 2000, vol. 13, 183–230.
8. S. Freud, *On Sexuality* (1977) [1905], 45.
9. Ibid., 49–50.
10. Ibid., 55.
11. Ibid, 57.
12. Ibid., 51, 55.
13. Ibid., 52.
14. On Livi, see Chapter 3.
15. Freud, *On Sexuality*, 52–4.
16. Ibid., 74.
17. R. Hall, *The Well of Loneliness* (1928).
18. S. Freud, 'Five Lectures on Psycho-analysis (1909)', in *The Standard Edition of the Complete Psychological Works of Sigmund Freud* (1957), vol. 11, 45–6.
19. For ruptures between the fields, see G. Bachelard, *The New Scientific Spirit* (1984); M. Foucault, *L'Archéologie du Savoir* (1969). For synchronic ruptures, see Mol, *The Body Multiple*.
20. As, for example, Laqueur has recently done with masturbation. See Laqueur, *Solitary Sex*.
21. T. J. Williams et al., 'Finger-Length Ratios and Sexual Orientation', *Nature*, 2000 (30 March), 404, 455–6.
22. American Psychiatric Association, *Diagnostic and Statistical Manual of Mental Disorders: DSM-IV* (1994), pp. 532–3.
23. Ibid., p. 533.
24. Faderman, *Surpassing the Love of Men*; Grosskurth, *Havelock Ellis*.
25. Faderman, *Surpassing the Love of Men*, 317–18.
26. L. Doan and J. Prosser, *Palatable Poison: Critical Perspectives on The Well of Loneliness* (2002).
27. D. Hamer et al., 'A Linkage between DNA Markers on the X Chromosome and Male Sexual Orientation', *Science*, 1993, vol. 261, no. 5119, 321–7.
28. S. LeVay, 'A Difference in Hypothalamic Structure between Heterosexual and Homosexual Men', *Science*, 199, vol. 253, no. 5023, 1034–7.
29. D. Nimmons, 'Sexual Brain', *Discover*, 1994 (March), vol. 5, no. 3, 64–7.
30. Benadusi, *Il nemico dell'uomo nuovo*; N. Milletti, 'Accuse innominabili', in N. Milletti and L. Passerini (eds.), *Fuori dalla norma* (2007), 135–70.
31. http://www.repubblica.it/2007/02/sezioni/politica/coppie-di-fatto-4/binetti-polemica/binetti-polemica.html, accessed 20 April 2010.
32. http://www.smh.com.au/world/child-abuse-is-a-gay-problem-says-vatican-20100414-se4y.html, accessed 20 April 2010.
33. Ibid.
34. For a powerful discussion of the limits of words in the process of defining, see J. Butler, *Antigone's Claim* (2000), 57–82.

Bibliography

Primary sources

Acton, W. (1857) *The Functions and Disorders of the Reproductive Organs, in Childhood, Youth, Adult Age, and Advanced Life, Considered in the Physiological, Social, and Moral Relations* (London: Churchill).

——(1870) [1857] *Prostitution, Considered in its Moral, Social and Sanitary Aspects* (London: Churchill).

Aletrino, A. (1901) 'La situation sociale de l'uraniste', *La scuola positiva*, vol. 11, 481–96.

Andriezen, W. L. (1899) 'On the Basis and Possibilities of a Scientific Psychology and Classification in Mental Disease', *Journal of Mental Science*, vol. 45, 257–90.

Andronico, C. (1882) 'Prostitute e delinquenti', *Archivio di psichiatria, scienze penali ed antropologia criminale*, vol. 3, 143–6.

Anon. (1836–7) 'On Prostitution in the City of Paris: Considered under the Heads of Public Hygiene, Morals, and Internal Police', *The Lancet*, vol. 28, 25 Mar., 1 Apr., 15 Apr., 22 Apr., 19 Aug., 16–23; 41–8; 106–11; 157–63; 755–63.

——(1867) 'Aberrations of the Sexual Instinct', *Medical Times and Gazette*, 9 Feb., 141–6.

——(1877) (review), ' "Ein Fall von Pseudohermaphodisie", by E. Hoffmann (1877)', *Rivista sperimentale di freniatria e medicina legale in relazione con l'antropologia e le scienze giuridiche e sociali*, vol. 3, 724–6.

——(1891) 'Discussion at the Quarterly Meeting of the Medico-Psychological Association, November 13, 1890', *Journal of Mental Science*, vol. 37, 200.

——(1894) 'Transaction – Psychological Section, Medico-Legal Society – October Meeting – Discussion of Ellis paper on Sexual Inversion', *The Medico-Legal Journal*, vol. 12, 345–7.

——(1896) 'Biblioteca antropologica-giuridica', *Archivio di psichiatria, scienze penali ed antropologia criminale*, vol. 17, 343.

——(1896) 'Collaboratori', *Archivio di psichiatria, antropologia criminale e scienze penali*, vol. 17, first page.

——(1896) (review) ' "The Development of Homosexuality" by Marc André Raffalovich', *Journal of Mental Science*, vol. 42, 156–60.

——(1897) (review) ' "Das Conträre Geschlechtsgefühl", von H. Ellis und J. A. Symonds translated by Dr. Hans Kurella' and ' "Uranisme et Unisexualité: Etude sur différentes manifestations de l'instinct sexuel", par Marc-André Raffalovich, 1896', *Journal of Mental Science*, vol. 43, 565–74.

——(1897) (review) '*Uranisme et Unisexualité: Etude sur sifférentes manifestations de l'instinct sexuel*, by Marc-André Raffalovich (1896)', *Journal of Mental Science*, vol. 43, 569–74.

——(1898) 'Charge of Publishing and Selling Obscene Literature', *British Medical Journal*, vol. 2, 5 Nov., 1466.

——(1898) 'Primo congresso di medicina legale', *Rivista mensile di psichiatria forense, antropologia criminale e scienze affini*, vol. 1, 333–68.

——(1900) 'The Question of Indecent Literature', *The Lancet*, 27 Jan., 250.

——(1902) 'The Sale of Pernicious Literature', *British Medical Journal*, 31 May, 1356–7.

——(1902) 'Sexual Psychology and Pathology', *British Medical Journal*, vol. 1, 8 Feb., 339–41.

——(1904) 'Necrologio di Pasquale Penta', *Archivio di psichiatria, neuropatologia, antropologia criminale e medicine legale*, vol. 25, 774.

Antonelli, J. (1905) *Medicina pastoralis in usum confessarium, professorum Theologiae moralis et curiarum ecclesiasticarum* (Rome: Pustet).

Baker, J. (1892) 'Some Points Connected with Criminals', *Journal of Mental Science*, vol. 38, 364–9.

Barnes, R. (1890) 'On the Correlations of the Sexual Functions and Mental Disorders of Women', *British Gynaecological Journal*, vol. 4, 390–406 and 'Discussion', 406–30.

Bell, W. Blair (1913) 'The Arris and Gale Lectures: The Genital Functions of the Ductless Glands in the Female', *The Lancet*, 22 Mar., 809–16.

——(1913) 'The Arris and Gale Lectures: The Genital Functions of the Ductless Glands in the Female', *The Lancet*, 5 Apr., 938–44.

——(1915) 'Hermaphroditism', *Liverpool Medico-Chirurgical Journal*, no. 35, 272–92.

——(1916) *The Sex Complex: A Study of the Relationships of the Internal Secretions to the Female Characteristics and Function in Health and Disease* (London: Baillière, Tindall and Cox).

——(1917) 'Disorders of Functions', in T. Watts Eden and C. Lockyer (eds.), *The New System of Gynaecology* (London: MacMillan and Co.), vol. 1, 287–415.

——(1920) 'The Correlation of Function: With Special Reference to the Organs of Internal Secretion and the Reproductive System', *British Medical Journal*, 12 June, 787–91.

——(1920) [1916] *The Sex Complex: A Study of the Relationships of the Internal Secretions to the Female Characteristics and the Function in Health and Disease* (London: Baillière, Tindall and Cox).

Bianchi, L. (1895) 'The Function of the Frontal Lobes', *Brain: A Journal of Neurology*, vol. 18, 497–522.

——(1904) 'Introduzione', in P. Penta, *La simulazione della pazzia e il suo significato antropologico, etnico, clinico e medico legale* (Naples: Perella).

——(1904) *Trattato di psichiatria* (Naples: Pasquale).

——(1905) [1904] 'Prefazione', in P. Penta, *La simulazione della pazzia* (Naples: Perella).

Binet, A. (1887) 'Le fétichisme dans l'amour', *Revue Philosophique*, vol. 24, 143–67, 252–74.

Bloch, I. (1910) [1907] *La vita sessuale dei nostri tempi nei suoi rapporti con la civiltà moderna* (Turin: Soc. Tipografica Nazionale).

Blundell, J. (1828–9) 'Lectures on the Diseases of Women and Children', *The Lancet*, 5 Sept., vol. 2, 707.

Bonucci, F. (1863) *Medicina legale delle alienazioni mentali* (Perugia: Cantucci).

Bouchereau, G. (1892) 'Nymphomania', in H. Tuke (ed.), *A Dictionary of Psychological Medicine* (London: J. & A. Churchill), vol. 2, 863–6.

Brown, F. W. S. (1915) *Sexual Variety and Variability among Women and their Bearing upon Social Reconstruction* (London: British Society for the Study of Sex Psychology).

Bucknill, J. C. and D. H. Tuke (1879) [1858] *A Manual of Psychological Medicine containing the Lunacy Laws, Aetiology, Statistics, Description, Diagnosis, Pathology, and Treatment of Insanity with an Appendix of Cases* (London: J. & A. Churchill).

Burnett, H. C. (1891) *Hospitals and Asylums of the World: Their Origin, History, Construction, Administration, Management, and Legislation* (London: J. & A. Churchill).

Burton, R. F. (1886) 'Terminal Essay. Section D: Pederasty', in *A Plain and Literal Translation of the Arabian Nights Entertainments, Now Intituled the Book of the Thousand Nights and a Night with Introduction Explanatory Notes on the Manners and Customs of Moslem Men and Terminal Essay upon the History of the Nights*, vol. 10 (London: Burton Club), 205–54.

Butler, J. (1987) [1870] 'An Appeal to the People of England', reproduced in S. Jeffreys (ed.), *The Sexuality Debates* (London: Routledge), 111–50.

Callari, I. (1903) 'Prostituzione e prostituta in Sicilia', *Archivio di psichiatria, scienze penali ed antropologia criminale*, vol. 24, 193–205.

Camera dei Deputati (1887) *Progetto del Codice penale per il Regno d'Italia e disegno di legge che ne autorizza la pubblicazione*, Relazione ministeriale, seduta del 22 novembre 1887), vol. 1 (Rome: Stamperia Reale), 213–14.

Cantarano, G. (1883) 'Contribuzione alla casuistica della inversione dell'istinto sessuale', *La psichiatria, la neurologia, e le scienze affini*, vol. 1, 201–16.

——(1890) 'Inversione e pervertimenti dell'istinto sessuale', *La psichiatria*, vol. 8, fasc. 3 and 4, 275–93.

Capaccini, F. lli (1896) 'Programma', *Archivio delle psicopatie sessuali*, iii–v.

——(1896) Letter, 28 Apr., Rome, attached to *Archivio delle psicopatie sessuali*.

Carpenter, E. (1948) [1896] *Love's Coming-of-Age: A Series of Papers on the Relation of the Sexes* (London: George Allen & Unwin Ltd).

Casper, J. L. (1852) 'Über Nothzucht und Päderastie und deren Ermittelung Seitens des Gerichtsarztes', *Vierteljahrschrift für gerichtliche öffentliche Medizin*, 1852, vol. 1, 21–78.

——(1859) [1852] *Manuale pratico di medicina legale* (Turin: Botta).

——(1865) [1852] *Handbook for the Practice of Forensic Medicine, Based upon Personal Experience* (London: Sydenham Society).

Celesia, P. (1900) 'Sull'inversione sessuale', *Archivio di psichiatria, scienze penali ed antropologia criminale*, vol. 21, 209–20.

Charcot, J. M. and V. Magnan (1882), 'Inversion du sens génital', *Archives de neurologie*, vol. 3, 53–60; concluded in vol. 4, 296–322.

——(1989) [1882] 'Inversione del senso genitale e altre perversioni genitali', in J. M. Charcot, *La donna dell'isteria* (Milan: Spirali/Vel).

Chavalier, J. (1893). *Una Maladie de la personalité. L'inversion sexuelle* (Lyon: A. Stork).

Chiarugi, V. (1793) *Della pazzia in genere, e in specie. Trattato medico-analitico con una centuria di osservazioni* (Florence: Luigi Carlieri).

——(1987) [1793] *On Insanity and its Classification* (USA: Science History Publications).

Churchill, F. (1835) *Outlines of the Principal Diseases of Females. Chiefly for the Use of Students* (Dublin: Martin Keene and Son).

——(1874) [1850] *On the Diseases of Women* (London: Longman).

Clevenger, S. V. (1881) 'Hunger: The Primitive Desire', *Science* (15 Jan.), vol. 2, 14.

Clouston, T. S. (ed.) (1873–4) 'David Skae, M.D., F.R.C.S.E. (by the late), The Morisonian Lectures on Insanity', *Journal of Mental Science*, vol. 19, 340–55, 491–507.

——(1883) *Clinical Lectures on Mental Diseases* (London: J. & A. Churchill).

——(1894) 'The Developmental Aspects of Criminal Anthropology', *Journal of the Anthropological Institute of Great Britain and Ireland*, vol. 33, 215–25.

——(1898) [1883] *Clinical Lectures on Mental Diseases* (London: J. & A. Churchill).

Copland, J. (1844–58) [1838] *Dictionary of Practical Medicine* (London: Longman, Brown, Green, and Longmans), vol. 1.

Crichton-Browne, J. (1860) 'Psychical Diseases of Early Life', *Journal of Mental Science*, vol. 6, 284–320.

Cristiani, A. (1896) 'Autopederastia in un alienato, affetto da follia periodica', *Archivio delle psicopatie sesuali*, 183–7.

Darwin, C. (1981) [1871] *The Descent of Man, and Selection in Relation to Sex* (Princeton: Princeton University Press).

Davidson, J. H. (1874) 'Remarks on Some of the Large Asylums of Italy', *Journal of Mental Science*, vol. 20, 410–15.

de' Liquori, A. (1842) [1760] *La vera sposa di Gesù Cristo cioè la monaca santa per mezzo delle virtù proprie d'una religiosa* (Bassano: Tipografia Remondini).

De Sanctis, S. (1911) *Trattato di medicina sociale. Patologia e profilassi mentale* (Milan: Vallardi).

De Watteville, A. (1885) (review) ' "Rivista di Filosofia Scientifica" by E. Morselli and "L'origine dei fenomeni psichici" by G. Sergi', *Brain: A Journal of Neurology*, vol. 8, 105–6.

Drysdale, G. (1882) [1854] *The Elements of Social Science or Physical, Sexual and Natural Religion: An Exposition of the True Cause and Only Cure of the Three Primary Social Evils: Poverty, Prostitution, and Celibacy* (London: E. Truelove).

Ellis, H. *Papers and Manuscripts*, British Library ADD 70524.

——*Papers and Manuscripts*, British Library ADD 70536.

——(1890) *The Criminal* (London: Walter Scott).

——(1890) 'The Study of the Criminal', *Journal of Mental Science*, vol. 36, 1–15.

——(1894) 'The Study of Sexual Inversion', *The Medico-Legal Journal*, vol. 12, 148–57.

——(1895) 'Sexual Inversion in Women', *Alienist and Neurologist*, vol. 16, 141–58.

——(1896) 'Nota sulle facoltà artistiche degli invertiti', *Archivio delle psicopatie sessuali*, 243–5.

——(1896) 'Sexual Inversion in Men', *Alienist and Neurologist*, vol. 17, 115–50.

——(1898) *A Note on the Bedborough Trial* (London: The University Press).

——(1898) 'The Question of Indecent Literature – to the Editors of "The Lancet" ', *The Lancet*, 26 Nov., 1431.

——(1899) [1894] *Man and Woman: A Study of Human Secondary Sexual Characters* (London: Walter Scott).

——(1901) [1890] *The Criminal* (London: Walter Scott).

——(1918) *The Erotic Rights of Women and the Objects of Marriage* (London: The British Society for the Study of Sex Psychology).

——(1932) *Views and Reviews: A Selection of Uncollected Articles, 1884–1932* (London: Desmond Harmsworth).

——(1934) [1894] *Man and Woman: A Study of Secondary and Tertiary Sexual Characters* (London: William Heinemann).

——(1940) *My Life* (London: William Heinemann).

Ellis, H. and J. A. Symonds (1897) *Sexual Inversion* (London: Wilson and MacMillan) (reprinted edition: Ayer Company Publisher 1994).

——(1901) [1897] *Studies in the Psychology of Sex: Sexual Inversion*, vol. 2 (Philadelphia: F. A. Davis Company).

——(1921) [1897] *Studies in the Psychology of Sex: Sexual Inversion*, vol. 2 (Philadelphia: F. A. Davis Company).

——(1924) [1897] *Studies in the Psychology of Sex: Sexual Inversion*, vol. 2 (Philadelphia: F. A. Davis Company).

Épaulard, A. (1901) *Vampirisme, nécrophilie, nécrosadisme, nécrophagie* (Lyon: A Stork & Co.).

Ferrari., G. C. (1896) (review) 'M. A. Raffalovich, *L'uranismo. Il processo Oscar Wilde*, Bocca, Torino, 1896; A. V. Schrenck Notzing, *La terapia suggestiva delle psicopatie sessuali*, Bocca, Torino, 1897; A. Niceforo, *Il gergo nei normali, nei degenerati e nei criminali*, Bocca, Torino, 1897', *Rivista sperimentale di freniatria e medicina legale delle alienazioni mentali*, vol. 22, 888–9.

Filippi, A. (n.d.) *Manuale di medicina legale conforme al nuovo codice penale* (Palermo-Catania: Vallardi).

——(1878) *Manuale di afrodisiologia civile, criminale e Venere forense* (Pisa: Nistri).

Flexner, A. (1914) *Prostitution in Europe* (New York: The Century Co.).

Forel, A. (1907) [1905] *La questione sessuale esposta alle persone colte* (Turin: Bocca).

——(1909) [1906] *Etica sessuale* (Turin: Bocca).

Forster, E. M. (1976) [1905] *Where Angels Fear to Tread* (Harmondsworth: Penguin).

——(2000) [1908] *A Room with a View* (New York: Penguin).

Frassinetti, G. (1862) [1859] *La monaca in casa* (Florence: Luigi Manuelli).

Freschi, F. (1855) *Manuale teorico pratico di medicina legale* (Milan: Volpato).

Freud, S. (1957) [1909] 'Five Lectures on Psycho-analysis (1909)', in *The Standard Edition of Complete Psychological Works of Sigmund Freud* (London: Hogarth Press), vol. 11, 3–56.

——(1977) [1905] *On Sexuality: Three Essays on the Theory of Sexuality and Other Works* (New York: Penguin Books), vol. 7.

Fronda, R. (1912) 'L'omosessualità nella donna', *Il manicomio*, vol. 17, 123–34.

Gandolfi, F. (1862–3) *Fondamenti di medicina forense analitica* (Rome: Capaccini).

Garnier, P. (1907) [1883] *L'onanismo a due in tutte le sue forme e conseguenze. Psicopatia sessuale* (Rome: Capaccini).

Gasquet, J. R. (1881) 'Italian Retrospect', *Journal of Mental Science*, vol. 26, 632–6.

——(1884) 'Italian Psychological Literature', *Journal of Mental Science*, vol. 29, 586–91.

Geddes, P. and J. A. Thomson (1889) *Evolution of Sex* (London: Walter Scott).

Gemelli, A. (1911) *Non Moechaberis. Disquisitiones Medicale in Usum Confessarioum. Quaestiones Theologiae Medico-Pastoralis* (Florence: Libreria Editrice Fiorentina).

——(1946) [1941] *La tua vita sessuale. Lettera ad uno studente universitario* (Milan: Vita e Pensiero).

Gock, H. (1875) 'Beitrag zur Kenntniss der conträren Sexualempfindung', *Archiv für Psychiatrie und Nervenkrankheiten*, vol. 5, 564–74.

Griesinger, W. (1868) 'Vortrag zur Eröffnung der psychiatrischen Clink', *Archiv für Psychiatrie und Nervenkrankheiten*, vol. 1, 636–54.

Hammond, W. A. (1882) 'The Disease of the Scythians (Morbus Feminarum) and Certain Analogous Conditions', *American Journal of Neurology and Psychiatry*, vol. 1, 339–55.

Hirschfeld, M. (2000) [1914] *Homosexuality of Men and Women* (New York: Prometheus Books).

Journal of Mental Science (Editors) (1899) 'The Bedborough Case', *Journal of Medical Science*, vol. 45, 122–3.

Kiernan, J. C. (1884) 'Perverted Sexual Instinct', *Chicago Medical Journal and Examiner*, vol. 48, 263–5.

Krafft-Ebing, R. von (1877) 'Über gewisse Anomalies des Geschlechtstriebs und die klinisch-forensich Verwertung derselben als eines wahrscheinlich funktionellen Degenerationszeichens des centralen Nervensystems', *Archiv für Psychiatrie und Nervenkrankheiten*, vol. 7, 291–312.

——(1889) [1886] *Le psicopatie sessuali con speciale considerazione alla inversione sessuale* (Turin: Bocca).

——(1893) [1892] *Psychopathia Sexualis with Special Reference to Contrary Sexual Instinct: A Medico Legal Study*, trans. C. Gilbert Chaddock (of the seventh German edition 1892) (Philadelphia and London: F. A. Davis).

——(1895) 'Zur Erklärung der conträren Sexualempfindung', *Jahrbücher für Psychiatrie und Neurologie*, vol. 13, 1–16.

——(1897) [1894] *L'inversione sessuale nell'uomo e nella donna. Omosessualità, evirazione, effeminatezza, metamorfosi sessuale, ermafroditismo, uranismo, viraginità, androginia, ginandria* (Rome: Capaccini).

——(1897) [1875] *Trattato di psicopatologia forense in rapporto alle disposizioni vigenti in Austria, Germania ed in Francia* (Turin: Bocca).

——(1931) [1886] *Psychopathia Sexualis with special reference to the Antipathic Sexual Instinct* (London: William Heinemann).

Krueg, J. (1881) 'Perverted Sexual Impulses', *Brain: A Journal of Neurology*, vol. 4, 368–76.

Kurella, H. (1896) 'Osservazioni sul significato biologico della bisessualità', *Archivio di psichiatria, scienze penali ed antropologia criminale*, vol. 17, 418–25.

——(1911) [1910] *Cesare Lombroso, a Modern Man of Science* (London: Rebman).

La Cara, A. (1924) [1902] *La base organica dei pervertimenti sessuali e la loro profilassi sociale* (Turin: Bocca).

Lancet, The (Editors) (1898) 'Editorial article', *The Lancet*, 19 Nov., 1344–5.

Laura, S. (1874) *Trattato di medicina legale* (Turin: Camilla e Bertolero).

Laycock, T. (1840) *A Treatise on the Nervous Diseases of Women; Comprising an Inquiry into the Nature, Causes, and Treatment of Spinal and Hysterical Disorders* (London: Longman).

Livi, C. (1875) 'Del metodo sperimentale in freniatria e medicina legale. Discorso che potrebbe servire a uso di programma', *Rivista sperimentale di freniatria e di medicina legale in relazione con l'antropologia e le scienze giuridiche e sociali*, vol. 1, 1–7.

Lombroso, C. (1873) 'Verzeni e Agnoletti', *Rivista di discipline carcerarie in relazione con l'antropologia, col diritto penale, colla statistica*, vol. 3, 193–213.

——(1874) 'Sul tatuaggio in Italia fra i delinquenti. Studio medico legale', *Archivio per l'antropologia e l'etnologia*, vol. 4, 393–400.

——(1876) *L'uomo delinquente studiato in rapporto alla antropologia, alla medicina legale ed alle discipline carcerarie* (Milan: Ulrico Hoepli).

——(1881) 'L'amore nei pazzi', *Archivio di psichiatria, antropologia criminale e scienze penali*, vol. 2, 1–32.

——(1883) 'Amori anomali e precoci nei pazzi', *Archivio di psichiatria, scienze penali ed antropologia criminale*, vol. 4, 17–25.

——(1883) 'Delitti di libidine', *Archivio di psichiatria, scienze penali, ed antropologia criminale*, vol. 4, 169–78, 320–48.

——(1884) (review) ' "Contribuzione alla casuistica della inversione dell'istinto sessuale" di G. Cantarano', *Archivio di psichiatria, scienze penali ed antropologia criminale*, vol. 5, 133–4.

——(1884) [1876] *L'uomo delinquente in rapporto all'antropologia, giurisprudenza ed alle discipline carcerarie. Delinquente nato e pazzo morale* (Turin: Bocca).

——(1884) 'Pazzia morale e delinquente nato', *Archivio di psichiatria, scienze penali ed antropologia criminale per servire allo studio dell'uomo alienato*, vol. 5, 17–22.

——(1885) 'Del tribadismo nei manicomi', *Archivio di psichiatria, scienze penali ed antropologia criminale*, vol. 4, 218–21.

——(1885) 'Ninfomania paradossa', *Archivio di psichiatria, scienze penali ed antropologia criminale*, vol. 4, 362–9.

——(1889) [1876] *L'uomo delinquente in rapporto all'antropologia, giurisprudenza ed alle discipline carcerarie. Delinquente nato e pazzo morale* (Turin: Bocca).

——(1890) 'Anomalie psichiche in Michelangelo e Virgilio', *Archivio di psichiatria, scienze penali ed antropologia criminale per servire allo studio dell'uomo alienato e delinquente*, vol. 11, 331–3.

——(1891) 'Due geni nevrotici femminili', *Archivio di psichiatria, scienze penali ed antropologia criminale*, vol. 12, 478–84.

——(1894) 'Prefazione', in E. Fornasari di Verce, *La criminalità e le vicende economiche d'Italia dal 1878 al 1890* (Turin: Bocca), v–xxii.

——(1903) 'La psicologia di una uxoricida tribade', *Archivio di psichiatria, scienze penali ed atropologia criminale*, vol. 24, 6–10.

——(1906) 'Du parallélisme entre l'homosexualité et la criminalité innée. VI Congrès International d'Anthropologie Criminelle, Turin 1906', *Archivio di psichiatria, neuropatologia, antropologia criminale e medicina legale*, vol. 27, 378–81.

Lombroso, C. and G. Ferrero (1893) *La donna delinquente, la prostituta e la donna normale* (Turin: Roux).

Luzenberger, A. (1896) 'Sul meccanismo dei pervertimenti sessuali e loro terapia, Comunicazione fatta al Congresso della Società Freniatrica Italiana tenuto in Firenze nell'ottobre 1896', *Archivio delle psicopatie sessuali*, 265–71.

Macnaughton Jones, H. (1884) *Practical Manual of Diseases of Women and Uterine Therapeutics for Students and Practitioners* (London: Baillière, Tindall and Cox).

——(1901) *Points of Practical Interest in Gynaecology* (London: Baillière, Tindall and Cox).

Magendie, M. (1836–7) 'Lectures of the Physiology of the Nervous System, delivered in 1836, in the College of France', *The Lancet*, vol. 2, 24 June 1837, 463–6.

Mantegazza, P. (1886) *Gli amori degli uomini: saggio di un'etnologia dell'amore* (Milan: Stampato in proprio).

——(1892) [1886] *Gli amori degli uomini: saggio di un'etnologia dell'amore* (Milan: Paolo Mantegazza).

——(1893) [1891] *Fisiologia della donna*, vol. 1 (Milan: Treves).

——(1893) (review) 'A. Moll, "Les perversions de l'instinct génital" 1893', *Archivio per l'antropologia e l'etnologia*, vol. 23, 464–5.

——(1893) (review) 'P. Penta, "I pervertimenti sessuali nell'uomo e Vincenzo Verzeni strangolatore di donne"', *Archivio per l'antropologia e l'etnologia*, vol. 23, 476.

——(1897) (review) 'A. Niceforo, "Le psicopatie sessuali acquisite e i reati sessuali"', *Archivio per l'antropologia e l'etnologia*, vol. 27, 440.

——(1897) (review) 'P. Penta, "L'origine e la patologenesi delle inversioni sessuali ecc."', *Archivio per l'antropologia e l'etnologia*, vol. 27, 433.

——(1935) [1886] *The Sexual Relations of Mankind* (New York: Eugenics Publishing).

Marchesini, G. and G. Obici (1903) [1898] *Le 'amicizie' di collegio. Ricerche sulle prime manifestazioni dell'amore sessuale* (Rome: Soc. editrice Dante Alighieri di Albrighi Segati e C.).

Martineau, L. (1896) [1884] *Le deformazioni vulvari ed anali prodotte dalla masturbazione, dal saffismo, dalla deflorazione e dalla sodomia* (Rome: Capaccini).

Masserotti, V. (1913) *Nel regno di Ulrichs* (Rome: Lux).

Maudsley, H. (1875) [1874] *La responsabilità nelle malattie mentali* (Milan: Dumolard).

——(1895) 'Criminal Responsibility in Relation to Insanity', *Journal of Mental Science*, vol. 51, 657–65.

Mercier, C. (1890) *Sanity and Insanity* (London: Walter Scott).

——(1905) *Criminal Responsibility* (Oxford: Clarendon Press).

——(1918) *Crime and Criminals Being the Jurisprudence of Crime, Medical, Biological and Psychological* (London: University of London Press).

Michéa, C. F. (1849) 'Des Déviations de l'appétit vénérien', *Union Medicale*, vol. 3, 338–9.

Moll, A. (1897) [1893] *Uranismo e prostituzione mascolina* (Rome: Capaccini).

Monaco, G. (1887) *Manuale di medicina legale* (Naples: G. Novenne).

Mongeri, L. (1907) *Patologia speciale delle malattie mentali* (Milan: Hoepli).

Montessori, M. (1912) *La morale sessuale nell'Educazione, in Atti del I Congresso Nazionale delle Donne Italiane*. Rome, 24–30 April, 1908 (Rome: Stabilimento Topografico della Societa Editrice Laziale), 272–81.

Moraglia, G. B. (1895) 'Nuove ricerche su criminali, prostitute e psicopatiche', *Archivio di psichiatria, scienze penali ed antropologia criminale per servire allo studio dell'uomo alienato e delinquente*, vol. 16, 309–20, 501–23.

Moreau de Tours, J. J. (1897) [1880] *Le aberrazioni del senso genesico* (Rome: Capaccini).

Morselli, E. (1885–9) *Manuale di semejotica delle malattie mentali* (Milan: Vallardi).

——(1903) 'Introduzione', in G. Marchesini and G. Obici, *Le 'amicizie' di collegio. Ricerche sulle prime manifestazioni dell'amore sessuale* (Rome: Soc. editrice Dante Alighieri di Albrighi Segati e C.), i–xli.

——(1944) *La psicanalisi* (Milan: Bocca).

Näcke, P. (1896) (review) 'M. A. Raffalovich, *Uranisme et unisexualité'*, *Archivio di psicopatie sessuali*, 168.

——(1908) 'Penta als einer der besten Kenner und Förderer der Sexualwissenschaft', *Zeitschrift für Sexualwissenschaft*, vol. 1, 74–81.

Needham, F. (1892) 'A Visit to Some Foreign Asylums', *Journal of Mental Science*, vol. 38, 222–7.

Neri, S. A. (1896) 'Pervertito, necrofiliaco, pederasta, masochista, etc.', *Archivio delle psicopatie sessuali*, 109–10.

Niceforo, A. (1897) *Il gergo nei normali, nei degenerati e nei criminali* (Turin: Bocca).

——(1897) *Le psicopatie sessuali acquisite e i reati sessuali* (Rome: Capaccini).

Nicolson, D. (1873) 'The Morbid Psychology of Criminals', *Journal of Mental Science*, vol. 19, 222–31, 398–409.

——(1874) 'The Morbid Psychology of Criminals', *Journal of Mental Science*, vol. 20, 20–37, 167–85, 527–51.

——(1875) 'The Morbid Psychology of Criminals', *Journal of Mental Science*, vol. 21, 18–31, 225–50.

Norman, C. (1892) 'Sexual Perversion', in H. Tuke (ed.), *A Dictionary of Psychological Medicine* (London: J. & A. Churchill), 1156–7.

Orsi, A. (1905) *La donna nuda. Saggio di psicologia del pudore* (Turin: Streglio).

——(1907) *Lussuria e castità* (Turin: Streglio).

Parent-Duchâtelet, A. J. B. (1835–6) *De la prostitution dans la ville de Paris* (Paris: Baillière).

——(1840) [1835–6] *On Prostitution in the City of Paris* (Haymarket: T. Burgess).

——(2008) [1835–6] *Prostitution in the City of Paris*, trans. R. Ridley-Smith (Wellington: The Translator)

Pellegrini, R. (1954) *Sessuologia* (Padua: CEDAM).

Penta, G. (1896) 'Un altro caso d'inversione sessuale in donna', *Archivio delle psicopatie sessuali*, 94–6.

Penta, P. (1890) *Vincenzo Verzeni (Lo strangolatore delle donne). Esame psichiatrico, La tribuna giudiziaria*, vol. 4 (extract).

——(1893) *I pervertimenti sessuali nell'uomo e Vincenzo Verzeni strangolatore di donne: studio biologico* (Naples: Luigi Pierro).

——(1896) 'Dei pervertimenti sessuali. Caratteri generali, origine e significato dimostrato collo autobiografie di Alfieri e di Rousseau e col dialogo "Gli amori" di Luciano', *Archivio delle psicopatie sessuali*, 1–7, 17–20.

——(1896) *Dei pervertimenti sessuali. Caratteri generali. Origine e significato dimostrati colle autobiografie di Alfieri e di Rousseau e col dialogo 'Gli amori' di Luciano* (Rome: Capaccini).

——(1896) 'Dei varii studii pubblicati sui pervertimenti sessuali dai primi sino ai più recenti dei nostri giorni', *Archivio delle psicopatie sessuali*, 8–14.

——(1896) (review) 'Havelock Ellis, *A Note on the Treatment of Sexual Inversion* in *Alienist and Neurologist*, 1896', *Archivio delle psicopatie sessuali*, 261–2.

——(1896) (review) 'W. L. Howard, "Sexual Perversion" in *Alienist and Neurologist*', *Archivio delle psicopatie sessuali*, 123–5.

——(1896) 'Influenza degli organi e delle funzioni sessuali sul modo di agire del sistema nervoso', *Archivio delle psicopatie sessuali*, 246–58, 263–73.

——(1896) 'L'origine e la patogenesi della inversione sessuale, secondo Krafft-Ebing e gli altri autori', *Archivio delle psicopatie sessuali*, 53–70.

——(1896) (review) 'E. S. Talbot and Havelock Ellis, "A Case of Developmental Degenerative Insanity, with Sexual Inversion, Melancholia following Removal of Testicles, Attempted Murder and Suicide", in *Journal of Mental Science*, April 1896', *Archivio delle psicopatie sessuali*, 191–4.

——(1898) (review) 'R. von Krafft-Ebing, *Psychopathia Sexualis*', *Rivista di psichiatria forense, antropologia criminale escienze affini*, vol. 1, 300–1.

——(1898) 'Sopra un caso di inversione sessuale', *Rivista di psichiatria forense, antropologia criminale escienze affini*, vol. 1, 109–15.

——(1900) 'In tema di pervertimenti sessuali. Documenti umani (lettere di amore tra individui dello stesso sesso)', *Rivista mensile di psichiatria forense, antropologia criminale e scienze affini*, vol. 3, 69–89.

——(1900) *Lezioni di psichiatria dettate nell'anno scolastico 1899–1900* (Naples), vols. 1 and 2, Manuscript.

——(1903) (review) ' "Jahrbuch für sexuelle Zwischenstufen mit besenderer Berucksochtigung der Homosexualität" 1899', *Rivista mensile di psichiatria forense, antropologia criminale e scienze affini*, vol. 2, 518–24.

——(1903) 'Parere medico-legale sulle condizioni psichiche del prete Pietro Paolo Potenza, accusato di duplice assassinio', *Rivista mensile di prichiatria forense, antropologia criminale e scienze affini*, vol. 6, 101–11, 253–62, 289–96, 325–49, 361–79.

——(1903) 'Un pedofilo fellatore', *Rivista mensile di psichiatria forense, antropologia criminale e scienze affini*, vol. 6, 477–518.

——(1904) *La simulazione della pazzia e il suo significato antropologico, etnico, Clinico e medico legale* (Naples: Perella).

Penta, P. and A. d'Urso (1896) 'Sopra un caso d'inversione sessuale in donna epilettica', *Archivio delle psicopatie sessuali*, 33–9.

Pouillet, T. (1897) [1880] *L'onanismo della donna* (Rome: Capaccini).

Raffalovich, M. A. (1896) 'Gli studi sulle psicopatie sessuali in Inghilterra', *Archivio delle psicopatie Sessuali*, 176–81.

——(1896) *L'uranismo, inversione sessuale congenita* (Turin: Bocca).

Riberio, L. (1940) *Omosessualità ed endocrinologia* (Milan: Bocca).

Rivista mensile di psichiatria forense (Editors) (1898) 'Programma', *Rivista mensile di psichiatria forense*, vol. 1, 1–3.

Rivotto-Peccei (1900) 'Sopra un caso di urninga femmina', *Archivio di psichiatria, scienze penali ed antropologia criminale*, vol. 21, 36–41.

Robertson, W. F. (1902) (review) '*Trattato di Psichiatria* by Prof. L. Bianchi', *Journal of Mental Science*, vol. 48, 558.

——(1905) (review) '*Trattato di psichiatria ad uso dei medici e degli studenti* by L. Bianchi', *Journal of Mental Science*, vol. 51, 411–13.

Routh, C. H. F. (1887) 'On the Etiology and Diagnosis, Considered specially from a Medico-legal Point of View, of those Cases of Nymphomania which Lead Women to Make False Charge against their Medical Attendants', *British Medical Journal*, vol. 2, 485–511.

——(1894) 'The Conservative Treatment of Disease of the Uterine Appendages', *British Gynaecological Journal*, vol. 10, 51–87.

Ryan, M. (1839) *Prostitution in London: with a Comparative View of that of Paris and New York* (London: Baillière).

Saporito, F. (1903) 'Sulla delinquenza militare', *Rivista mensile di psichiatria forense, antropologia criminale e scienze affini*, vol. 6, 44–68, 125–84.

Savage, G. H. (1878) 'Case of Malformation of Genitalia in Insanity', *Journal of Mental Science*, vol. 24, 459–61.

——(1881) 'Moral Insanity', *Journal of Mental Science*, vol. 27, 147–55.

——(1884) 'Case of Sexual Perversion in a Man', *Journal of Mental Science*, vol. 30, 390–1.

——(1886) [1882] *Insanity and Allied Neuroses: Practical and Clinical* (London: Cassel).

Savage, G. and C. Mercier (1896) 'Insanity of Conduct', *Journal of Mental Science*, vol. 42, 1–17.

Schrenck Notzing, A. V. (1897) [1892] *La terapia suggestiva delle psicopatie sessuali* (Turin: Bocca).

Senato del Regno (1888) *Relazione della commissione speciale che autorizza il Re a pubblicare il Codice penale per il Regno d'Italia*, s.i.t., 183–5.

Sergi, G. (1899) *Le degenerazione umane* (Milan: Dumolard).

Shaw, J. C. and G. N. Ferris (1883) 'Perverted Sexual Instinct (Conträre Sexualempfindung: Westphal; Inversione dell'instincto sessuale, Arrigo Tamassia; Inversion du sens génital: Charcot et Magnan', *Journal of Nervous and Mental Disease*, vol. 10, 185–204.

Sighele, S. (1892) 'La coppia criminale', *Archivio di psichiatria, scienze penali ed antropologia criminale*, vol. 13, 519–41.

——(1893) *La coppia criminale: psicologia degli amori morbosi* (Turin: Bocca).

Smith, H. (1877) *Practical Gynaecology: A Handbook of the Diseases of Women* (London: J. & A. Churchill).

Stura, F. (1904) *Le miserie di Venere* (Turin: Renzo Stroglio e C.).

Sullivan, W. C. (1893) 'Notes on a Case of Acute Insanity with Sexual Perversion', *Journal of Mental Science*, vol. 39, 225–6.

Symonds, J. A. (1883) *A Problem in Greek Ethics* (London: Privately Printed).

——(1896) *A Problem in Modern Ethics Being an Inquiry into the Phenomenon of Sexual Inversion. Addressed Especially to Medical Psychologists and Jurists* (London: Charles R. Dawes Ex Libris).

——(1901) *A Problem in Greek Ethics Being an Inquiry into the Phenomenon of Sexual Inversion Addressed Especially to Medical Psychologists and Jurists* (London: Privately Printed).

——(1969) [1885–93] *The Letters of John Addington Symonds, vol. 3, 1885–1893*, ed. H. M. Schuller and R. L. Peters (Detroit: Wayne State University Press).

——(1984) *The Memoirs of John Addington Symonds*, ed. P. Grosskurth (London: Hutchinson).

Tait, L. (1889) *Diseases of Women and Surgery* (Leicester: Richardson & Company).

Tamassia, A. (1875) (review) ' "Ein Fall von Gewaltsamer Unnatürlicher Nothzuch" di Reimann, 1875', *Rivista sperimentale di freniatria e di medicina legale in relazione con l'antropologia e le scienze giuridiche e sociali*, vol. 1, 489.

——(1877) 'La pazzia morale', *Rivista sperimentale di freniatria e di medicina legale in relazione con l'antropologia e le scienze giuridiche e sociali*, vol. 3, 160–2.

——(1878) 'Sull'inversione dell'istinto sessuale', *Rivista sperimentale di freniatria e di medicina legale in relazione con l'antropologia e le scienze giuridiche e sociali*, vol. 4, 97–117.

Tanzi, E. (1905) *Trattato delle malattie mentali* (Milan: Soc. Ed. Libraria).

——(1911) *Psichiatria Forense* (Milan: Vallardi).

Tardieu, A. (1898) [1858] *I delitti di libidine. Oltraggi al pudore, stupri e attentati al pudore, pederastia e sodomia* (Rome: Capaccini).

Taylor, A. S. (1905) [1865] *The Principles and Practice of Medical Jurisprudence*, 5th edn., ed. F. J. Smith (London: J. & A. Churchill).

Thomson, J. B. (1870) 'The Psychology of Criminals', *Journal of Mental Science*, vol. 16, 321–50.

Tonini, F. (1862) *Igiene e fisiologia del matrimonio ossia storia naturale e medica dello stato coniugale e della igiene speciale dei coniugi nelle diverse loro fasi e del neonato* (Milan: Gaetano Brigola).

Trevelyan, G. M. (1907) *Garibaldi's Defence of the Roman Republic* (London and New York: Longmans).

——(1909) *Garibaldi and the Thousand* (London and New York: Longmans).

——(1911) *Garibaldi and the Making of Italy: June–November 1860* (London and New York: Longmans).

Tuke, H. (ed.) (1892) *A Dictionary of Psychological Medicine* (London: J. & A. Churchill).

Una tribade (1914) *Tribadismo, saffismo, clitorismo* (Florence: Pensiero).

Dr Urquhart (1891) 'Case of Sexual Perversion', *Journal of Mental Science*, vol. 37, 94–6.

Venturi, S. (1892) *Le degenerazioni psico-sessuali nella vita degli individui e nella storia della società* (Turin: Bocca).

Westphal, C. F. O. (1869) 'Die Conträre Sexualempfindung: Symptom eines neuropathischen (psychopathischen) Zustandes', *Archiv für Psychiatrie und Nervenkrankheiten*, vol. 2, 73–108, in M. A. Lombardi-Nash (2006) *Sodomites and Urnings: Homosexual Representations in Classic German Journals* (New York: Harrington Park Press), 87–120.

Willard, E. O. G. (1867) *Sexology as the Philosophy of Life: Implying Social Organization and Government* (Chicago: Walsh).

Williamson, T. (1896) 'On the Symptomatology of Gross Lesions (Tumours and Abscesses) Involving the Prae-Frontal Region of the Brain', *Brain: A Journal of Neurology*, vol. 19, 346–65.

Wise, M. (1883) 'Case of Sexual Perversion', *Alienist and Neurologist*, vol. 4, 87–91.

Zacchia, P. (1673–74) [1621–35] *Quaestionum medico-legalium* (Frankfurt: Lugduni).

Zuccarelli, A. (1888) *Inversione congenita dell'istinto sessuale in due donne* (Naples: Stabilimento Tocco e C.).

Secondary sources

Aldrich, R. (1993) *The Seduction of the Mediterranean: Writing, Art, and Homosexual Fantasy* (London: Routledge).

Alexander, S. (1994) *Becoming a Woman and Other Essays in 19th and 20th Century Feminist History* (London: Virago Press).

American Psychiatric Association (1994) *Diagnostic and Statistical Manual of Mental Disorders: DSM-IV* (Washington: American Psychiatric Association).

Anderson, P. (1995) *When Passion Reigned: Sex and the Victorians* (New York: Basic Books).

Augstein, H. F. (1996) 'J. C. Prichard's Concept of Moral Insanity: A Medical Theory of the Corruption of Human Nature', *Medical History*, vol. 40, 311–43.

Babini, V. P. (1986), 'L'infanticida tra letteratura medica e letteratuta giuridica', in P. Rossi (ed.), *L'età del positivismo* (Bologna: Il Mulino), 453–74.

——(1989) 'Organicismo e ideologie nella psichiatria italiana dell'Ottocento', in F. M. Ferro (ed.), *Passioni della mente e della storia* (Milan: Vita e Pensiero), 331–50.

——(1996) *La questione dei frenastenici. Alle origini della psicologia scientifica in Italia (1870–1910)* (Milan: F. Angeli).

——(1999) 'Un altro genere. La costruzione scientifica della "natura femminile"', in A. Bugio (ed.), *Nel nome della razza. Il razzismo nella storia d'Italia 1870–1945* (Bologna: Il Mulino), 475–89.

——(2007) 'In the Name of Father: Gina and Cesare Lombroso', in V. P. Babini and R. Simili (eds.), *More than Pupils: Italian Women in Science at the Turn of the 20th Century* (Florence: Olschki), 75–106.

——(2009) *Liberi Tutti. Manicomi e psichiatri in Italia: una storia del Novecento* (Bologna: Il Mulino).

Babini, V. P. and L. Lama (2000) *Una 'donna nuova'. Il femminismo scientifico di Maria Montessori* (Milan: Angeli).

Babini, V. P., M. Cotti, F. Minuz, and A. Tagliavini (1982) *Tra sapere e potere. La psichiatria italiana nella seconda metà dell'Ottocento* (Bologna: Il Mulino).

Babini, V. P., F. Minuz, and A. Tagliavini (1986) *La donna nelle scienze dell'uomo* (Milan: Angeli).

Bachelard, G. (1984) *The New Scientific Spirit*, trans. Arthur Goldhammer (Boston: Beacon Press).

Basaglia, F. (1981) *Scritti 1953–1968* (Turin: Einaudi).

Bauer, H. (2009) *English Literary Sexology Translations of Inversion, 1860–1930* (Basingstoke: Palgrave Macmillan).

——(2009) 'Theorizing Female Inversion: Sexology, Discipline, and Gender at the Fin de Siècle', *Journal of the History of Sexuality*, Vol. 18, 84–102.

Beauvoir, S. de (1993) [1949] *The Second Sex* (London: Everyman's Library).

Beccalossi, C. (2009) 'The Origin of Italian Sexological Studies: Female Sexual Inversion ca. 1870–1900', *Journal of the History of Sexuality*, vol. 18, 103–20.

——(2010) 'Nineteenth-Century European Psychiatry on Same-Sex Desires: Pathology, Abnormality, Normality and the Blurring of Boundaries', *Psychology & Sexuality*, vol. 1, 226–38.

——(2012) 'Female Same-Sex Desires: Conceptualizing a Disease in Competing Medical Fields in Nineteenth-Century Europe', *Journal of the History of Medicine and Allied Sciences* (forthcoming).

Bellassai, S. (2004), *La mascolinità contemporanea* (Rome: Carocci).

Benadusi, L. (2005) *Il nemico dell'uomo nuovo. L'omosessualità nell'esperimento totalitario fascista* (Milan: Feltrinelli).

Berrios, G. E. and H. Freeman (eds.) (1991) *150 Years of British Psychiatry (1841–1991)* (London: Gaskell).

Berrios, G. E. and N. Kennedy (2002) 'Erotomania: A Conceptual History', *History of Psychiatry*, vol. 13, 381–400.

Bland, L. (1995) *Banishing the Beast: English Feminism and Sexual Morality, 1885–1914* (London: Penguin).

Bland, L. and L. Doan (eds.) (1998) *Sexology in Culture: Labelling Bodies and Desires* (Cambridge: Polity Press).

Bloch, M. (1967) [1966] 'A Contribution towards a Comparative History of European Societies', in *Land and Work in Mediaeval Europe* (London: Routledge).

Bock Berti, G. (1992) 'Sulla "formazione" medico-legale di Arrigo Tamassia (1849–1917)', *Rivista di storia della medicina*, vol. 23, 37–43.

Borris, K. and G. Rousseau (eds.) (2008) *The Sciences of Homosexuality in Early Modern Europe* (London: Routledge).

Brady, S. (2005) *Masculinity and Male Homosexuality in Britain, 1861–1913* (Basingstoke: Palgrave Macmillan).

Brand, C. P. (1957) *Italy and the English Romantics: The Italian Fashion in Early Nineteenth-Century England* (Cambridge: Cambridge University Press).

Brecher, E. (1969) *The Sex Researchers* (Boston: Little, Brown).

Bulfaretti, L. (1975) *Cesare Lombroso* (Turin: UTET).

Bullough, V. L. (1974) 'Homosexuality and the Medical Model, *Journal of Homosexuality*, vol. 1, 99–110.

——(1989) 'The Physician and Research into Human Sexual Behavior in Nineteenth-Century Germany', *Bulletin for the History of Medicine*, vol. 63, 247–67.

Burgio, A. (ed.) (1999) *Nel nome della razza:Il razzismo nella storia d'Italia, 1870–1945* (Bologna: Il Mulino).

Butler, J. (2000) *Antigone's Claim: Kinship Between Life and Death* (New York: Columbia University Press).

Buttafuoco, A. (1985) *Le mariuccine. Storia di un'istituzione laica. L'asilo Mariuccia* (Milan: Angeli).

——(1988) *Cronache femminili: temi e momenti della stampa emancipazionista in Italia dall'Unità al Fascismo* (Arezzo: Dipartimento di studi storico-sociali e filosofici Università degli Studi di Siena).

——(1989) ' "In servitù regine". Educazione ed emancipazione nella stampa politica femminile', in S. Soldani (ed.), *L'educazione delle donne. Scuole e modelli di vita femminile nell'Italia dell'Ottocento* (Milan: Angeli), 363–91.

Butterfield, H. (1951) *The Whig Interpretation of History* (New York: Scribner).

Buzard, J. (1993) *The Beaten Track: European Tourism, Literature and the Ways to 'Culture', 1800–1918* (Oxford: Oxford University Press).

Caine, B. (1997) *English Feminism, 1780–1920* (Oxford: Oxford University Press).

Canguilhem, G. (1998) [1966] *Il normale e il patologico* (Turin: Einaudi).

Castle, T. (1993) *The Apparitional Lesbian: Female Homosexuality and Modern Culture* (New York: Columbia University Press).

Cavalli Pasini, A. (1986) 'Ruolo e figura femminili nella pubblicistica e nella letteratura popolare', in P. Rossi (ed.), *L'età del positivismo* (Bologna: Il Mulino), 405–38.

Chamberlain, J. E. and S. L. Gilman (eds.) (1985) *Degeneration: The Dark Side of Progress* (New York: Columbia University Press 1895).

Chauncey, G. (1982–3) 'From Sexual Inversion to Homosexuality: Medicine and the Changing Conceptualization of Female Deviance', *Salmagundi*, vol. 58–9, 114–46.

——(1994) *Gay New York: Gender, Urban Culture and the Makings of the Gay Male World, 1890–1940* (New York: Basic Books).

Cocks, H. G. (2003) *Nameless Offences: Homosexual Desire in the Nineteenth Century* (London: I. B. Tauris Publishers).

——(2006) 'Religion and Spirituality', in H. G. Cocks and M. Houlbrook (eds.), *The Modern History of Sexuality* (Basingstoke: Palgrave Macmillan).

Colley, L. (1992) *Britons: Forging the Nation, 1707–1837* (New Haven: Yale University Press).

Colombo, G. (1975) *La scienza infelice: Il museo di antropologia criminale di Cesare Lombroso* (Turin: Boringhieri).

Cook, H. (2004), *The Long Sexual Revolution: English Women, Sex and Contraception, 1800–1975* (Oxford: Oxford University Press).

Cook, M. (2003) *London and the Culture of Homosexuality, 1885–1914* (Cambridge: Cambridge University Press).

Copley, A. (1989) *Sexual Moralities in France, 1780–1980: New Ideas on the Family, Divorce and Homosexuality* (London: Routledge).

Corbin, A. (1990) [1978] *Women for Hire: Prostitution and Sexuality in France after 1850* (Cambridge, MA: Harvard University Press).

Cosmacini, G. (1987) *Storia della medicina e della sanità in Italia. Dalla peste europea alla guerra mondiale, 1348–1918* (Milan: Laterza).

Cott, N. F. (1978) 'Passionlessness: An Interpretation of Victorian Sexual Ideology, 1790–1850', *Signs*, vol. 4, 219–36.

Crozier, I. (2000) 'Havelock Ellis, Eonism and the Patient's Discourse; or, Writing a Book about Sex', *History of Psychiatry*, vol. 11, 125–54.

——(2000) 'Taking Prisoners: Havelock Ellis, Sigmund Freud, and the Construction of Homosexuality, 1897–1951', *Social History of Medicine*, vol. 13, 447–66.

——(2001) 'The Medical Construction of Homosexuality and its Relation to the Law in Nineteenth-Century England', *Medical History*, vol. 45, 61–82.

——(2004) 'Philosophy in the English Boudoir: Havelock Ellis, "Love and Pain", in the Sexological Discourses on Algophilia', *Journal of the History of Sexuality*, vol. 13, 275–305.

——(2005) 'All the Appearances Were Perfectly Natural: The Anus of the Sodomite in Nineteenth-Century Medical Discourse', in C. E. Forth and I. Crozier (eds.), *Body Parts: Critical Explorations in Corporeality* (Lanham, MD: Lexington Books), 65–84.

——(2008) 'Introduction', in H. Ellis and J. A. Symonds, *Sexual Inversion* (Basingstoke: Palgrave Macmillan).

——(2008) 'Nineteenth-Century British Psychiatric Writing on Homosexuality before Havelock Ellis: The Missing Story', *Journal for the History of Medicine and Allied Sciences*, vol. 63, 65–102.

——(2008) 'Pillow Talk: Credibility, Trust and the Sexological Case History', *History of Science*, vol. 46, 375–404.

Dall'Orto, G. (1984) *Leggere Omosessuale* (Turin: Edizioni Gruppo Abele).

——(1988) 'La "tolleranza repressiva" dell'omosessualità. Quando un atteggiamento legale diviene tradizione', in Arci Gay Nazionale (ed.), *Omosessuali e stato* (Bologna: Cassero), 37–57.

David, M. (1966) *La psicoanalisi nella cultura italiana* (Turin: Boringhieri).

Davidson, A. I. (1987) 'How to Do the History of Psychoanalysis: A Reading of Freud's "Three Essays on the Theory of Sexuality" ', *Critical Inquiry*, vol. 13, 252–77.

——(1987) 'Sex and the Emergence of Sexuality', *Critical Inquiry*, vol. 14, 16–48.

——(1990) 'Closing up the Corpses: Diseases of Sexuality and the Emergence of the Psychiatric Style of Reasoning', in G. Boolos (ed.), *Meaning and Method: Essays in Honor of Hilary Putnam* (Cambridge: Cambridge University Press), 295–326.

——(2001) *The Emergence of Sexuality: Historical Epistemology and Formation of Concepts* (Cambridge, MA: Harvard University Press).

Davidson, R. (2000) *Dangerous Liaisons: A Social History of Venereal Disease in Twentieth-Century Scotland* (Amsterdam: Rodopi).

Davidson, R. and L. Hall (eds.) (2001) *Sex, Sin and Suffering: Venereal Disease and European Society since 1870* (London: Routledge).

Davie, N. (2005) *Tracing the Criminal: The Rise of Scientific Criminology in Britain, 1860–1918* (Turner, Oxford: Bradwell Press).

Davis, J. A. (2000) 'Introduction: Italy's Difficult Modernization', in J. A. Davis (ed.), *Italy in the Nineteenth Century, 1796–1900* (Oxford: Oxford University Press), 1–24.

Davis, K. (1965) *Human Society* (New York: Macmillan).

Dean, C. J. (1996) *Sexuality and Modern Western Culture* (New York: Twayne).

D'Emilio, J. and E. B. Freedman (1990) 'Problems Encountered in Writing the History of Sexuality: Sources, Theory and Interpretation', *Journal of Sex Research*, vol. 27, 481–95.

De Giorgio, M. (1992) *Le italiane dall'Unità ad oggi. Modelli culturali e comportamenti sociali* (Rome: Laterza).

De Peri, F. (1984) 'Il medico e il folle: introduzione psichiatrica, sapere scientifico e pensiero medico fra Otto e Novecento', in F. Della Peruta (ed.), *Storia d'Italia. Annali 7* (Turin: Einaudi), 1054–1140.

Doan, L. (1998) 'Acts of Female Indecency: Sexology Intervention in Legislating Lesbianism', in L. Bland and L. Doan (eds.), *Sexology in Culture: Labelling Bodies and Desires* (Cambridge: Polity Press), 199–213.

——(2001) *Fashioning Sapphism: The Origin of the Modern Lesbian Culture* (New York: Columbia University Press).

Doan, L. and J. Prosser (2002) *Palatable Poison: Critical Perspectives on* The Well of Loneliness (New York: Columbia University Press).

Dolza, D. (1990) *Essere figlie di Lombroso: due donne intellettuali tra '800 e '900* (Milan: Angeli).

Donoghue, E. (1993) *Passions between Women: British Lesbian Culture, 1668–1801* (London: Scarlet).

Dörner, K. (1981) [1969] *Madmen and the Bourgeoisie: A Social History of Insanity and Psychiatry* (Oxford: Basil Blackwell).

Dowbiggin, I. A. (1991). *Inheriting Madness: Professionalization and Psychiatric Knowledge in Nineteenth-Century France* (Berkeley: University of California Press).

Dreger, A. D. (1998) *Hermaphrodites and the Medical Invention of Sex* (Cambridge, MA: Harvard University Press).

Duggan, C. (2007) 'Gran Bretagna e Italia nel Risrgimento', in A. M. Banti and P. Ginsborg (eds.), *Storia d'Italia, Il Risorgimento* (Turin: Einaudi), 777–96.

Duggan, L. (1990) 'Review Essay: From Instincts to Politics: Writing the History of Sexuality in the U.S.', *Journal of Sex Research*, vol. 27, 95–109.

——(1993) 'The Trials of Alice Mitchell: Sensationalism, Sexology and the Lesbian Subject in Turn-of-the-Century America', *Signs*, vol. 18, 791–814.

——(2000) *Sapphic Slashers: Sex, Violence and American Modernity* (Durham: Duke University Press).

Eder, F. X., L. A. Hall, and G. Hekma (eds.) (1999) *Sexual Cultures in Europe: National Histories* (Manchester: Manchester University Press).

Faderman, L. (1991) *Odd Lovers and Twilight Girls: A History of Lesbian Life in Twentieth Century America* (New York: Columbia University Press).

——(1991) [1981] *Surpassing the Love of Men: Friendships between Women from the Renaissance to the Present* (London: Women's Press).

Finnegan, F. (1979) *Poverty and Prostitution: A Study of Victorian Prostitutes in York* (Cambridge: Cambridge University Press).

Fish, F. (1965) 'David Skae, M.D., F.R.C.S., Founder of the Edinburgh School of Psychiatry', *Medical History*, vol. 9, 36–53.

Foldy, M. S. (1997) *The Trials of Oscar Wilde: Deviance, Morality, and Late-Victorian Society* (New Haven and London: Yale University Press).

Forrester, J. (1996) 'If p, then what? Thinking in Cases', *History of the Human Sciences*, vol. 9, 1–25.

Foucault, M. (1969) *L'Archéologie du Savoir* (Paris: Gallimard).

——(1998) [1976] *The Will to Knowledge, History of Sexuality, Volume 1: An Introduction*, trans. Robert Hurley (Harmondsworth: Penguin Books).

——(2003) [1999] *Abnormal: Lectures at the Collège de France, 1974–1975* (London: Verso).

Fout, J. C. (ed.) (1990) *Forbidden History: The State, Society, and the Regulation of Sexuality in Modern Europe* (Chicago: University of Chicago Press).

Friedli, L. (1987) ' "Passing Women": A Study of Gender Boundaries in the Eighteenth Century', in G. Rousseau and R. Porter (eds.), *Sexual Underworlds of the Enlightenment* (Manchester: Manchester University Press), 234–60.

Frigessi, D. (2003) *Cesare Lombroso* (Turin: Einaudi).

Gadebusch Bondio, M. C. (1996) 'La tipologizzazione della donna deviante nella seconda metà dell'800', in M. Beretta, F. Mondella, and M. T. Monti (eds.), *Per una storia critica della scienza* (Bologna: Cisalpino), 283–314.

Gargano, C. (2007) *Capri pagana: Uranisti ed amazzoni tra Ottocento e Novecento* (Capri: La Conchiglia)

Garland, D. (1997) [1994] 'Of Crimes and Criminals: The Development of Criminology in Britain', in M. Maguire, R. Morgan, and R. Reiner (eds.), *The Oxford Handbook of Criminology* (Oxford: Oxford University Press), 11–56.

Gattei, G. (1984) 'La sifilide: medici e poliziotti intorno alla "Venere politica" ', in F. Della Peruta (ed.), *Storia d'Italia, Annali 7* (Turin: Einaudi), 741–98.

Gay, P. (1984) *The Bourgeois Experience, from Victoria to Freud: Education of the Senses*, vol. 1 (Oxford: Oxford University Press).

Gervasoni, M. (1997) ' "Cultura della degenerazione" tra socialismo e criminologia alla fine dell'Ottocento in Italia', *Studi Storici*, vol. 3, 1087–1119.

Giacanelli, F. (1975) 'Appunti per una storia della psichiatria italiana', in K. Dörner, *Il borghese e il folle. Storia sociale della psichiatria* (Bari: Laterza), v–xxxii.

——(1986) 'Psichiatria e storiografia', *Psicoterapia e scienze umane*, vol. 20, 81–93.

Gibson, M. (1986) *Prostitution and the State of Italy, 1860–1915* (New Brunswick: Rutgers University Press).

——(1995) [1986] *Stato e prostituzione in Italia, 1860–1915* (Milan: Il Saggiatore).

——(2002) *Born to Crime: Cesare Lombroso and the Origins of Biological Criminology* (Westport: Praeger).

——(2004) 'Labelling Women Deviant: Heterosexual Women, Prostitutes and Lesbians in Early Criminological Discourse', in P. Willson (ed.), *Gender, Family and Sexuality: The Private Sphere in Italy, 1860–1945* (Basingstoke: Palgrave Macmillan), 89–104.

——(2006) 'Cesare Lombroso and the Italian Criminology', in P. Becker and R. F. Wetzell (eds.), *Criminals and Their Scientists: The History of Criminology in International Perspective* (Cambridge: Cambridge University Press), 137–58.

Gibson, M. and N. H. Rafter (eds.) (2006) 'Introduction', in C. Lombroso, *Criminal Man* (Durham: Duke University Press), 1–41.

Gieryn, T. F. (1999) *Cultural Boundaries of Science: Credibility on the Line* (Chicago and London: University of Chicago Press).

Gilman, S. L. (1985) *Difference and Pathology: Stereotypes of Sexuality, Race, and Madness* (Ithaca: Cornell University Press).

——(1989) *Sexuality: An Illustrated History* (New York: Wiley).

Glover, E. (1949) 'Victorian Ideas of Sex', in BBC (ed.), *Ideas and Beliefs of the Victorians: An Historic Revaluation of the Victorian Age* (London: Sylvan Press), 358–64.

Goldstein, J. (2001). *Console and Classify: The French Psychiatric Profession in the Nineteenth Century* (Chicago: University of Chicago Press).

Gould, S. J. (1977) *Ontogeny and Phylogeny* (Cambridge, MA: Belknap Press).

——(1981) *The Measure of the Man* (London: Penguin Books).

Gowing, L. (2006) 'Lesbians and Their Like in Early Modern Europe, 1500–1800', in R. Aldrich (ed.), *Gay Life and Culture* (London: Thames and Hudson), 126–43.

Graziosi, M. (1993) 'Infirmitas sexus. La donna nell'immaginario penalistico', *Democrazia e diritto*, vol. 33, 99–143.

Groneman, C. (1994) 'Nymphomania: The Historical Construction of Female Sexuality', *Signs*, vol. 19, 337–67.

Grosskurth, P. (1980) *Havelock Ellis: A Biography* (London: Allen Lane).

——(ed.) (1984) *The Memoirs of John Addington Symonds* (London: Hutchinson).

Guarnieri, P. (1986) *Individualità difformi: la psichiatria antropologica di Enrico Morselli* (Milan: Claudio Angeli).

——(1988) 'Between Soma and Psyche: Morselli and Psychiatry in Late-Nineteenth-Century Italy', in W. F. Bynum, R. Porter and M. Shepherd (eds.), *The Anatomy of Madness: Essays in the History of Psychiatry* (London: Tavistock Publications), vol. 3, 102–24.

——(1991) 'Alienists on Trial: Conflict and Convergence between Psychiatry and Law (1876–1913)', *History of Science*, vol. 29, 393–410.

Gutmann, P. (2006) 'On the Way to Scientia Sexualis: "On the relation of the sexual system to the psyche in general and to cretinism in particular" (1826) by Joseph Häussler', *History of Psychiatry*, vol. 17, 45–53.

Guy, J. M. (ed.) (2007) *The Complete Works of Oscar Wilde* (Oxford: Oxford University Press).

Halberstam, J. (1998) *Female Masculinity* (Durham, NC: Duke University Press).

Hall, L. (1990) 'Forbidden by God, Despised by Men: Masturbation, Medical Warnings, Moral Panic, and Manhood in Great Britain, 1850–1950', in J. C. Fout (ed.), *Forbidden History: The State, Society, and the Regulation of Sexuality in Modern Europe* (Chicago: University of Chicago Press), 293–315.

——(1994) ' "The English Have Not Hot Water Bottles": the Morganatic Marriage Between Sexology and Medicine since William Acton', in R. Porter and M. Teich (eds.), *Sexual Knowledge, Sexual Science* (Cambridge: Cambridge University Press), 350–66.

——(1995) ' "Disinterested Enthusiasm for Sexual Misconduct": The British Society for the Study of Sex Psychology, 1913–47', *Journal of Contemporary History*, vol. 30, 665–86.

——(1998) 'Feminist Reconfigurations of Heterosexuality in the 1920s', in L. Bland and L. Doan (eds.), *Sexology in Culture: Labelling Bodies and Desires* (Cambridge: Polity Press), 135–49.

——(2000) 'Malthusian Mutations: The Changing Politics and Moral Meaning of Birth Control in Britain', in B. Dolan (ed.), *Malthus, Medicine and Morality: 'Malthusianism' after 1798* (Amsterdam: Rodopi,), 141–64.

——(2000) *Sex, Gender and Social Change in Britain since 1880* (Basingstoke: Macmillan).

——(2004) 'Hauling Down the Double Standard: Feminism, Social Purity and Sexual Science in Late Nineteenth-Century Britain', *Gender & History*, vol. 16, 36–56.

Halperin, D. M. (1990) *One Hundred Years of Homosexuality and Other Essays on Greek Love* (New York: Routledge).

——(2002) *How to Do the History of Homosexuality* (Chicago: University of Chicago Press).

Hamer, D. et al. (1993) 'A Linkage between DNA Markers on the X Chromosome and Male Sexual Orientation', *Science*, vol. 261, no. 5119, 321–7.

Hansen, B. (1989) 'American Physicians' Earliest Writings about Homosexuals, 1880–1900', *The Milbank Quarterly*, vol. 67, suppl. 1, 99–108.

Harris, K. (1984) *Sexual Ideology and Religion: The Representation of Women in the Bible* (Brighton: Wheatsheaf).

Hekma, G. (1989) 'A History of Sexology: Social and Historical Aspects of Sexuality', in J. Bremmer (ed.), *From Sappho to De Sade: Moments in the History of Sexuality* (London: Routledge), 173–93.

——(1994) 'A Female Soul in a Male Body: Sexual Inversion as Gender Inversion in Nineteenth Century Sexology', in G. Herdt (ed.), *Third Sex, Third Gender: Beyond Dimorphism in Culture and History* (New York: Zone Books), 213–39.

Horn, D. G. (2003) *The Criminal Body: Lombroso and the Anatomy of Deviance* (New York: Routledge).

——(2006) 'Making Criminologists: Tools, Techniques, and the Production of Scientific Authority', in P. Becker and R. F. Wetzell (eds.), *Criminals and Their Scientists: The History of Criminology in International Perspective* (Cambridge: Cambridge University Press), 317–36.

Howard, J. J. (1978) 'The Civil Code of 1865 and the Origins of the Feminist Movement in Italy', in B. Boyd Caroli, R. F. Harney, and L. F. Tomasi (eds.), *The Italian Immigrant Woman in North America* (Proceedings of the Tenth Annual Conference of the American Italian Historical Association held in Toronto, Ontario (Canada), 28–29 October 1977 in conjunction with the Canadian Italian Historical Association (Toronto: The Multicultural History Society of Ontario), 14–24.

Hutter, J. (1993) 'The Social Construction of Homosexuals in the Nineteenth Century: The Shift from the Sin to the Influence of Medicine on Criminalizing Sodomy in Germany', *Journal of Homosexuality*, vol. 24, 73–93.

Isabella, M. (2009) *Risorgimento in Exile: Italian Émigrés and the Liberal International in the Post-Napoleonic Era* (Oxford: Oxford University Press).

Jackson, L. A. (2010) 'The Regularisation of Sexuality in the Age of Empire', in C. Beccalossi and I. Crozier (eds.), *Sexuality in the Age of Empire* (Oxford: Berg).

Jackson, M. (1994) *The Real Facts of Life: Feminism and the Politics of Sexuality, c.1850–1940* (London: Taylor and Francis).

Jacob, M. C. (1999), 'Science Studies after Social Constructionism: The Turn toward the Comparative and the Global', in V. E. Bonnell and L. Hunt (eds.), *Beyond the Cultural Turn: New Directions in the Study of Society and Culture* (Berkeley: University of California Press), 95–120.

Jay, M. and M. Neve (eds.) (1999) *1900: A Fin-de-Siècle Reader* (London: Penguin Books).

Jeffreys, S. (1985) *The Spinster and Her Enemies: Feminism and Sexuality* (London: Pandora).

Jones, K. (1991) 'The Culture of the Mental Hospital', in G. E. Berrios and H. Freeman (eds.), *150 Years of British Psychiatry (1841–1991)* (London: Gaskell), 17–28.

——(1991) 'Law and Mental Health: Sticks or Carrots', in G. E. Berrios and H. Freeman (eds.), *150 Years of British Psychiatry (1841–1991)* (London: Gaskell), 89–102.

Jordan, M. D. (1997) *The Invention of Sodomy in Christian Theology* (Chicago: University of Chicago Press).

Jordanova, L. (1980) 'Natural Facts: A Historical Perspective on Science and Sexuality', in C. McCormack and M. Strathern (eds.), *Nature, Culture and Gender* (Cambridge: Cambridge University Press), 42–69.

——(1989) *Sexual Visions: Images of Gender in Science and Medicine between the Eighteenth and the Twentieth Century* (London: Harvester Wheatsheaf).

——(1993) 'Gender and the Historiography of Science', *British Journal for the History of Science*, vol. 26, 469–83.

Katz, J. N. (1995) *The Invention of Heterosexuality* (New York: Penguin).

Kerr, J. M. M., R.W. J. Miles, and H. Phillips (eds.) (1954) *Historical Review of British Obstetrics and Gynaecology, 1800–1950* (Edinburgh: E. & S. Livingstone Ltd).

Kertzer, D. I. (2000) 'Religion and Society, 1789–1892', in J. A. Davis (ed.), *Italy in the Nineteenth Century, 1796–1900* (Oxford: Oxford University Press), 181–205.

Knight, F. (1997) ' "Male and Female He Created Them": Men, Women and the Question of Gender', in J. Wolffe (ed.), *Religion in Victorian Britain*, vol. 5, *Culture and Empire* (Manchester: Manchester University Press), 23–58.

Koehler, P. (1999) 'The Evolution of British Neurology in Comparison with Other Countries', in F. C. Rose (ed.), *A Short History of Neurology: The British Contribution, 1660–1910* (Oxford: Butterworth Heinemann), 58–74.

Kuhn, T. S. (1977) *The Essential Tension: Selected Studies in Scientific Tradition and Change* (Chicago: University of Chicago Press).

Laing, R. D. (1960) *The Divided Self: A Study of Sanity and Madness* (London: Tavistock).

Laqueur, T. (1990) *Making Sex: Body and Gender from the Greeks to Freud* (Cambridge, MA: Harvard University Press).

——(2003) *Solitary Sex: A Cultural History of Masturbation* (New York: Zone Books).

Latour, B. (1987) *Science in Action: How to Follow Scientists and Engineers through Society* (Milton Keynes: Open University Press).

Latour, B. and S. Woolgar (1979) *Laboratory Life: The Social Construction of Scientific Facts* (London: Sage Publications).

Lawrence, C. (1985) 'Incommunicable Knowledge: Science, Technology and the Clinical Art in Britain, 1850–1914', *Journal of Contemporary History*, vol. 20, 503–20.

——(1994) *Medicine in the Making of Modern Britain, 1700–1920* (London: Routledge).

LeVay, S. (1991) 'A Difference in Hypothalamic Structure between Heterosexual and Homosexual Men, *Science*, vol. 253, no. 5023, 1034–7.

Lynch, M. (1985) ' "Here is adhesiveness": From Friendship to Homosexuality', *Victorian Studies*, vol. 29, 67–96.

MacIver, R. M. and C. H. Page (1964) [1950] *Society: An Introductory Analysis* (London: Macmillan).

Macrelli, R. (1981) *L'indegna schiavitù: Anna Maria Mozzoni e la lotta contro la prostituzione di stato* (Rome: Editori Riuniti).

Mahood, L. (1990) *The Magdalenes: Prostitution in the Nineteenth Century* (London: Routledge).

Manacorda, M. A. (1989) 'Istruzione ed emancipazione della donna nel Risorgimento. Riletture e considerazioni', in S. Soldani (ed.), *L'educazione delle donne. Scuole e modelli di vita femminile nell'Italia dell'Ottocento* (Milan: Angeli), 1–33.

Mangoni, L. (1985) *Una crisi di fine secolo. La cultura italiana e la Francia fra Otto e Novecento* (Turin: Einaudi).

Marcus, S. (1996) *The Other Victorians: A Study of Sexuality and Pornography in Mid-Nineteenth-Century England* (London: Weidenfeld and Nicolson).

Marcus, S. (2002) 'Comparative Sapphism', in M. Cohen and C. Dever (eds.), *The Literary Channel: The Inter-National Invention of the Novel* (Princeton: Princeton University Press), 251–85.

Martucci, P. (2002) *Le piaghe d'Italia. I lombrosiani e i grandi crimini economici nell'Europa di fine Ottocento* (Milan: Angeli).

Mason, M. (1994) *The Making of Victorian Sexuality* (Oxford: Oxford University Press).

Mayer, A. J. (1981) *The Persistence of the Old Regime* (New York: Pantheon Books).

McIntosh, M. (1968) 'The Homosexual Role', *Social Problems*, vol. 16, 182–92.

McLaren, A. (1999) *Twentieth-Century Sexuality: A History* (Oxford: Blackwell Publishers).

Meadows, J. (2004) *The Victorian Scientist: The Growth of a Profession* (London: British Library).

Medvei, V. C. (1982) *A History of Endocrinology* (Lancaster: MPT Press).

——(1993) *The History of Clinical Endocrinology* (Carnforth: Parthenon).

Meer, T. van der (2007) 'Sodomy and its Discontents: Discourse, Desire, and the Rise of a Same-Sex Proto-Something in the Early Modern Dutch Republic', *Historical Reflections*, vol. 33, 41–67.

Milletti, N. (1994) 'Analoghe sconcezze. Tribadi, saffiste, invertite e omosessuali: categorie e sistemi sesso/genere nella rivista di antropologia criminale fondata da Cesare Lombroso (1880–1949)', *DWF*, vol. 4, 50–122.

——(2007) 'Accuse innominabili. Lesbiche e confino di polizia durante ul fascismo', in N. Milletti and L. Passerini (eds.), *Fuori dalla norma. Storie lesbiche nell'Italia della prima metà del Novecento* (Turin: Rosenberg & Sellier), 135–70.

Minuz, F. (1986) 'La norma femminile nell'antropologia', in P. Rossi (ed.), *L'età del positivismo* (Bologna: Il Mulino), 439–52.

Mol, A. (2002) *The Body Multiple: Ontology in Medical Practice* (Durham and London: Duke University Press).

Money, J. (1986) *Capri: Island of Pleasure* (London: Hamish Hamilton).

Moran, L. J. (1996) *The Homosexual(ity) of Law* (London: Routledge).

Morants, R. (1974) 'The Lady and Her Physician', in M. S. Hartmann and L. Banner (eds.), *Clio's Consciousness Raised: New Perspectives in the History of Women* (New York: Harper Torchbooks), 38–53.

Mort, F. (1987) *Dangerous Sexualities: Medico-Moral Politics in England since 1830* (London: Routledge).

Moscucci, O. (1990) *The Science of Woman: Gynaecology and Gender in England, 1800–1929* (Cambridge: Cambridge University Press).

Mosse, G. L. (1984) [1982] *Sessualità e nazionalismo. Mentalità borghese e rispettabilità* (Rome: Laterza).

Nead, L. (1988) *Myths of Sexuality* (Oxford: Oxford University Press).

Neve, M. (1997) 'The Influence of Degenerationist Categories in Nineteenth-Century Psychiatry, with Special Reference to Great Britain', in Y. Kawakita, S. Sakai, and Y. Otsuka (eds.), *History of Psychiatric Diagnosis* (Tokyo: Ishiyaku EuroAmerica), 141–63.

Newton, E. (1984) 'The Mythic Mannish Lesbian: Radclyffe Hall and the New Woman', *Signs*, vol. 9, 557–75.

Nimmons, D. (1994) 'Sexual Brain', *Discover*, March, vol. 5, no. 3, 64–7.

Nottingham, C. (1999) *The Pursuit of Serenity: Havelock Ellis and the New Politics* (Amsterdam: Amsterdam University Press).

Nye, R. A. (1984) *Crime, Madness and Politics in Modern France: The Medical Concept of National Decline* (Princeton: Princeton University Press).

——(1991) 'The History of Sexuality in Context: National Sexological Tradition', *Science in Context*, vol. 4, 387–406.

——(1993) 'The Medical Origins of Sexual Fetishism', in E. Apter and W. Pietz (eds.), *Fetishism as Cultural Discourse* (Ithaca: Cornell University Press), 13–30.

O'Connor, M. (1998) *The Romance of Italy and the English Political Imagination* (New York: St. Martin's Press).

Oliari, E. (2006) *L'omo delinquente. Scandali e delitti gay dall'Unità a Giolitti* (Rome: Prospettiva).

Oosterhuis, H. (2000) *Stepchildren of Nature: Krafft-Ebing, Psychiatry and the Making of Sexual Identity* (Chicago: University of Chicago Press).

Oppenheim, J. (1991) *Shattered Nerves: Doctors, Patients, and Depression in Victorian England* (Oxford: Oxford University Press).

Oram, A. (2006) 'Cross-Dressing and Transgender', in H. G. Cocks and M. Houldbrook (eds.), *The Modern History of Sexuality* (Basingstoke: Palgrave Macmillan), 263–6.

Oram, A. and A. Turnbull (eds.) (2001) *The Lesbian History Sourcebook: Love and Sex between Women in Britain from 1780 to 1970* (London: Routledge).

Oudshoorn, N. (1994) *Beyond the Natural Body: An Archaeology of Sex Hormones* (London: Routledge).

Park, K. (1997) 'The Rediscovery of the Clitoris: French Medicine and the Tribade, 1570–1620', in D. Hillman and C. Mazzio (eds.), *The Body Parts: Fantasies of Corporeality in Early Modern Europe* (New York: Routledge), 171–94.

Paul, D. B. (1995) *Controlling Human Heredity, 1865 to the Present* (Atlantic Highlands: Humanities Press).

Pearsall, R. (1983) [1969] *The Worm in the Bud: The World of Victorian Sexuality* (Harmondsworth: Penguin).

Peel, J. (1986) *William Blair-Bell: Father and Founder* (London: The Royal College of Obstetricians and Gynaecologists).

Pemble, J. (1987) *The Mediterranean Passion: Victorians and Edwardians in the South* (Oxford: Clarendon Press).

——(2000) 'Art, Disease, and Mountains', in J. Pemble (ed.), *John Addington Symonds: Culture and the Demon Desire* (Basingstoke: Macmillan), 1–21.

Pfister, M. and R. Hertel (eds.) (2008) *Performing National Identity: Anglo-Italian Cultural Transactions* (Amsterdam: Rodopi).

Pick, D. (1988) ' "Terrors of the night": *Dracula* and "Degeneration" in the Late Nineteenth Century', *Critical Quarterly*, vol. 30, 71–87.

——(1989) *Faces of Degeneration: A European Disorder, c.1848–1918* (Cambridge: Cambridge University Press).

Poggi, S. (1987) *Il positivismo* (Rome-Bari: Laterza).

Pomata, G. (2001) 'Menstruating Men: Similarity and Difference of the Sexes in Early Modern Medicine', in V. Finucci and K. Brownlee (eds.), *Generation and Degeneration: Tropes of Reproduction in Literature and History from Antiquity through Early Modern Europe* (Durham: Duke University Press), 109–52.

Porter, R. and L. Hall (1995) *The Facts of Life: The Creation of Sexual Knowledge in Britain, 1650–1950* (New Haven and London: Yale University Press).

Porter, R. and M. Teich (eds.) (1994) *Sexual Knowledge, Sexual Science: The History of Attitudes to Sexuality* (Cambridge: Cambridge University Press).

Porter, T. M. (1986) *The Rise of Statistical Thinking, 1820–1900* (Princeton: Princeton University Press).

Pritchard, D. (2001) *Oscar Wilde* (New Lanark: Geddes & Grosset).

Pyenson, L. (2002) 'Comparative History of Science', *History of Science*, vol. 40, 1–33.

Rafter, N. H. (1997) *Creating Born Criminals* (Urbana: University of Illinois Press).

Rapp, D. (1990) 'The Early Discovery of Freud by the British General Educated Public, 1912–1919', *Social History of Medicine*, vol. 3, 217–43.

Renvoize, E. (1991) 'The Association of Medical Officers of Asylums and Hospitals for the Insane, the Medico-Psychological Association, and their Presidents', in G. E. Berrios and H. Freeman (eds.), *150 Years of British Psychiatry* (London: Gaskell), 29–78.

Riall, L. (2009) *Risorgimento: The History of Italy from Napoleon to Nation-State* (Basingstoke: Palgrave Macmillan).

Richards, G. (2000) 'Putting Britain on the Couch: The Popularisation of Psychoanalysis in Britain, 1918–1940', *Science in Context*, vol. 13, 183–230.

Robinson, P. (1976) *The Modernisation of Sex: Havelock Ellis, Alfred Kinsey, William Masters, and Virginia Johnson* (London: Paul Elek).

Rosa, A. A. (1975) *Storia d'Italia. Dall'unità ad oggi*, vol. 4 (Turin: Einaudi).

Rosario, V. A. (1997) *The Erotic Imagination: French Histories of Perversity* (New York and Oxford: Oxford University Press).

——(ed.) (1997) *Science and Homosexualities* (New York: Routledge).

Rossi-Doria, A. (1999) 'Antisemitismo e antifemminismo nella cultura positivistica', in A. Burgio (ed.), *Nel nome della razza* (Bologna: Il Mulino), 455–73.

——(ed.) (2003) *A che punto è la storia delle donne in Italia* (Rome: Viella).

Rowold, K. J. (1996) ' "The Academic Woman": Minds, Bodies and Education in Britain and Germany, 1860–1914', PhD Thesis, University College London.

Russett, C. E. (1989) *Sexual Science: The Victorian Construction of Womanhood* (Cambridge, MA: Harvard University Press).

Sarogni, E. (2004) [1995] *La donna italiana, 1861–2000. Il lungo cammino verso i diritti* (Milan: Il Saggiatore).

Schiebinger, L. (1989) *The Mind Has No Sex? Women in the Origins of Modern Science* (Cambridge, MA: Harvard University Press).

Scott, J. W. (1986) 'Gender: A Useful Category of Historical Analysis', *American Historical Review*, vol. 91, 1053–75.

Scull, A. (1975) *Museums of Madness: The Social Organisation of Insanity in 19th Century England* (Ann Arbor: University Microfilms International).

Scull, A., C. MacKenzie, and N. Hervey (1996) *Masters of Bedlam: The Transformation of the Mad-Doctoring Trade* (Princeton: Princeton University Press).

Secord, J. A. (2004) 'Knowledge in Transit', *Isis*, vol. 95, 654–72.

Sedgwick, E. K. (1990) *Epistemology of the Closet* (Berkeley: University of California Press).

Sengoopta, C. (2000) 'The Modern Ovary: Constructions, Meanings, Uses', *History of Science*, vol. 38, 425–88.

——(2000) *Otto Weininger: Sex, Science and Self in Imperial Vienna* (Chicago: Chicago University Press).

——(2006) *The Most Secret Quintessence of Life: Sex, Glands, and Hormones, 1850–1950* (Chicago: University of Chicago Press).

Shorter, E. (1982) *A History of Women's Bodies* (Harmondsworth: Penguin).

——(2005) *A Historical Dictionary of Psychiatry* (Oxford: Oxford University Press).

Showalter, E. (1987) *Female Malady: Women, Madness, and English Culture, 1830–1980* (New York: Penguin Books).

Smith, F. B. (1976) 'Labouchère's Amendment to the Criminal Law Amendment Act', *Historical Studies*, vol. 17, 165–75.

Smith-Rosenberg, C. (1975) 'The Female World of Love and Ritual', *Signs*, vol. 1, 1–29.

——(1985) *Disorderly Conduct: Visions of Gender in Victorian America* (New York: Knopf).

——(1989) 'Discourses of Sexuality and Subjectivity: The New Woman, 1870–1936', in M. Duberman, M. Vicinus, and G. Chauncey (eds.), *Hidden From History: Reclaiming the Gay and Lesbian Past* (New York: New American Library), 264–80.

Soldani, S. (ed.), (1989) *L'educazione delle donne e modelli di vita femminile nell'Italia dell'Ottocento* (Milan: Angeli).

Spongberg, M. (1997) *Feminizing Venereal Disease* (New York: New York University Press).

Steakley, J. (1975) *The Homosexual Emancipation Movement in Germany* (New York: Arno Press).

——(1997) ' "Per Scientiam ad justitiam": Magnus Hirschfeld and the Sexual Politics of Innate Homosexuality', in V. A. Rosario (ed.), *Science and Homosexualities* (New York: Routledge).

Stearns, C. Z. and P. N. Stearns (1985) 'Victorian Sexuality: Can We Do It Better?', *Journal of Social History*, vol. 19, 625–34.

Stepansky, P. E. (1980) 'A Footnote to the History of Homosexuality in Britain: Havelock Ellis and the Bedborough Trial of 1898', in *Essays in the History of Psychiatry* (tenth Anniversary Supplementary Volume to the *Psychiatric Forum*), 72–102.

Stone, L. (1990) *Road to Divorce: England, 1530–1987* (Oxford: Oxford University Press).

Strachey, L. (1918) *Eminent Victorians* (London: Chatto & Windus).

Sulloway, F. J. (1992) [1979] *Freud, the Biologist of the Mind: Beyond the Psychoanalytic Legend* (Cambridge, MA: Harvard University Press).

Summers, A. (1999) ' "The Constitution Violate": The Female Body and the Female Subject in the Campaigns of Josephine Butler', *History Workshop Journal*, vol. 48, 1–15.

Szasz, T. S. (1960) 'The Myth of Mental Illness', *American Psychologist*, vol. 15, 113–18.

——(1980) *Sex by Prescription* (New York: Doubleday).

Tagliavini, A. (1982) '"La scienza psichiatrica": La costruzione del sapere nei congressi della Società Italiana di Freniatria (1874–1907)', in V. P. Babini, M. Cotti, F. Minuz, and A. Tagliavini, *Tra sapere e potere. La psichiatria italiana nella seconda metà dell'Ottocento* (Bologna: Il Mulino), 77–134.

——(1985) 'Aspects of the History of Psychiatry in Italy in the Second Half of the Nineteenth Century', in W. F. Bynum, R. Porter, and M. Shepherd (eds.), *The Anatomy of Madness: Essays in the History of Psychiatry* (London: Tavistock Publications), vol. 2, 175–96.

——(1986) 'La "mente femminile" nella psichiatria dell'Ottocento', in P. Rossi (ed.), *L'età del positivismo* (Bologna: Il Mulino), 475–91.

Tanteri, M. and E. Trimboli (1999–2000) 'Il processo Verzeni', in L. Condò, M. Cotugno, C. Giardiello, G. Malerba, F. Ottaviani and L. Saletti (eds.), *Modelli, giudizi e pregiudizi:materiali per una storia di fine secolo, seminars 1999–2000, Dipartimento di storia moderna e contemporanea, Università degli di Studi di Rome 'La Sapienza'* (available online http://w3.unirome1.it/dsmc/ricerca/materiali_fine_secolo.htm, accessed July 2007).

Tasca, L. (2004) 'Il "senatore erotico". Sesso e matrimonio nell'antropologia di Paolo Mantegazza', in B. P. F. Wanrooij (ed.), *La mediazione matrimoniale in Europa fra Otto e Novecento* (Fiesole-Roma: Georgetown University), 295–322.

Temkin, O. (1977) *The Double Face of Janus, and Other Essays in the History of Medicine* (Baltimore: Johns Hopkins University Press).

Terry, J. (1995) 'Anxious Slippages between "Us" and "Them": A Brief History of the Scientific Search for Homosexual Bodies', in J. Terry and J. Urla (eds.), *Deviant Bodies: Critical Perspectives on Difference in Science and Popular Culture* (Bloomington: Indiana University Press), 129–69.

——(1999) *An American Obsession: Science, Medicine and Homosexuality in Modern Society* (Chicago and London: University of Chicago Press).

Tomaselli, S. (ed.) (1995) *Mary Wollstonecraft: A Vindication of the Rights of Men with A Vindication of the Rights of Women* (Cambridge: Cambridge University Press).

Tomes, N. (1994) 'Feminist Histories of Psychiatry', in M. S. Micale and R. Porter (eds.), *Discovering the History of Psychiatry* (Oxford: Oxford University Press), 348–83.

Traub, V. (2002) *The Renaissance of Lesbianism in Early Modern England* (Cambridge: Cambridge University Press).

——(2007) 'The Present Future of Lesbian Historiography', in G. E. Haggerty and M. McGarry (eds.), *A Companion to Lesbian, Gay, Bisexual, Transgender, and Queer Studies* (Oxford: Blackwell), 124–45.

Trumbach, R. (1977) 'London's Sodomites: Homosexual Behavior and Western Culture in the Eighteenth Century', *Journal of Social History*, vol. 11, 1–33.

——(1991) 'London's Sapphists: From Three Sexes to Four Genders in the Making of Modern Culture', in J. Epstein and K. Straub (eds.), *Body Guards: The Cultural Politics of Gender Ambiguity* (London: Routledge), 112–41.

——(1998) *Sex and the Gender Revolution: Heterosexuality and the Third Gender in the Enlightenment London*, vol. 1 (Chicago: University of Chicago Press).

Vicinus, M. (ed.) (1972) *Suffer and Be Still: Women in the Victorian Age* (Bloomington: Indiana University Press).

——(1989) 'They Wonder to Which Sex I Belong: The Historical Roots of the Modern Lesbian Identity', in D. Altman, C. Vance, and M. Vicinus (eds.), *Homosexuality, Which Homosexuality?* (Amsterdam: Dekker and Shorer), 467–97.

——(2004) *Intimate Friends: Women who Loved Women, 1778–1928* (Chicago: University of Chicago Press).

Villa, R. (1985) *il deviante e i suoi segni: Lombroso e la nascita dell'antropologia criminale* (Milan: Angeli).

Wahl, E. S. (1999) *Invisible Relations: Representations of Female Intimacy in the Age of Enlightenment* (Stanford: Stanford University Press).

Walkowitz, J. R. (1980) *Prostitution and Victorian Society: Women, Class and State* (Cambridge: Cambridge University Press).

——(1993) 'Dangerous Sexualities', in G. Duby and M. Perrot (eds.), *A History of Women: Emerging Feminism from Revolution to World War* (Cambridge, MA: Harvard University Press), vol. 4, 369–98.

Wanrooij, B. P. F. (1987) '"La carne vedova", immagini della sessualità femminile', *Belfagor*, vol. 42, 454–66.

——(1988) 'La passione svelata: sessualità, crimine ed educazione in Italia tra Ottocento e Novecento', *Sanità, scienza e storia*, vol. 5, 393–428.

——(1990) *La Storia del pudore* (Venice: Marsilio).

——(1999) 'Italy: Sexuality, Morality and Public Authority', in F. X. Eder, L. Hall, and G. Hekma (eds.), *Sexual Cultures in Europe* (Manchester: Manchester University Press), 114–37.

——(2004) 'The History of Sexuality in Italy (1860–1945)', in P. Willson (ed.), *Gender, Family and Sexuality: The Private Sphere in Italy, 1860–1945* (Basingstoke: Palgrave Macmillan), 173–91.

Waters, C. (2006) 'Sexology', in H. G. Cocks and M. Houlbrook (eds.), *Palgrave Advances in the Modern History of Sexuality* (Basingstoke: Palgrave Macmillan), 41–63.

Weeks, J. (1979) 'Movements of Affirmations: Sexual Meanings and Homosexual Identities', *Radical History Review*, vol. 20, 164–79.

——(1981) *Sex, Politics and Society: The Regulation of Sexuality since 1800* (London: Longman).

——(1982) 'Foucault for Historians', *History Workshop Journal*, vol. 14, 106–19.

——(1985) *Sexuality and its Discontents: Meanings, Myths and Modern Sexualities* (London: Routledge & Kegan Paul).

——(2000) *Making Sexual History* (Cambridge: Polity Press).

——(2005) 'Remembering Foucault', *Journal of the History of Sexuality*, vol. 14, 186–201.

Wetzell, R. F. (2000) *Inventing the Criminal: A History of German Criminology, 1880–1945* (Chapel Hill: University of North Carolina Press).

Williams, T. J. et al. (2000) 'Finger-Length Ratios and Sexual Orientation', *Nature* 404 (30 March), 455–6.

Willson, P. (2004) 'Introduction: Gender and the Private Sphere in Liberal and Fascist Italy', in P. Willson (ed.), *Gender, Family and Sexuality: The Private Sphere in Italy, 1860–1945* (Basingstoke: Palgrave Macmillan), 1–19.

Wolffe, J. (2000) *Great Deaths: Grieving, Religion, and Nationhood in Victorian and Edwardian Britain* (Oxford: Oxford University Press).

Index